Remote Sensing and Visualization Methods: Monitoring, Modeling, Simulations and Interaction of Forest Resource—Volume I

Remote Sensing and Visualization Methods: Monitoring, Modeling, Simulations and Interaction of Forest Resource—Volume I

Topic Editors

Huaiqing Zhang
Juan Suárez-Minguez
Qi Chen
Yunsheng Wang
Safa Tharib
Hua Sun
Weipeng Jing
Huaguo Huang
Ting Yun
Meili Wang

MDPI

Basel • Beijing • Wuhan • Barcelona • Belgrade • Novi Sad • Cluj • Manchester

Topic Editors

Huaiqing Zhang
Chinese Academy of Forestry
Beijing
China

Juan Suárez-Minguez
Northern Research Station
Roslin
UK

Qi Chen
University of Hawai'i at Mānoa
Honolulu
USA

Yunsheng Wang
Finnish Geospatial Research Institute
Masala
Finland

Safa Tharib
Buckinghamshire New University
Buckinghamshire
UK

Hua Sun
Central South University of Forestry & Technology
Changsha
China

Weipeng Jing
Northeast Forestry University
Harbin
China

Huaguo Huang
Beijing Forestry University
Beijing
China

Ting Yun
Nanjing Forestry University
Nanjing
China

Meili Wang
Northwest A&F University
Xi'an
China

Editorial Office
MDPI AG
Grosspeteranlage 5
4052 Basel, Switzerland

This is a reprint of the Topic, published open access by the journals *Applied Sciences* (ISSN 2076-3417), *Forests* (ISSN 1999-4907), *Remote Sensing* (ISSN 2072-4292), *Sensors* (ISSN 1424-8220) and *ISPRS International Journal of Geo-Information* (ISSN 2220-9964), freely accessible at: https://www.mdpi.com/topics/58YHQZ7DBC.

For citation purposes, cite each article independently as indicated on the article page online and as indicated below:

Lastname, A.A.; Lastname, B.B. Article Title. *Journal Name* **Year**, *Volume Number*, Page Range.

Volume I
ISBN 978-3-7258-4721-1 (Hbk)
ISBN 978-3-7258-4722-8 (PDF)
https://doi.org/10.3390/books978-3-7258-4722-8

Set
ISBN 978-3-7258-4719-8 (Hbk)
ISBN 978-3-7258-4720-4 (PDF)

Contents

remote sensing

MDPI

Article

A Novel Scheme about Skeleton Optimization Designed for ISTTWN Algorithm

Jie Yang [1,2], Xiaorong Wen [1,2,*], Qiulai Wang [3], Jin-Sheng Ye [3], Yanli Zhang [4] and Yuan Sun [1,2]

1 Co-Innovation Center for the Sustainable Forestry in Southern China, Nanjing Forestry University, Nanjing 210037, China
2 College of Forestry, Nanjing Forestry University, Nanjing 210037, China
3 Guangdong Forestry Survey and Planning Institute, 338 Guangshanyilu, Guangzhou 510520, China
4 Arthur Temple College of Forestry and Agriculture, Stephen F. Austin State University, Nacogdoches, TX 75962, USA
* Correspondence: wxr9872@njfu.edu.cn

Abstract: The ISTTWN algorithm overcame the defect of separating the production process of skeleton points and skeleton lines in tree branch point cloud skeleton extraction and improved the accuracy of the extracted initial skeletons, but the skeletons need further optimization. In the existing skeleton optimization, it is difficult to see the stump adjustment, and most of the bifurcation optimization and skeleton smoothness adopt fitting. Based on the characteristics of the initial skeletons extracted by the ISTTWN algorithm, this research optimizes the skeleton from four aspects. An algorithm for the stump adjustment for reconstructing the stump based on the layer and hierarchical relationship and an algorithm for the bifurcation optimization based on the local branch point cloud and cosine correlation are proposed, and an existing pruning method and a skeleton smoothing method are used. The results show that the skeleton optimization method proposed or used in this research has a high computational efficiency in general and can ultimately retain the necessary skeleton lines. In a visual analysis, the optimized skeleton is obviously much more natural and more in line with the actual topology of trees. In the quantitative analysis, the completeness, accuracy and effectiveness reached 97.82%, 95.72% and 89.47%, respectively. In this study, in addition to the existing tree parameters extracted by the skeleton or generalized cylinder model, the generated skeleton is used to extract the branch attributes. The R^2 of the deflection angle of the branch tip, distance from branch tip and branch length are about 0.897, 0.986 and 0.988, respectively, which illustrates that their models are very good. This research can further expand the application of the skeleton.

Keywords: skeleton extraction; skeleton optimization; tree skeleton; tree point cloud

1. Introduction

As the shapes of different trees in nature are various, making and visualizing 3D tree models in the field of Computer Graphics is always a hot research topic and has been developed for at least half a century. This research has been similarly affecting forestry in recent years, attempting to find the possibility to replace the method of traditional forest investigation [1]. The way of combining an advanced surveying and mapping technique with novel information technology has gradually been widely used and is steadily developed to improve the visual effect and the reality of 3D tree models. In addition, in forestry, it can also be used to obtain more accurate existing parameters from measured single trees or the whole measured standing forest without a manual measurement, and some novel and meaningful parameters can be further proposed to describe the form not at the scene, know the actual situation and help make forecasts and plans [2,3].

LiDAR technology has been developed for one decade, and it uses lots of spatial points (these constitute the point cloud) to approximately describe the surface of objects in the measuring zone. If a LiDAR scanner is used to scan a tree and irrelevant parts in the tree point cloud are filtered out, a tree point cloud model, one type of tree model, will be presented. Some parameters that only rely on point information, such as tree heights, the DBH, the crown width and the clear bole height (CBH), can be obtained directly or indirectly by this kind of tree model [4]. Apart from that, many parameters cannot be obtained, such as branch length, branching angle and tree volume [5]. This is because this type of tree model is unorganized and has noise. In addition, it also has a disadvantage of huge storage. Therefore, it usually plays a role in the input or source data of other types of tree models. The generalized cylinder model is one of the commonly used types and it uses cylinders or truncated cones to fit the tree branch point cloud. The axes of these cylinders or truncated cones correspond to the lines on the tree skeleton. It can be asserted that the visual effect of one tree model and the accuracy of the subsequent extraction of the parameters highly depend on the effect of the skeleton extraction. After a branch model is produced, the models of the leaves will be virtually added by a certain algorithm [6]. With the model of trees, many studies not only extracting tree parameters can be carried out. For example, Huang et al. [7] tried to retrieve aerodynamic parameters in rubber tree forests by modeling trees with TLS point clouds.

The existing algorithms for extracting the skeleton from a tree branch point cloud can be divided into three major categories—algorithms based on Voxel Space, algorithms based on Point Clouds Contraction and algorithms based on a Geometrical Characteristic (hereinafter referred to as the VS, PCC and GC algorithms, respectively). Shi [8] pointed out that, in general, GC algorithms are next to PCC algorithms in computational efficiency but are the best or equally excellent among the three in dealing with small offshoots, the model topology and overall effects, as well as dealing with adjacent bifurcations and central deviation. In addition, he concluded that GC algorithms have the relatively best performance among them. In fact, there is still a frequently used algorithm named the Space Colonization Algorithm (SCA) [9] that does not belong to any of the categories. However, as it is hard to control the type of trunk bifurcations [10], the SCA is mainly used to simulate the small offshoots [11] through the leaf point cloud. Alternatively, it can be applied to extracting the tree skeleton with sparse points on the part of the branch from the tree point clouds obtained by ALS scanning [12]. As the Unmanned Aerial Vehicle (UAV) technique has been widely used with its low cost, high resolution and portability [13], there will be more research using ALS point clouds to extract tree skeletons and model trees [14]. Apart from the SCA, in general, the TLS point clouds are the data source of the mentioned algorithms for a skeleton extraction.

The reason why GC algorithms can be better in keeping the original appearance of trees is that the common principle of the GC algorithms is to cluster the branch point cloud by utilizing the similarity of the local features. According to the chosen feature or features, different GC algorithms can be divided into three categories—layering by the similarity of the shortest-path distance from each point to the root point (or layering by the shortest-path distances), such as Xu et al.'s [15]; layering by heights, such as Gao et al.'s [16]; and layering by spatial orientation and the radius of the branch (or layering by the tree branching geometry), such as You's [17] and the TreeQSM [18]. It should be noticed that algorithms that adopt layering by tree branching geometry integrate one of the other two categories (where the height's rising direction is not necessarily along the positive direction of the z-axis) and introduce extra information obtained from the tree branch point cloud to distinguish the points, which brings complexity for programming.

The initial outputs, produced by any algorithms for extracting tree skeletons from branch point clouds, exist inaccurately in many detailed positions and have a poor local visual effect. So, the procedure of skeleton optimization (or skeleton refinement) is imperative. It contains many aspects.

One, the breakpoint connection. Breakpoints make the skeleton unconnected, and the reconstructed tree model will form multiple parts. Xu et al. [15] used the existing trend of the skeleton and angle constraint to ensure the connection between the breakpoint and the main skeleton. Shi [8] pointed out that the Arterial Snake Model (ASM) proposed by Li et al. [19] can be used to deal with breakpoints by the fitting, growth merger of the arterial snakes. Zhou [20] proposed that a Cardinal Spline can be used to interpolate the vacancy caused by a breakpoint. Sun et al. [21] proposed an algorithm for using the change in adjacent angles formed by skeleton lines near the breakpoints to decide which skeleton points should connect to the breakpoint. Zhang et al. [22] defined a part of a regularizer for penalizing the positions of skeleton points in the different blocks belonging to the two joints, which is used to fix the breakpoints.

Two, the bifurcation optimization. In the initial skeleton, the lines at the bifurcations are always sharp and do not go through the center of the local point cloud. He et al. [23] pointed out that a bifurcation node, its parent node, its child nodes on all sub-branches and the tangent vectors formed by these nodes can construct a Hermite curve to fit and interpolate some transition points. Linvy et al. [24] used a Branch-Structure Graph (BSG) to differentiate different branches and the constructed local generalized cylinder models to compute an overlap rate between two cylinders in order to recognize which edge should be removed. He [25] proposed that after a reliable point found by the RANdom SAmple Consensus (RANSAC) is identified, its subsequent 16 points can use parabola fitting to estimate the suitable bifurcation point. Wang [26] adopted a rule to find a suitable connection where the angle between a parent branch and a sub-branch is no more than 45 degrees. Nouri et al. [27] reused and modified the idea of GC algorithms layering by axial heights to reconstruct local skeletons at bifurcations. Xu et al. [28] gave an algorithm for modifying the local skeleton at a bifurcation on the basis of the direction-first principle and combined with the idea of the shortest path. Fu et al. [29] adopted a way of refinement that, firstly, executed K-Means Clustering with several chosen initial cluster centers at a bifurcation, and the corresponding local point set as inputs and clusters can be obtained; then, similar to the approximation method, the local skeleton points and lines can be reconstructed by repeatedly excluding recognizable points on branches from a cluster and recomputing the centroid of the rest points by Cylindrical Prior Constraint (CPC) optimization.

Three, the skeleton smoothness. The initial skeleton lines in the part of each branch are unnatural and have excessive changes in their direction. Mei et al. [30] applied a Laplacian-based contraction which belongs to the PCC algorithms into the skeleton smoothness by modifying the parameter of controlling the force of a contraction in the matrix equation in order to enable it to adjust the smoothness. Li et al. [31] provided a simple method of directly averaging a continuous three points into the new middle one. Chaundhury et al. [32] pointed out that a B-spline curve can be used to resample each branch part of the tree skeleton, and they proposed Stochastic Modeling—a modeling technology for controlling the termination of skeleton smoothness according to the Degree Of Difference (DOD) between each part of the tree point cloud and the corresponding skeleton. Xu et al. [33] used a method by introducing 3D coordination differences which are used instead of directly averaging three points and a coefficient to balance the smoothness result, and after several rounds of iteration, the skeleton can be obviously more natural in the visual.

In addition, the pruning twigs, filtering isolated skeleton points and merging too close points are also frequently used according to the requirement of a study [34]. According to the feature of an algorithm, the skeleton optimization scheme can be selected from above, or additional methods can be designed. For example, Li et al. [35] designed an overall translation for the skeletons produced by a similar implement as AdTree.

We proposed an algorithm to extract the branch skeleton from the tree point cloud in the way of an Incomplete Simulation of Tree Transporting Water and Nutrients, which was named as the ISTTWN algorithm in short—a new GC algorithm layering by the shortest-path distances to simulate the transporting process of water and nutrients in a tree [36]. Compared to the previous various algorithms for extracting the tree skeleton from the branch point cloud, this method overcomes the defect that the process of producing skeleton points is out of sync with the process of connecting skeleton lines. In addition, when producing a new skeleton point and forming the skeleton line related to it at the same time, the result showed that the produced initial skeletons have a more accurate topology and cost a relatively short time. In this research, we will give some methods of skeleton optimization suitable for the ISTTWN algorithm. All the methods follow the common principles that any skeleton points obtained by fitting, which does not rely on point cloud information, will not be produced, and that the existing stable skeleton points cloud will not be removed. Based on this principle, our contributions in this research are as follows:

(1) We proposed a new aspect of skeleton optimization, which was named as the stump adjustment, and designed an algorithm for reconstructing the stump based on the layer and hierarchical relationship.
(2) We proposed an algorithm for bifurcation optimization based on the local branch point cloud and cosine correlation.
(3) After additionally adopting pruning twigs and Xu et al.'s [33] skeleton smoothness, we verified the differences between unoptimized skeletons and optimized skeletons and tested the optimized skeletons to extract some branch attributes and evaluate the accuracy.

2. Materials and Methods

Figure 1 is the flow chart about all the steps to obtain an optimized skeleton from a tree branch point cloud.

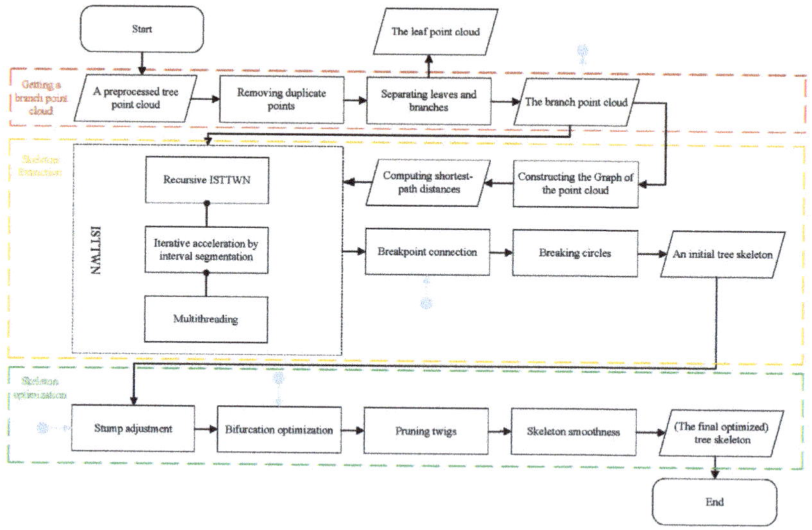

Figure 1. The overall workflow.

2.1. Source and Preprocessing of Experimental Data

2.1.1. Data Source

We used three tree point clouds. Two of them were randomly chosen from the public tree point cloud dataset provided by Seidel et al.'s [37] research, including a European Ash (*Fraxinus excelsior*) and a European Oak (*Quercus robur* L.). In addition, the remaining one was randomly chosen from the preprocessed point clouds of sample plots of American black poplar (*Populus deltoides*) scanned by RIEGL VZ-400i Terrestrial LiDAR Scanner (TLS) with scanning mode Panorama40. The three tree point clouds are shown in Figure 2. All point clouds executed the removal of duplicate points.

(a) European ash (b) European oak (c) American black poplar

Figure 2. Tree point clouds.

2.1.2. Separating Leaves and Branches

According to optical properties of LiDAR and considering that TLS is usually placed on a tripod on the ground, the point cloud of a tall tree scanned by TLS has high resolution at the trunk and low resolution on the crown. This makes it difficult to form points at petioles. Therefore, the parts of leaves can relatively be clearly differentiated with the parts of branches. We adopted a method similar to region growing by choosing an initial point at the trunk, adding it into the set that represents branches and continuing to expand the set with all unadded neighbor points of each point in it until there are no more points added in. At this time, points still not added into the set are all thought of as points on leaves. This method is implemented by a kd-tree and its radius search to complete separation [38], where radius parameter $r = 10\rho$ (ρ is the average resolution) and there must be no duplicate points in the input tree point cloud. This method is suitable for any kind of tree point clouds as it only uses spatial coordinate information. Certainly, in order to improve accuracy, it is suggested that trees should be best scanned in leafless seasons. Figure 3 shows three tree point clouds' results of separating leaves and branches. In this research, we only used the branch point clouds.

(a) European ash (b) European oak (c) American black poplar

Figure 3. Separation of leaves and branches.

2.1.3. Computing the Shortest-Path Distances

Graph theory algorithms can be used to deal with point cloud when points in a point cloud are regarded as vertices and lines between each point and its neighbor points within a fixed certain range are regarded as edges and the length of each line is regarded as the edge weight. As points in the branch point cloud are all on branches, such a range should be chosen appropriately to ensure the graph is connected. In this research, we firstly computed an initial search radius calculated by equation

$$\rho_{\max} = \max_{\boldsymbol{p} \in P} \min_{\boldsymbol{q} \in P - \{\boldsymbol{p}\}} \|\boldsymbol{p} - \boldsymbol{q}\|_2, \quad (1)$$

where $\boldsymbol{p}$ and $\boldsymbol{q}$ represent certain spatial points, $\rho_{\max}$ is the maximum of all spacing between the points and their 1-nearest neighbor points in a point cloud. Such initial search radius is taken as cluster tolerance to cluster the branch point cloud (in the form of set) P using Euclidean Cluster Extraction (ECE). Then, we constructed a new undirected complete graph

G with cluster indices as vertices, minimum cluster spacing between any two clusters as edge weights and computed Minimum Spanning Tree (MST) T of graph G. Finally, we set such range as the longest edge weight in T. This method can save memory for storing huge amounts of edge information to the maximum extent.

If such a graph could be regarded as an undirected connected graph, we set the shortest-path distance between a source vertex and all other vertices would be easily obtained by using algorithms solving the Single-Source Shortest Path (SSSP) problem. Point corresponding to the source vertex is chosen as the lowest one at the stump. The algorithm we used was the queue optimized Bellman–Ford algorithm.

2.2. Skeleton Extraction

The principle of ISTTWN algorithm can be described as a process that by using the shortest-path distances and the specified distance that water and nutrients are transported each turn (or step length), one part of points to produce skeleton points and the rest points to continue transporting can be decided. In addition, a clustering algorithm (here, use ECE with $eps = \rho_{\max}, MinPts = 1$) can help distinguish whether there are different offshoots or not, then the produced skeleton points can connect with the skeleton point in the same offshoot in previous turn.

It is obvious that the algorithm is recursive. As recursion is very slow, we tried some methods to improve it, including using iterative acceleration by interval segmentation and using multithreading. These improvements indeed decrease the cost of time and memory usage. However, the cluster object in improved ISTTWN algorithm is different from that in original ISTTWN algorithm, as the former is all the unexplored points in the whole branch point cloud, while the latter is just those in a local part. For improved ISTTWN algorithm, on the positive side, this difference can solve the issue that two too close branches are erroneously interpreted into one branch in a certain range. The range is a sphere which radius is equal to the product of step length multiplied by the number of steps in each interval, in other words, interval length. We wished skeletons could be as detailed as possible, so we let step length equal to the minimum value $\rho_{\max}$; therefore, in this research, we would give the method of stump adjustment, so we let the number of steps in each interval equal to the minimum value 2. On the negative side, it produces some breakpoints. These breakpoints exist in alternate positions between intervals and influence a small range of topology, so we gave a simple way of breakpoint connection as follows, where a bin (or a branch segment) is the point set of a local branch near a skeleton point, and initially, each skeleton point is centroid of a bin (it should be noted that the plural of "bin" is bin-s, and a bin means the point set consisting of bin-s of the same layer according to definitions given by Xu et al. [15]).

1. Because skeleton lines are stored by ordered pairs, they can be regarded as directed edges in graph theory. Therefore, the indegree and outdegree of vertex corresponding to each skeleton point can be computed. Apart from the same vertex as the source vertex, any vertices where indegree is 0 are vertices corresponding to breakpoints.
2. For each breakpoint, use a kd-tree and k-Nearest-Neighbors (k-NN) search to find 3 neighbor points for which corresponding vertices are not on the tree, with the root being the vertex corresponding to the breakpoint. Compute the number of intersections between bin of breakpoints and bin of each neighbor point. We regarded that a neighbor point that the bin of which has the maximum number of intersections and the number is not equal to 0 should be the same point as the breakpoint. Then, the two points and their corresponding bin-s are required to be merged. If all numbers are equal to 0, directly connect the nearest one to the breakpoint.

This method of breakpoint connection is not optimized, but the following bifurcation optimization and skeleton smoothness can further optimize the skeleton as well. At last, although there are usually no circles in the skeleton, the process of breaking circles is needed in case of potential bad situations. Due to particularity of the produced skeleton, it can be simply described that in directed graph formed by the skeleton, if a vertex where indegree is more than 1, only one edge ending at it should be reserved, and such an edge should have the most comprehensive information branch information.

2.3. Skeleton Optimization

2.3.1. Stump Adjustment

In principle, if a branch is near horizontal and many points on its surface are in the same section of heights, the skeleton points of the branch produced by GC algorithms layering by heights will abruptly shift and not distribute according to the growth direction any more. In addition, there is hardly any tree in which all the segments of the branches are almost straight up. Unless there is reconstruction, the results caused by the defect can no longer be further refined even after optimization. Compared with layering by heights, layering by shortest-path distances can avoid this defect and its result has lower central deviation in general. However, as the shortest-path graph formed by the branch point is in the shape that spreads from the source vertex, the closer the distance is, the more radial the shape can be. Because the source point is on the surface of trunk, the skeleton point closer to the source point is more offset from the center of the trunk. In addition, when the step length is smaller, the center deviation here is more obvious. Point cloud at the stump could be complete or incomplete, not only affected by the process of filtering the ground points but also by undergrowth vegetation during scanning. If the point cloud at the stump is incomplete, the stump would not seem to be cut from a plane visually. When the step length is small, the interval of the branch point cloud near the stump might be clustered erroneously into more than one cluster and subsequently some unwanted branches would probably be produced. That is to say, the skeleton has terrible local distortion at the stump.

You [17] proposed a method to process the branch point cloud at the stump in advance so that a horizontal section can be found, and points below the section in the branch point cloud should be removed by repeatedly calculating the variance of lines connecting the centroids of adjacent horizontal layers below a certain height. Essentially, the method does not solve the problem of center deviation at the stump actually. Other than that, we had not found any other attempts to solve this problem.

We proposed a method to handle the issue without making any changes to the input branch point cloud. This method is processed after the initial skeleton is produced, so it belongs to skeleton optimization, and we named the process of reconstructing the local skeleton at the stump as stump adjustment. Reconstruction means redoing skeleton extraction. It contains three problems: which algorithm for skeleton extraction should be used, which part of the branch point cloud should be chosen as an input and how to replace the original corresponding part with the reconstruction result.

(1) About the chosen algorithm for skeleton extraction.

The way of layering by heights can work relatively normally when producing the skeleton close to the ground because there are not too many bifurcations near the ground in tall trees compared to their crowns. If one input of ISTTWN algorithm is changed from shortest-path distances into heights, the program will run but keep making the wrong progress, as Figure 4 shows. Xu et al.'s [15] algorithm, which used MST algorithm to obtain skeleton lines, can also use the input of heights (similar to Gao's [16] algorithm) but is easy to create an incorrect topology as MST only guarantees the minimum sum of weights rather than the match between topology and the point cloud. Figure 5 shows a wrong result if Xu et al.'s [15] algorithm is directly used to reconstruct a stump. It is illustrated that algorithms should consider the existence of some branches with abnormal growth direction and of some tree species with ground branching characteristics.

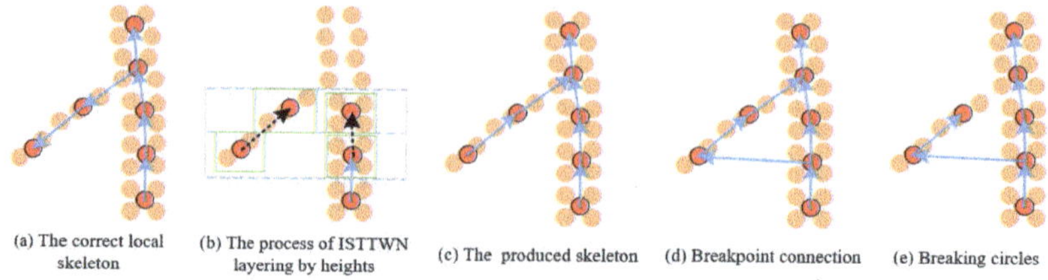

Figure 4. The problem in the skeleton produced by ISTTWN algorithm layering by heights.

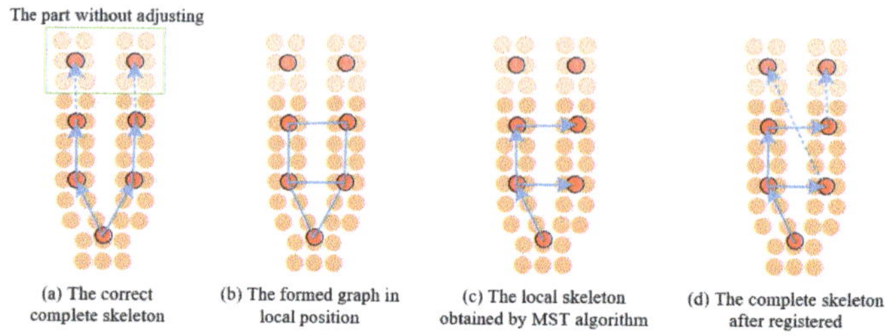

Figure 5. The problem in the skeleton produced by Xu et al.'s [15] algorithm layering by heights.

We modified the creation of the graph in Xu et al.'s [15] algorithm according to hierarchical relationship, which is similar to that which TreeQSM uses but more strict than it as we also used the bin information to prevent the unreasonable connection among adjacent layers, and this can satisfy the requirements, as is shown in Algorithm 1.

Algorithm 1 Graph formed according to hierarchical relationship

Input: Skeleton point sequence $\{q_n\}_{n=1}^{|Q|}$, where $|Q|$ is the size of skeleton point set; Bin sequence $\{B_n\}_{n=1}^{|Q|}$ corresponding to each skeleton point in turn; Layer number sequence $\{L_n\}_{n=1}^{|Q|}$ corresponding to each skeleton point in turn;
Output: Undirected connected graph G';

$eps := r_{\min} + \varepsilon$ ▷ $r_{\min}$ is the smallest search radius that can make the graph formed by point cloud connected, and ε is infinitesimal as the feasible region of radius search generally does not calculate the boundary
Initialize G', where $G'.V = \{v_1, v_2, \cdots, v_{|Q|}\}$
for $i := 1 \to |Q|$ **do**
  Find all vertices that the layer number of corresponding skeleton points is $L_i - 1$ and constitute set U
  foreach vertex $v_j \in U$ **do**
    if minimum cluster interval between B_i and B_j is less than eps **then**
      $G'.E \leftarrow G'.E \cup \{(v_j, v_i)\}$
    end if
  end for
end for

There are relatively fewer circles in the created graph, and at this time, the accuracy of MST algorithm is obviously better. Because the graph is undirected, edges in the graph need to be oriented to obtain the skeleton.

(2) About the chosen part of the branch point cloud.

In nature, some trees have prop roots and a few of their stumps are especially far from the ground. In order to cater to these trees, we chose a maximum datum height near breast height (1.3 m). Certainly there are some small trees, and the maximum datum height is found near 1.3 m/5 m * tree height when tree height is less than 5 m. In addition, the datum maximum height should be an integral multiple of the set layer height (when layering by heights, step length in the algorithm for skeleton extraction is called as layer height) in avoidance of the uneven height of each layer. The point cloud below this datum height is used as the input to reconstruct the local skeleton.

(3) About the replacement method.

Initially, the current datum height is the maximum datum height. When the current datum height is specified, the direction of local part of branches should be identified. However, such as willow trees, some branches bend over to the earth and their point cloud should be filtered out. In the rest trunk point cloud, for ensuring the reconstructed part can find a common part of the initial skeleton to be connected, the number of bin within the range of adding a layer height to the current datum height should be the same as the number of bin within the range of the current datum height minus a layer height. In addition, a bin in the lower layer should connect to a bin in the higher layer without repetition. The direction of local skeleton lines across the current datum height should also be from low to high. If not making all these conditions tenable, the current datum height will be itself minus a layer height and be retried until it is equal to the lowest height of point cloud plus layer height. Otherwise, reconstruction will be executed. Remove all the skeleton points and lines related to them from the initial skeleton lines, and piece it to the reconstructed local skeleton with the disconnected positions corresponding to each other.

It is obvious that the algorithm for stump adjustment we proposed reconstructs the stump based on layer and hierarchical relationship. It is worth noting that there are four major reasons for reconstruction failure, and stump adjustment can work normally by avoiding them:

(1) The layer height is too large.
(2) The tree species could be shrub.
(3) The point cloud quality is bad.
(4) Ground normalization has not been performed.

2.3.2. Bifurcation Optimization

We proposed an algorithm for bifurcation optimization based on local branch point cloud and cosine correlation. The principle is based on the characteristics of local branch point cloud covariance matrix. Let c_i be the centroid of bin S_i and construct the covariance matrix

$$M_i = \frac{1}{|S_i|} \sum_{p \in S_i} (p - c_i)(p - c_i)^{\mathrm{T}}. \tag{2}$$

The eigenvector corresponding to the maximum eigenvalue of matrix M_i describes the growth direction of the local branch [39]. It should be noted that such eigenvector has ambiguity, in other words, the growth direction may be along the positive or negative direction of the eigenvector.

Let the growth directions of a branch and one of its offshoots at the bifurcation and actual direction of local skeleton lines be v, v', τ, τ', respectively. The directions of skeleton lines are completely along the growth directions of the tree if and only if

$$v = k\tau, v' = k'\tau', \tag{3}$$

where $k \neq 0, k' \neq 0$. That is to say, angle $\langle v, \tau \rangle = 0^\circ$ or 180° and angle $\langle v', \tau' \rangle = 0^\circ$ or 180°. Using the formula of the cosine value of the angle between two vectors, another form of conclusion can be derived as follows:

$$\left| \frac{\tau v}{\|\tau\|_2 \|v\|_2} \cdot \frac{\tau' v'}{\|\tau'\|_2 \|v'\|_2} \right| = 1. \tag{4}$$

However, this formula is extremely difficult to hold. So, bifurcation optimization is to find a relatively best connection, in other words, to solve optimization problem

$$\underset{\tau,\tau'}{\arg\max}\left|\frac{\tau v}{\|\tau\|_2\|v\|_2}\cdot\frac{\tau' v'}{\|\tau'\|_2\|v'\|_2}\right|. \tag{5}$$

This is a method based on cosine similarity.

When the bifurcation node, its parent node and one of its son nodes are, respectively, $B(x,y,z), A(x_1,y_1,z_1), C(x_2,y_2,z_2)$, the target is to find a B to maximize the cosine similarity. In avoidance of abrupt line trend, B should be in a feasible region D, and the problem can be changed into finding the solution that makes the function

$$f(x,y,z) = \frac{[v_1(x-x_1)+v_2(y-y_1)+v_3(z-z_1)]^2[v'_1(x_2-x)+v'_2(y_2-y)+v'_3(z_2-z)]^2}{(v_1^2+v_2^2+v_3^2)(v'^2_1+v'^2_2+v'^2_3)\left[(x-x_1)^2+(y-y_1)^2+(z-z_1)^2\right]\left[(x_2-x)^2+(y_2-y)^2+(z_2-z)^2\right]} \tag{6}$$

reach the maximum value in the feasible region $(x,y,z) \in D$, where the absolute value is eliminated by means of square. However, this problem is difficult to solve because the feasible region is not always in the same shape or can be expressed in the same form, and when using Lagrange multiplier approach, it is very difficult to invert the parameters by making all derivative functions obtained after the derivation of different variables equal to 0. Although the approximate solution of the equation can be found by sampling the feasible region in a certain way, the position of B is not always fixed. Sometimes, when the parent node of B, its grandparent node, ..., even a point on lines ending at these points is B, a better solution can be found. With the change in B, if the shape of the feasible region maintains a single form of expression, it is likely to produce an overfitting phenomenon, that is, to find a point that maximizes the value of the equation but has abrupt connections.

In ISTTWN algorithm, a skeleton point is the centroid of a bin. The number of bin-s ECE clusters points into is decided by whether the points can be clearly divided. This feature illustrates that the line between a bifurcation node and its parent node basically goes through the center of the local branch and the production of the bifurcation point is always late. So, we adopted a method to find the relatively most suitable bifurcation point among test points on the broken lines of the local parent branch instead of reconstructing these lines. Figure 6 shows the complete process.

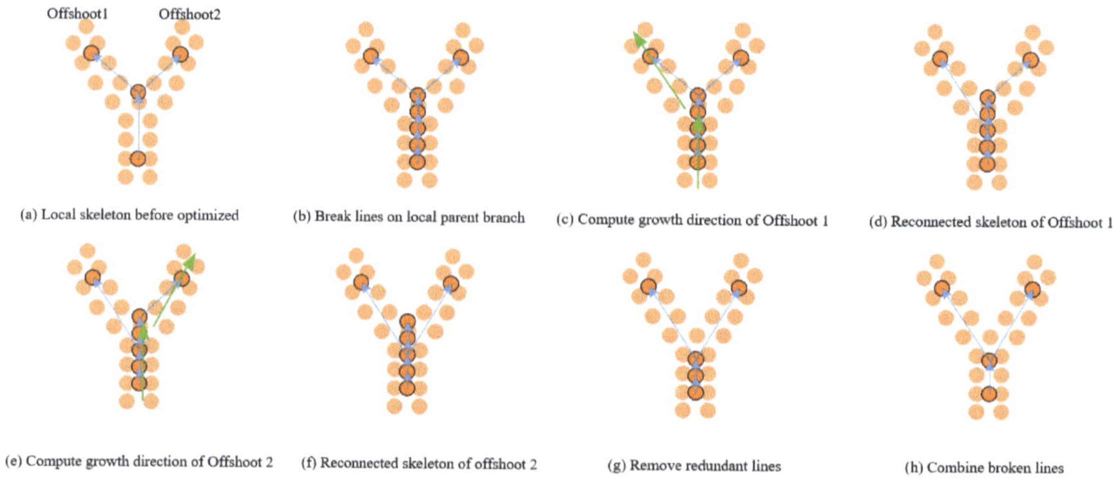

Figure 6. Bifurcation optimization.

Here are two problems that need to be further discussed.

(1) How are bin-s of three continuous skeleton points divided into two bin-s and then substituted into the equation?

Similarly, let the three skeleton points be A, B, C. Construct plane Π whose normal vector is $\overrightarrow{AC}$ and passes through point B. Use plane Π to divide the point cloud constituted by bin-s of A, B, C into two bin-s.

(2) What is the bin of a broken point?

The bin of a broken point can derive from bin-s of points on both ends of the original line, and the range should be within the range of plus or minus one step length along the direction of the original skeleton line in avoidance of unceasingly accumulating bin-s when encountering too many continuous bifurcations.

2.3.3. Pruning Twigs

As *MinPts* is set to 1, there are too many small twigs in the skeleton affected by noise points. The following two types of branches are regarded as twigs to be pruned:

(1) If the end point is located in a branch at the top of the tree and the number of its bin is less than the newly specified minimum cluster size, remove the skeleton line connected with the current end point and remove the end point and its bin;

(2) If the skeleton lines that represent one branch at the end of the tree top are less than the specified minimum number of skeleton lines of the branch, the branch shall be removed.

If a branch meets the conditions of (1) and (2) at the same time, it shall be processed according to (2). In this research, the minimum cluster size and the minimum number of skeleton lines of the branch are 4 and 2, respectively.

2.3.4. Skeleton Smoothness

We adopted the method proposed by Xu et al. [33] to smooth the skeleton. For each node $p_{(2)}$ not at both ends of a branch, find its parent node $p_{(1)}$ and its unique son node $p_{(3)}$, and after the k-th iteration,

$$p_{(2)}^{(k)} = \frac{\lambda}{2} p_{(1)}^{(k-1)} + (1-\lambda) p_{(2)}^{(k-1)} + \frac{\lambda}{2} p_{(3)}^{(k-1)}, \tag{7}$$

where the nodes are points in the initial skeleton when $k = 0$, λ is a coefficient to balance the smoothness result. In this research, the maximum k was set as 5 and λ was set as 0.1.

3. Results

3.1. Computational Performance

We tested each part when each experimental material was used as an input to obtain the final optimized skeleton. The basic information about the three tree point clouds, parameters or outputs is shown in Table 1. The running time of each part and the change in the skeleton lines are shown in Figure 7. The number of the identified branches is shown in Table 2.

Table 1. Basic information.

Tree Species	Size of Original Tree Point Cloud	Number of Duplicate Points	ρ	Size of Branch/Leaf Point Cloud	ρ_{max} of Branch Point Cloud	Number of Edges in Constructed Undirected Graph
European ash	97,895	0	0.016659 m	94,021/3874	0.166197 m	14,582,170
European oak	836,636	4	0.005627 m	819,556/17,076	0.056221 m	164,547,719
American black poplar	565,698	288,680	0.011883 m	274,688/2330	0.118744 m	70,006,006

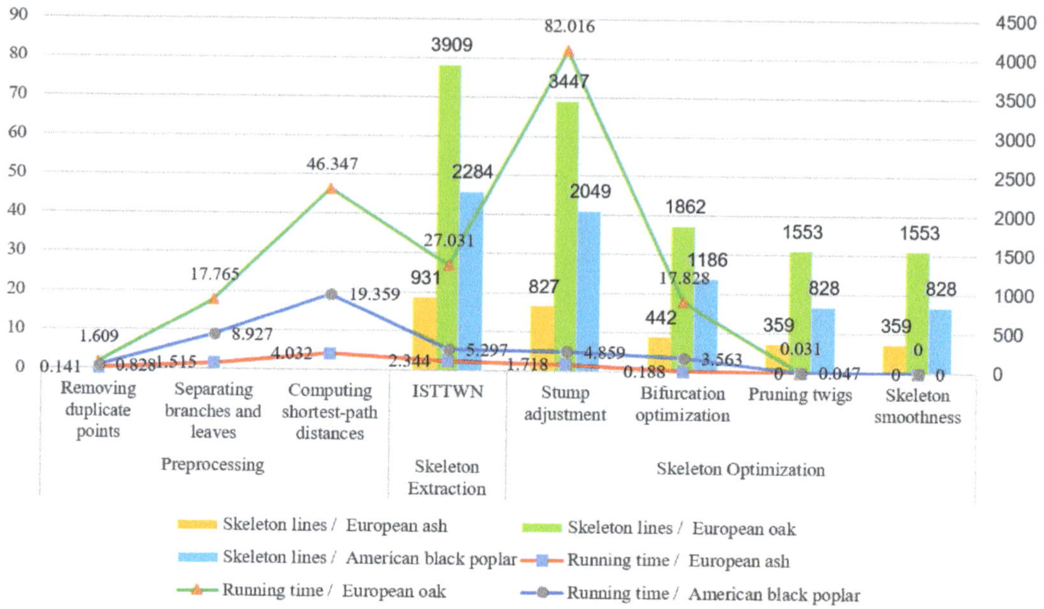

Figure 7. The running time of each part and the change in skeleton lines.

Table 2. The change in the identified branches

Tree Species	ISTTWN	Stump Adjustment	Bifurcation Optimization	Pruning Twigs	Skeleton Smoothness
European ash	244	138	183	67	67
European oak	1036	518	674	257	257
American black poplar	705	508	731	268	268

The result shows that the processing time is positively proportional to the size of the point cloud in general. As the branch point cloud of the European oak has a relatively sky-high cloud density, the quantity of the points near the earth is quite a lot and it burdens the process of the stump adjustment, which leads to a sudden increase in the time requirement. The time in extracting the other two point clouds is almost the same as the following optimizing process. In addition, the reason why the skeleton lines decrease significantly after the bifurcation optimization is that the algorithm designed for combining broken lines in Figure 6h also combines two normal skeleton lines with a too small included angle in order to retain the necessary topology. The great changes in the number of branches after the stump adjustment will be explained after the visualization in the following section.

3.2. Visual Analysis

The skeletons before optimization and after optimization are shown in Figure 8. Three details were selected on the skeleton of each experimental material. In addition, the skeleton and point cloud are superimposed at these detailed positions to visually check the matching degree between them.

It can be seen from Figure 8 that before the optimization, the distortion of the initial skeletons at the stump is large. The center deviation is almost unavoidable when the skeletons are fine enough, and sometimes unreasonable branches exist, which causes severe topological errors. After the optimization, the unreasonable branches are eliminated successfully, and the corresponding skeleton points are judged as to whether they are removed or retained according to the connection relation. However, the center deviation seems to decrease more or less. It should be illustrated that there is a certain inclination in the point cloud of the European oak; therefore, the result after the stump adjustment is relatively worse than the others visually, while such a result indeed reaches the best in the vertical direction. However, like the point cloud of the American black poplar, the points at the trunk with a near vertical distribution and completely no group part can be processed very well and have a good visual effect.

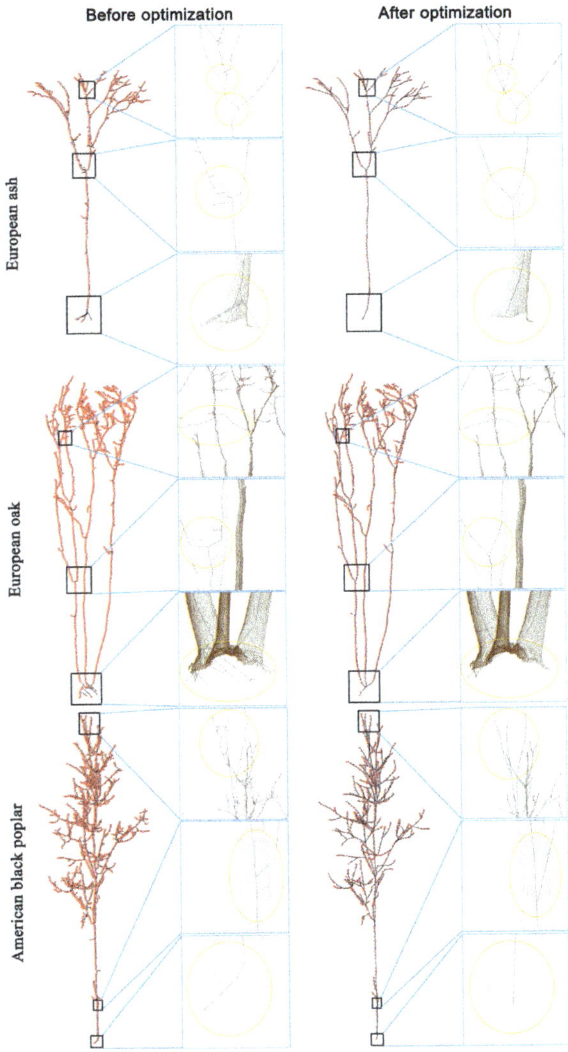

Figure 8. Comparison of skeletons before optimization and after optimization.

At bifurcations, although skeletons before the optimization are corresponding to the approximate growth direction, the lines do not go through the center of the bin-s, which bring a stiff felling. The optimized skeletons in the same place are obviously close to the center.

The process of pruning twigs and skeleton smoothness can make skeletons tidier, more beautiful and more natural on the condition that the skeleton lines match the bin-s.

In all, after the optimization in the four aspects we proposed or used, the visual effect of the skeleton is improved, and the errors in the initial skeletons are corrected, which means that the usability of the skeleton is enhanced.

3.3. Application

As the process of obtaining and modifying skeletons follows a basic rule that the fitting is completely unused, these skeletons can be regarded as the true portrayal of the topological structure of trees. Therefore, some parameters about branches of a tree can be extracted. In this research, we computed the branching angle [40] (or bifurcation angle [41], BA), the deflection angle of the branch tip [42] θ which relates to the size of the branch angle [43] Θ formed with the vertical direction of the trunk, the distance from the branch tip (or Branch Chord Length [44], BCL) and the branch length (BL) of each first-order branch. Figure 9 gives intuitive definitions of the relevant concepts.

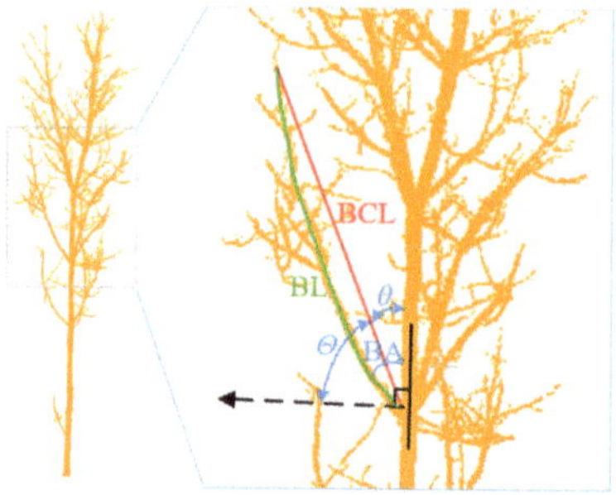

Figure 9. Some attributes of a branch.

We used evaluation indexes, including the Root Mean Squared Error $RMSE$, Mean Absolute Error MAE, Mean Absolute Percentage Error $MAPE$ and coefficient of determination (R-squared) R^2, to evaluate the accuracy. The following is the definition formula of each index:

$$RMSE = \sqrt{\frac{1}{n}\sum_{i=1}^{n}(x_i - \hat{x}_i)^2}, \tag{8}$$

$$MAE = \frac{1}{n}\sum_{i=1}^{n}|x_i - \hat{x}_i|, \tag{9}$$

$$MAPE = \frac{100\%}{n}\sum_{i=1}^{n}\left|\frac{x_i - \hat{x}_i}{x_i}\right|, \tag{10}$$

$$R^2 = 1 - \frac{\sum_{i=1}^{n}(x_i - \hat{x}_i)^2}{\sum_{i=1}^{n}(x_i - \bar{x})^2}, \tag{11}$$

where x_i is a real measured certain attribute value of the i-th branch, $\hat{x}_i$ is a certain attribute value of the i-th branch obtained by the skeleton and $\bar{x}$ is the mean value of a certain attribute value of all the branches.

The scatter plot of each branch attribute is constructed with the real values as the x-axis and the corresponding values obtained from the skeleton as the y-axis, as shown in Figure 10.

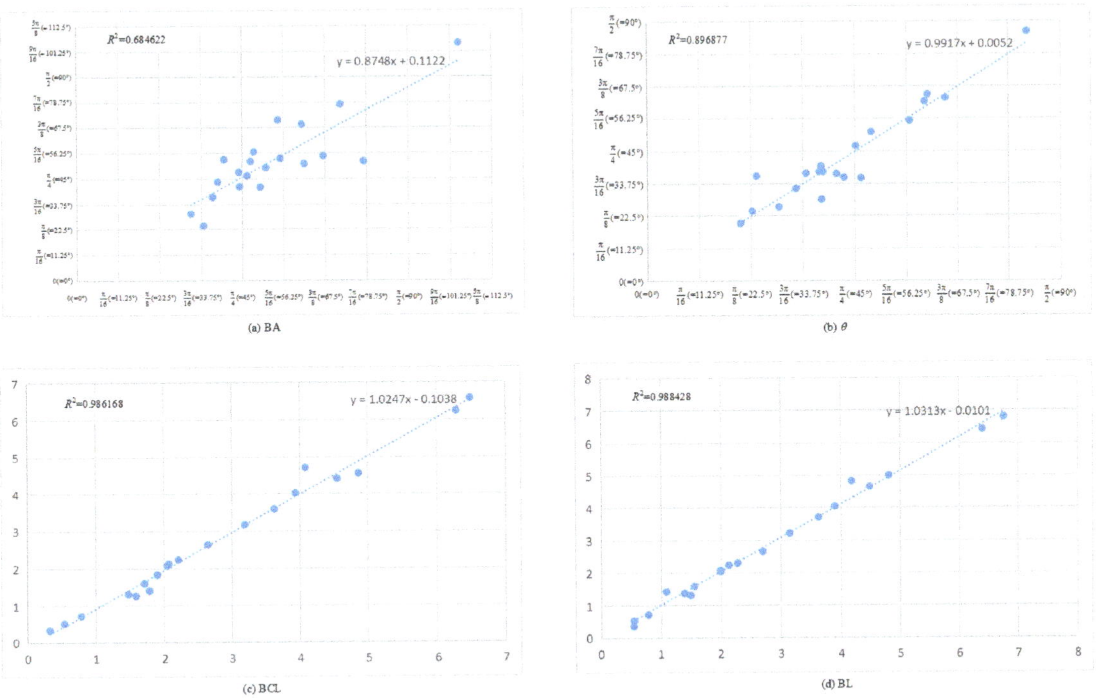

Figure 10. The scatter plot of each branch attribute.

Table 3. Accuracy of attribute values of branches.

Evaluation Indexes	BA	θ	BCL	BL
RMSE	0.164641 (9.43323°)	0.086456 (4.95359°)	0.203494 m	0.192263 m
MAE	0.124162 (7.11395°)	0.061605 (3.52970°)	0.129652 m	0.125959 m
MAPE	13.513%	9.4927%	5.6421%	6.9631%
R^2	0.684622	0.896877	0.986168	0.988428

Table 3 shows the accuracy of the attribute values of the branches. Among them, the BA and θ use the angle system, which are converted to the radian system before calculating the evaluation indexes.

The R^2 of the BA is relatively the lowest, and its *MAE* is about 7°. This is because measuring the BA manually is measuring the inner edge of two branches with a protractor, while the BA extracted from the skeleton uses almost the centerlines of the local branches. In addition, the size of the protractor affects the accuracy, while the line lengths of the skeleton similarly are affected. Therefore, the BA itself is relatively not a strictly defined quantity. As we used the parameter that makes the skeleton fine enough, the extracted results would be generally more accurate if the BA could be considered as the included angle of the centerlines.

The R^2 of the others illustrates that these extracted quantities are basically corresponding to the measured *MAPE*s, indicating that the models are very good. However, it should be noticed that the skeleton of an offshoot always includes a short part on its parent offshoot, and points at treetops are sporadic, which makes the skeleton points at treetops probably deviate the distance up to one step length. These potential errors propagate and cause the $RMSE$ and MAE of the BL to be beyond one step length, and affect the θ and BCL. So, as the skeleton is good at expressing distances and angles, these quantities can be extracted well, but they still have room for further improvement.

4. Discussion

4.1. Quantitative Analysis and Comparison

We used the three matrices proposed by Xu et al. [33] to evaluate the skeletons themselves quantitatively, including the completeness A_p, correctness A_r and effectiveness A_e. A_p is equal to the proportion that the union of the local branch point cloud where all the skeleton points are located covers the whole branch point cloud; A_r is equal to the proportion of skeleton points inside the space formed by the local branch point cloud; and A_e is equal to the proportion of skeleton points very close to the center of the local branch point cloud. It should be noted that the local branch point cloud where the skeleton points are located no longer uses the foregoing bin-s but is obtained from the surrounding sphere again. A_p measures how much the skeleton represents the branch point cloud. A_r measures whether the skeleton conforms to the general trend of the branches. And A_e measures whether the skeleton nearly passes through the center of the branch on the basis of A_r. The average results of the three values obtained by the selected trees are compared with the average results obtained by theirs. The results are shown in Table 4.

Table 4. Quantitative analysis.

Matrix	AdTree [45]	Xu et al. [33]	This Research (Before Optimization)	This Research (After Optimization)
Average A_p	81.10%	86.99%	97.84%	97.82%
Average A_r	99.21%	99.49%	95.20%	95.72%
Average A_e	79.26%	36.41%	86.22%	89.47%

It can be seen that due to the good feature that the ISTTWN algorithm makes full use of the branch point cloud, the requirement of completeness and effectiveness can be guaranteed in the initial skeletons. However, before the skeleton optimization, the skeleton bifurcates with hysteresis, so this value is slightly low. After the optimization, the skeleton lines decrease obviously, and after the combination, part of the skeleton lines become longer. So, it is normal for A_p to decrease slightly. A higher A_p means that our method is more complete in the expression of the canopy details. However, due to the sparse canopy points, it is easy to cause the wrong identification and classification. A_r will be affected by this factor, resulting in a situation slightly inferior to the other two. When the total of the skeleton points as the denominator decreases, the augment of A_r and A_e reveals that the skeleton is going to be finer. Therefore, the results of both the visual analysis and quantitative analysis show that the skeleton optimization proposed in this study is indeed effective and necessary, and it can improve the results of the ISTTWN algorithm.

4.2. Prospect

In terms of data source, wind disturbance during scanning and severe occlusion of the obtained point cloud will cause errors in tree reconstruction [46]. The corresponding processing technology needs further development.

There are some possible improvements in the complete process of skeletonization. Firstly, a better and highly compatible algorithm for separating the branches and leaves should be developed. Secondly, a dynamic method to construct the shortest-path distances is required as currently the paths constructed by a fixed search radius at different density positions are not always reasonable. Thirdly, as the previous research described, a more efficient allocation method of multithreading can further accelerate the execution. Lastly, research from the aspect of various skeleton optimizations allows for endless topics.

There was some research about more tree parameters extracted from skeletons or a subsequent generalized cylinder model, which are meaningful to forestry, for example, the inclination angle IA [47], azimuth [43] and above-ground biomass [48]. Including the parameters we discussed before, these concepts need to be more clearly defined mathematically and universally acknowledged when skeleton extraction is used. In addition, limited to the current extraction of the main trunk skeleton only referring to angles [49], the results of the branch display at all levels and the parameter extraction of any trees are not available with complete automation and a more suitable extraction method should be further explored.

5. Conclusions

In this research, we proposed and used several methods of skeleton optimization based on the results of the ISTTWN algorithm and the principle of not using fitting. Specifically, it includes the following aspects.

(1) Different from the existing aspects of skeleton optimization, a stump adjustment is proposed according to the feature at the stump of the tree skeletons layering by the shortest-path distances. An algorithm for reconstructing the stump based on the layer and hierarchical relationship is designed correspondingly.
(2) From the aspect of bifurcation optimization, an algorithm based on the local branch point cloud and cosine correlation is proposed.
(3) Pruning twigs and Xu et al.'s [33] skeleton smoothness are adopted in the process of optimization.

With this optimization scheme, the skeletons can be neat, natural and fit with branch point clouds in the visualization. Optimized skeletons are used to extract some branch attributes and evaluate the accuracy. Relatively, the deflection angle of the branch tip, distance from branch tip and branch length can be accurately extracted with strong correlations ($R^2 = 0.897$, $R^2 = 0.986$ and $R^2 = 0.988$, respectively). The availability of the skeletons has been improved, which reveals that this research can help further explore the application scope of tree skeletons.

Author Contributions: Conceptualization, J.Y.; methodology, J.Y. and X.W.; software, J.Y.; formal analysis, J.Y.; data curation, J.Y., Q.W., J.-S.Y. and Y.S.; writing—original draft preparation, J.Y.; writing—review and editing, J.Y., X.W. and Y.Z.; visualization, J.Y.; supervision, X.W. and Y.Z. All authors have read and agreed to the published version of the manuscript.

Funding: This research was funded by the Forestry Innovation Foundation of Guangdong Province (No. 2021KJCX001) and the Priority Academic Program Development of Jiangsu Higher Education Institutions (PAPD).

Data Availability Statement: Data are available upon request from the section editors.

Acknowledgments: This work was funded by Project 2021KJCX001, supported by the Forestry Innovation Foundation of Guangdong Province and Priority Academic Program Development of Jiangsu Higher Education Institutions (PAPD). J. Y. is thankful for the patience and support from Yuxin Duanmu who cooperated with him to write the paper in English and correct it and for the helpful discussion from Jun Wang from the College of Chemical Engineering, Nanjing Tech University of China.

Conflicts of Interest: The authors declare no conflict of interest.

References

1. Guan, X. Research on the Modeling and Analysis of the Tree-dimensional Point Cloud Data of Trees. Master's Thesis, Central South University of Forestry & Technology, Changsha, China, 2016.
2. Sun, C.; Huang, C.; Zhang, H.; An, F.; Wang L.; Yun T. Individual Tree Crown Segmentation and Crown Width Extraction From a Heightmap Derived From Aerial Laser Scanning Data Using a Deep Learning Framework. *Front Plant Sci.* **2022**, 13, 914974. [CrossRef] [PubMed]

3. Xue, X.; Jin, S.; An, F.; An, F.; Zhang, H.; Fan, J.; Eichhorn, M.P.; Jin C.; Chen B.; Jiang L.; et al. Shortwave Radiation Calculation for Forest Plots Using Airborne LiDAR Data and Computer Graphics. *Plant Phenomics.* **2022**, *2022*, 9856739. [CrossRef] [PubMed]
4. Wang, G.; Li, Q.; Yang, X.; Wang, C. Real Scene Modeling of Forest Based on Airborne LiDAR Data. *J. Beijing Univ. Civ. Eng. Archit.* **2021**, *37*, 39–46. [CrossRef]
5. Li, Y.; Su, Y.; Zhao, X.; Yang, M.; Hu, T.; Zhang, J.; Liu, J.; Liu, M.; Guo, Q. Retrieval of tree branch architecture attributes from terrestrial laser scan data using a Laplacian algorithm. *Agric. For. Meteorol.* **2020**, *284*, 107874. [CrossRef]
6. Ji, J. The broad-leaved tree leaves reconstruction and deformation based on laser-point-clouds. Master's Thesis, Nanjing Forestry University, Nanjing, China, 2015. [CrossRef]
7. Huang, Z.; Huang, X.; Fan, J.; Eichhorn, M.; An, F.; Chen, B.; Cao, L.; Zhu, Z.; Yun, T. Retrieval of Aerodynamic Parameters in Rubber Tree Forests Based on the Computer Simulation Technique and Terrestrial Laser Scanning Data. *Remote Sens.* **2020**, *12*, 1318. [CrossRef]
8. Shi, Y. *3D Simulation Theory and Technology of Botanic Tree Based on Scattered Point Cloud*; Tongji University Press: Shanghai, China, 2018.
9. Runions, A.; Lane, B.; Prusinkiewicz, P. Modeling Trees with a Space Colonization Algorithm. In *EurographicsWorkshop on Natural Phenomena*; The Eurographics Association; Prague, Czech Republic, 2007. [CrossRef]
10. Yang, H.; Wang, Y.; Wang, Z.; Zhang, Z. 3D Tree-modeling Approach Based on Competition over Space Resources. *Comput. Sci.* **2019**, *46*, 38–41.
11. Zhang, Y.; Yu, W.; Zhao, X.; LYU, Y.; Feng, W.; Li, Z.; Hu, S. Interactive tree segmentation and modeling from ALS point clouds. *J. Graph.* **2021**, *42*, 599–607. [CrossRef]
12. Hu, S.; Li, Z.; Zhang, Z.; He, D.; Wimmer, M. Efficient tree modeling from airborne LiDAR point clouds. *Comput. Graph.* **2017**, *67*, 1–13. [CrossRef]
13. Zheng, J.; Fu, H.; Li, W.; Wu, W.; Yu, L.; Yuan, S.; Tao, W.Y.W.; Pang, T.K.; Kanniah, K.D. Growing status observation for oil palm trees using Unmanned Aerial Vehicle (UAV) images. *ISPRS J. Photogramm. Remote Sens.* **2021**, *173*, 95–121. [CrossRef]
14. Zhou, C.; Yang, G.; Liang, D.; Yang, X.; Xu, B. An Integrated Skeleton Extraction and Pruning Method for Spatial Recognition of Maize Seedlings in MGV and UAV Remote Images. *IEEE Trans. Geosci. Remote Sens.* **2018**, *56*, 4618–4632. [CrossRef]
15. Xu, H.; Gossett, N.; Chen, B. Knowledge and heuristic-based modeling of laser-scanned trees. *ACM Trans. Graph.* **2007**, *26*, 19. [CrossRef]
16. Gao, S.; Zhang, H.; Liu, M.; He, Q.; Luo, L. Contour extraction of point cloud data for tree branches. *J. Zhejiang A F Univ.* **2013**, *30*, 648–654. [CrossRef]
17. You. Stem Form Measurement Based on Point Cloud Data. Ph.D. Thesis, Chinese Academy of Forestry, Beijing, China, 2016.
18. Raumonen, P.; Kaasalainen, M.; Åkerblom, M.; Kaasalainen, S.; Kaartinen, H.; Vastaranta, M.; Holopainen, M.; Disney, M.; Lewis, P. Fast Automatic Precision Tree Models from Terrestrial Laser Scanner Data. *Remote Sens.* **2013**, *5*, 491–520. [CrossRef]
19. Li, G.; Liu, L.; Zheng, H.; Mitra, N. Analysis, Reconstruction and Manipulation using Arterial Snakes. *ACM Trans. Graph.* **2010**, *29*, 152. [CrossRef]
20. Zhou, J. Research on Skeleton Extraction from Tree Point Cloud via k-Nearest-Neighbors-based Contraction. Master's Thesis, Chongqing University, Chongqing, China, 2020.
21. Sun, J.; Wang, P.; Li, R.; Zhou, M.; Wu, Y. Fast Tree Skeleton Extraction Using Voxel Thinning Based on Tree Point Cloud. *Remote Sens.* **2022**, *14*, 2558. [CrossRef]
22. Zhang, Y.; Shen, B.; Wang, S.; Kong, D.; Yin, B. L0-regularization-based skeleton optimization from consecutive point sets of kinetic human body. *ISPRS J. Photogramm. Remote Sens.* **2018**, *143*, 124–133. [CrossRef]
23. He, G.; Yang, J.; Behnke, S. Research on geometric features and point cloud properties for tree skeleton extraction. *Pers. Ubiquitous Comput.* **2018**, *22*, 903–910. [CrossRef]
24. Livny, Y.; Yan, F.; Olson, M.; Chen, B.; Zhang, H.; El-Sana, J. Automatic Reconstruction of Tree Skeletal Structures from Point Clouds. *ACM Trans. Graph.* **2010**, *29*, 151. [CrossRef]
25. He, L. Reseaches on 3D reconstruction of fruit tree's trunk and its dynamic characteristics fo vibratory harvesting. Ph.D. Thesis, Zhejiang Sci-Tech University, Hangzhou, China, 2014.
26. Wang, B. Studies on automatic 3D reconstruction techniques of trees based on terrestrial lidar point clouds. Master's Thesis, Univerity of Electronic Science and Technology of China, Chengdu, China, 2015.
27. Nouri, A.; Autrusseau, F.; Bourcier, R.; Gaignard, A.; l'Allinec, V.; Menguy, C.; Veziers, J.; Desal, H.; Loirand, G.; Redon, R. 3D bifurcations characterization for intra-cranial aneurysms prediction. In Proceedings of the Medical Imaging: Image Processing, San Diego, CA, USA, 19–21 February 2019. [CrossRef]
28. Xu, J.; Shan, J.; Wang, G. Hierarchical Modeling of Street Trees Using Mobile Laser Scanning. *Remote Sens.* **2020**, *12*, 2321. [CrossRef]
29. Fu, L.; Liu, J.; Zhou, J.; Zhang, M.; Lin, Y. Tree Skeletonization for Raw Point Cloud Exploiting Cylindrical Shape Prior. *IEEE Access* **2020**, *8*, 27327–27341. [CrossRef]
30. Mei, J.; Zhang, L.; Wu, S.; Zhen, W.; Zhang, L. 3D tree modeling from incomplete point clouds via optimization and L1-MST. *Int. J. Geogr. Inf. Sci.* **2016**, *31*, 1–23. [CrossRef]
31. Li, R.; Bu, G.; Wang, P. An Automatic Tree Skeleton Extracting Method Based on Point Cloud of Terrestrial Laser Scanner. *Int. J. Opt.* **2017**, *2017*, 1–11. [CrossRef]
32. Chaudhury, A.; Godin, C. Skeletonization of Plant Point Cloud Data Using Stochastic Optimization Framework. *Front. Plant Sci.* **2020**, *11*, 773. [CrossRef] [PubMed]

33. Xu, S.; Li, X.; Yun, J.; Xu, S. An Effectively Dynamic Path Optimization Approach for the Tree Skeleton Extraction from Portable Laser Scanning Point Clouds. *Remote Sens.* **2022**, *14*, 94. [CrossRef]
34. Yang, M.; Wu, E. Self-Adapting Algorithm of 3D Art-Designing Tree Skeleton Extraction. *J. Front. Comput. Sci. Technol.* **2012**, *6*, 1039–1048. [CrossRef]
35. Li, J.; Wu, H.; Xiao, Z.; Lu, H. 3D modeling of laser-scanned trees based on skeleton refined extraction. *Int. J. Appl. Earth Obs. Geoinf.* **2022**, *112*, 102943. [CrossRef]
36. Yang, J.; Wen, X.; Wang, Q.; Ye, J.S.; Zhang, Y.; Sun, Y. A Novel Algorithm Based on Geometric Characteristics for Tree Branch Skeleton Extraction from LiDAR Point Cloud. *Forests* **2022**, *13*, 1534. [CrossRef]
37. Seidel, D.; Annighöfer, P.; Thielman, A.; Seifert, Q.; Thauer, J.H.; Glatthorn, J.; Ehbrecht, M.; Kneib, T.; Ammer, C. Predicting Tree Species From 3D Laser Scanning Point Clouds Using Deep Learning. *Front. Plant Sci.* **2021**, *12*, 635440. [CrossRef]
38. Zhang, D.; Yan, R.; Yun, T.; Xue, L.; Ruan, H. The 3D reconstruction of tree branches from point cloud based on terrestrial laser scanner. *J. For. Eng.* **2016**, *1*, 107–114. [CrossRef]
39. Zhao, Y.; Guo, H. A Tree Branch Skeleton Extraction Approach Based on Point Cloud. *Comput. Digit. Eng.* **2016**, *44*, 1333–1337. [CrossRef]
40. Thurlbeck, A.; Horsfield, K. Branching angles in the bronchial tree related to order of branching. *Respir. Physiol.* **1980**, *41*, 173–181. [CrossRef] [PubMed]
41. Uylings, H. Optimization of diameters and bifurcation angles in lung and vascular tree structures. *Bull. Math. Biol.* **1977**, *39*, 509–520. [CrossRef] [PubMed]
42. King, D.; Loucks, O.L. The theory of tree bole and branch form. *Radiat. Environ. Biophys.* **1978**, *15*, 141–165. [CrossRef] [PubMed]
43. Harper, G.J. Quantifying branch, crown and bole development in Populus tremuloides Michx. from north-eastern British Columbia. *For. Ecol. Manag.* **2008**, *255*, 2286–2296. [CrossRef]
44. Dong, L.; Liu, Z.; Bettinger, P. Nonlinear mixed-effects branch diameter and length models for natural Dahurian larch (Larix gmelini) forest in northeast China. *Trees* **2016**, *30*, 1191–1206. [CrossRef]
45. Du, S.; Lindenbergh, R.; Ledoux, H.; Stoter, J.; Nan, L. AdTree: Accurate, Detailed, and Automatic Modelling of Laser-Scanned Trees. *Remote Sens.* **2019**, *11*, 2074. [CrossRef]
46. Ai, M.; Yao, Y.; Hu, Q.; Wang, Y.; Wang, W. An Automatic Tree Skeleton Extraction Approach Based on Multi-View Slicing Using Terrestrial LiDAR Scans Data. *Remote Sens.* **2020**, *12*, 3824. [CrossRef]
47. Wu, B.; Zheng, G.; Chen, Y.; Yu, D. Assessing inclination angles of tree branches from terrestrial laser scan data using a skeleton extraction method. *Int. J. Appl. Earth Obs. Geoinf.* **2021**, *104*, 102589. [CrossRef]
48. Fan, G.; Nan, L.; Dong, Y.; Su, X.; Chen, F. AdQSM: A New Method for Estimating Above-Ground Biomass from TLS Point Clouds. *Remote Sens.* **2020**, *12*, 3089. [CrossRef]
49. Huang, Z. Tree skeleton reconstruction and wind resistance analysis based on laser point cloud. Master's Thesis, Nanjing Forestry University, Nanjing, China, 2020. [CrossRef]

remote sensing

MDPI

Article

Improving GEDI Forest Canopy Height Products by Considering the Stand Age Factor Derived from Time-Series Remote Sensing Images: A Case Study in Fujian, China

Xiaocheng Zhou [1,*], Youzhuang Hao [1], Liping Di [2], Xiaoqin Wang [1], Chongcheng Chen [1], Yunzhi Chen [1], Gábor Nagy [3] and Tamas Jancso [3]

[1] Key Laboratory of Spatial Data Mining and Information Sharing of Ministry of Education, National & Local Joint Engineering Research Center of Satellite Geospatial Information Technology, Fuzhou University, Fuzhou 350108, China

[2] Center for Spatial Information Science and Systems (CSISS), George Mason University, Fairfax, VA 22030, USA

[3] Institute of Geoinformatics, Alba Regia Technical Faculty, Obuda University, 8000 Székesfehérvár, Hungary

* Correspondence: zhouxc@fzu.edu.cn; Tel.: +86-0591-63179578

Abstract: Forest canopy height plays an important role in forest resource management and conservation. The accurate estimation of forest canopy height on a large scale is important for forest carbon stock, biodiversity, and the carbon cycle. With the technological development of satellite-based LiDAR, it is possible to determine forest canopy height over a large area. However, the forest canopy height that is acquired by this technology is influenced by topography and climate, and the canopy height that is acquired in complex subtropical mountainous regions has large errors. In this paper, we propose a method for estimating forest canopy height by combining long-time series Landsat images with GEDI satellite-based LiDAR data, with Fujian, China, as the study area. This approach optimizes the quality of GEDI canopy height data in topographically complex areas by combining stand age and tree height, while retaining the advantage of fast and effective forest canopy height measurements with satellite-based LiDAR. In this study, the growth curves of the main forest types in Fujian were first obtained by using a large amount of forest survey data, and the LandTrendr algorithm was used to obtain the forest age distribution in 2020. The obtained forest age was then combined with the growth curves of each forest type in order to determine the tree height distribution. Finally, the obtained average tree heights were merged with the GEDI_V27 canopy height product in order to create a modified forest canopy height model (MGEDI_V27) with a 30 m spatial resolution. The results showed that the estimated forest canopy height had a mean of 15.04 m, with a standard deviation of 4.98 m. In addition, we evaluated the accuracy of the GEDI_V27 and the MGEDI_V27 using the sample dataset. The MGEDI_V27 had a higher accuracy (R^2 = 0.67, RMSE = 2.24 m, MAE = 1.85 m) than the GEDI_V27 (R^2 = 0.39, RMSE = 3.35 m, MAE = 2.41 m). R^2, RMSE, and MAE were improved by 71.79%, 33.13%, and 22.53%, respectively. We also produced a forest age distribution map of Fujian for the year 2020 and a forest disturbance map of Fujian for the past 32 years. The research results can provide decision support for forest ecological protection and management and for carbon sink analysis in Fujian.

Keywords: GEDI; canopy height; forest age; LiDAR; time-series remote sensing; Fujian

1. Introduction

Forests, as the main part of the ecosystem, are the largest carbon reservoir in terrestrial ecosystems and make important contributions to global carbon sinks [1–3]. In order to mitigate climate change, and to protect forest ecosystems, we need accurate access to large-scale forest structure parameters for monitoring and managing forest resources [4]. Among the many forest parameters, forest canopy height is both a visual representation of forest resources and an important attribute for estimating forest biomass and carbon stock [5,6]. At present, the accuracy of large-scale forest canopy height estimation is still limited, especially in mountainous and hilly areas [7,8]. Therefore, the rapid and accurate acquisition of large-scale forest canopy height and the analysis of the spatial pattern of the forest height can provide scientific and reasonable decision support to policy makers for forestry carbon sink trading, dynamic monitoring, and the assessment of forest carbon neutrality [9].

The traditional methods to obtain forest canopy height over large areas are time consuming and labor intensive, are difficult for surveyors to perform in areas with a complex terrain, and are dangerous. The satellite-based light detection and ranging (LiDAR) technology has the ability to actively acquire the global surface and target 3D information, and it can play an important role in vegetation height measurement, sea surface height measurement, and global climate monitoring, thus providing a new way to acquire forest canopy height over large areas [10,11]. In December 2018, the National Aeronautics and Space Administration (NASA) launched the global ecosystem dynamics investigation (GEDI) LiDAR altimetry mission. The GEDI is equipped with the world's first multi-beam linear laser altimeter for high-resolution forest vertical structure measurement, which is mainly used for the precise measurement of forest canopy height, vertical structure, and ground elevation in tropical and temperate regions. Compared with the early full-waveform laser altimeter ICESat (2003–2009), the GEDI provides denser coverage of observations and more accurate estimates [12,13]. However, the GEDI data still have limitations in their application. Firstly, the data are footprint-level and thus cannot cover the ground comprehensively, and they are sparse in subtropical mountainous regions compared with other regions. Secondly, they are missing time series information, which is not suitable for tracking data information through time. These shortcomings can be effectively remedied if the GEDI data are combined with time-series optical images.

For forest canopy height estimation, some scholars have used the spectral variation of long-time-series remote sensing images and have extracted the time series features for regression analysis in order to obtain the latest canopy height [14–16]. These studies have confirmed that time series features such as forest disturbance time, disturbance intensity, and spectral indices before and after the disturbance all have good correlation with the canopy height. In LiDAR applications, the mainstream approach for the large-scale mapping of forest canopy height is to use spatially continuous satellite imagery to extrapolate the vegetation attributes that are derived from satellite-based or airborne LiDAR [17,18]. With the availability of high-resolution satellite imagery and a new generation of on-board LiDAR data, the accuracy of regional and global forest canopy height mapping has been greatly improved compared to previous studies.

Potapov et al. [19] proposed a machine learning algorithm with bagged regression trees to map a global canopy height product (GEDI_V27) at a 30 m resolution for 2019, using the 95th percentile relative height metric from GEDI_2A footprint data as the dependent variable. Compared with the GEDI validation data (RMSE = 6.6 m; MAE = 4.45 m; R^2 = 0.62) and the available airborne LiDAR (RMSE = 9.07 m; MAE = 6.36 m, R^2 = 0.61), the GEDI_V27 product is a more stable canopy height model (CHM) that is obtained by machine learning inversion using a large number of spectral, textural, and phenological features; however, the accuracy of GEDI_V27 product is lower in subtropical complex mountainous regions, making it difficult to use the product for subsequent studies in such regions. The GEDI_V27 product mainly has the following deficiencies: (1) GEDI is easily affected by topography and thick cloud cover when collecting canopy height data, which leads to a large instability in the accuracy of the data that is collected in subtropical complex mountainous regions [20–22]. Therefore, during the inversion process, the value of the dependent variable RH_95 in the mountainous region has errors, which leads to the fact that no matter how high the model accuracy is, the results that are obtained still have large errors. (2) This product extrapolates satellite-based LiDAR footprint data using optical images, and the generated canopy height product still has the problem of potential saturation [23]. (3) The product is based on a global study area, and it is difficult to take into account all of the regions during the sample selection for model calibration. As a result, the accuracy of the regions with a small sample size is lower [24–26]. Based on the above-mentioned problems, we improved GEDI forest canopy height products in 2020 in Fujian province by considering the stand age factor derived from long time-series Landsat data from 1986 to 2020 and a large amount of local data in the paper, which reduces the influence of complex terrain and effectively reduces the problem of data saturation. Finally, it should be noted that the research region of this study is Fujian province, and the GEDI collected data in Fujian from April 2019 to September 2020, therefore, the forest age and sample set in this study were limited to this time period.

There is a very close correlation between forest stand age and its tree height, especially for immature forests, with a very clear linear relationship [27,28]. In this paper, therefore, we use long time series Landsat satellite data to produce a forest age distribution map and then combine the growth curves of different forest types in order to estimate the tree height. Finally, the tree height results are merged with the satellite-based LiDAR forest canopy height product GEDI_V27 in order to create a modified forest canopy height distribution map. The correlated features, such as spectral and texture features in the forest stand images, are usually adopted for remote sensing estimation of forest age [29]. Numerous remote sensing data have been used in order to establish the relationship between stand age and optical, LiDAR, and microwave parameters, including MODIS, ICESat, SPOT-5, and SAR [30–33]. Although the traditional remote sensing methods can be used to produce thematic maps of forest age, the spectral saturation of mature forests and the frequent areas of land change can reduce the accuracy of the forest age estimation models [34,35]. In addition, most forest age products that are estimated by traditional remote sensing have low spatial resolution and limited time points, and thus cannot capture forest disturbance and restore the history sufficiently in a space or time domain [33,36]. With Landsat time-series data archived and freely available, forest disturbance information, such as forest fires, logging, and afforestation, have been detected at a 30 m resolution over the past 35 years, allowing researchers to obtain the time point and spatial location of the origin of most of the existing forest stands for forest stand age estimation [37,38]. In this paper, LandTrendr, which is a long-time-series forest disturbance and restoration detection algorithm that was proposed by Kennedy et al. [39], was used to estimate forest age. This algorithm extracted the change process track of the spectral index pixel by pixel, for which potential vertices were used as segmentation nodes, and the node positions and spectral values provided effective information that could explain the forest structure. The trajectory curves before and after it represent the continuous changes or the stable states of the forest pixels in the spectral trajectory. Through this method, the forest disturbance and restoration time nodes were obtained in order to infer the starting time of afforestation, so as to obtain the distribution map of the existing forest age.

In order to reduce the error of forest canopy height estimation in mountainous regions, time series satellite remote sensing was adopted in order to estimate the forest stand age factor and the forest canopy height, which, by merging the forest canopy height that is derived from stand age, can reduce the uncertainty of exogenous factors in the process of data acquisition by satellite-based LiDAR and improve the accuracy of the GEDI canopy height models in mountainous regions. This study had the following four main objectives: (1) construct the growth curves of needleleaf, broadleaf, and mixed forests in Fujian, (2) estimate the forest age based on long time series Landsat satellite data, (3) merge the estimated tree height with the GEDI canopy height model to develop a modified forest canopy height model, and (4) evaluate and compare the accuracy of the modified canopy height product in the paper with the GEDI_V27 canopy height product.

2. Materials and Data

2.1. Study Area

The study area of this paper is Fujian, which is situated in Southeastern China. Fujian covers an area of 124,000 km^2 and has a subtropical maritime monsoon climate, with mountains and hills accounting for more than 80% of the province's area (Figure 1a,b). Fujian is rich in forest resources and is a key collective forest area in Southern China, with a forest coverage rate of over 66%. The forest types are mainly needleleaf, broadleaf, and mixed (Figure 1c). The forest types cover map of Fujian was produced by using 2020 Sentinel satellite images (with a spatial resolution of 10 m) and China's Gaofen series satellites (with a spatial resolution of 1–2 m), combined with the field survey data, using the random forest method. In order to understand the coverage of the GEDI original footprint data in Fujian, all of the GEDI level 2A datasets in Fujian were downloaded from the Land Processes Distributed Active Archive Center [40]. The time span of the GEDI Fujian data collection is from April 2019 to September 2020. There are 345 GEDI track files in HDF5 format. The quality filtering and the data visualization were completed using the pyGEDI package [41] in order to obtain 1,375,019 footprint data. The data collection of the GEDI in the forest region of Fujian was relatively sparse, and the available footprints per square kilometer were from 0 to 124, which partly explains the low accuracy of the GEDI_V27 in the Fujian region.

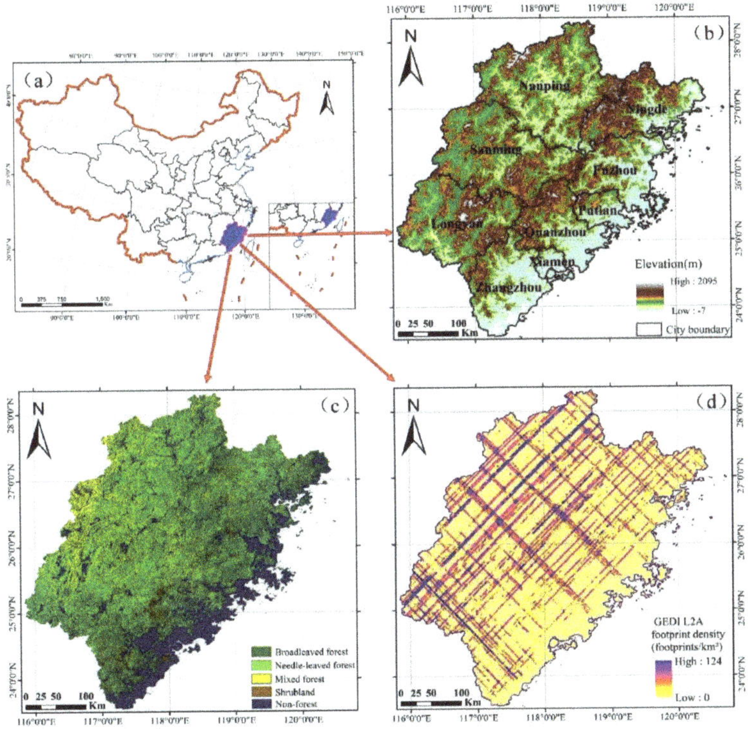

Figure 1. Study area. (**a**) The geographical location of Fujian, (**b**) a DEM of the study area, (**c**) forest types in the study area, and (**d**) a density distribution map of GEDI_2A footprint data in 1 km^2 of the study area.

2.2. Landsat Time-Series Imagery and Preprocessing

The time-series Landsat dataset that has been used in this study includes the surface reflectance products of Landsat-5, Landsat-7, and Landsat-8, corresponding to the sensors TM, ETM+, and OLI, respectively. The data that were acquired from the time range 1986–2020 were processed on the GEE platform, with the specific Landsat satellite parameters, which are shown in Table 1.

Table 1. Used Landsat satellite parameters.

Satellite	Sensor	Acquisition Years	Bands (μm)	
Landsat-5	TM	1984–2012	B1-Blue (0.45–0.52)	B2-Green (0.52–0.60)
			B3-Red (0.63–0.69)	B4-NIR (0.76–0.90)
			B5-SWIR (1.55–1.75)	B6-LWIR (10.40–12.50)
			B7-SWIR (2.08–2.35)	
Landsat-7	ETM+	1999–Ongoing	B1-Blue (0.45–0.52)	B2-Green (0.52–0.60)
			B3-Red (0.63–0.69)	B4-NIR (0.76–0.90)
			B5-SWIR (1.55–1.75)	B6-LWIR (10.40–12.50)
			B7-SWIR (2.08–2.35)	B8-Pan (0.52–0.9)
Landsat-8	OLI	2013–Ongoing	B1-Coastal (0.43–0.45)	B2-Blue (0.45–0.51)
			B3-Green (0.53–0.60)	B4-Red (0.63–0.68)
			B5-NIR (0.85–0.89)	B6-SWIR1 (1.56–1.66)
			B7-SWIR2 (2.11–2.29)	B8-Pan (0.50–0.68)
			B9-Cirrus (1.36–1.39)	B10-LWIR (10.06–11.19)

2.3. Field Survey Dataset

The field survey dataset mainly includes UAV data, forest plot inventory data, and field measurement datasets (Figure 2). The UAV data that were used in this study include LiDAR data and visible light images from 2020 to 2021. The distribution of the data includes Baisha forestry farms in the city of Fuzhou, Jiangle County in Sanming, and the Hu Boliao reserve in Zhangzhou.

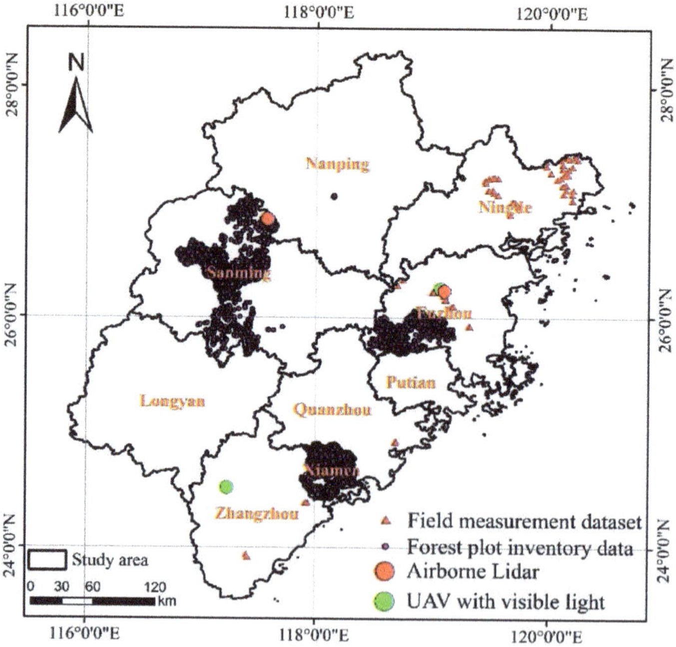

Figure 2. Distribution of field survey sample data in the study area.

The forest plot inventory data were divided into a plot according to the stand conditions, the stand factors, and the harvesting methods that were consistent with the stand. The forest plot data include information on the average tree height, the average diameter at breast height, the denseness, and the stocking volume. The forest plot inventory data that were used in this study include 66,413 forest plots in Mingxi County in 2008, Xiamen in 2009, Jiangle County in 2015, 2019, and 2021, the Wanmulin reserve in Nanping city in 2016,

Yongtai County in 2021, and Yongan County in 2021. The field measurement data were collected from October 2020 to December 2021. The average tree heights in a rectangular sample plot with a side length of 30 m were measured and recorded, with measuring tools such as a tape measure, a breast height ruler, a laser range altimeter, and the HUACE i70 GNSS receiver.

3. Methods

The technical flow of this study is shown in Figure 3. The first part of the research involved fitting curves between the average forest height from inventory plots and the forest age estimated from Landsat time series. Second, the forest age was estimated using time series Landsat satellite data. The time nodes of forest disturbance and restoration were obtained based on the LandTrendr algorithm using long-time-series Landsat data, and Fujian was divided into disturbance areas and non-disturbance areas. The forest stand age can be calculated in a disturbance area based on the recovery node; forest stand ages in the non-disturbance area were labeled as greater than the maximum forest age in the disturbance areas. Third, the canopy height model and the accuracy evaluation were inverted. The average tree height was calculated based on the obtained forest age factor and the growth curves of the different forest types in Fujian, and it was merged with the GEDI_V27 to obtain the canopy height model of Fujian in 2020. The sample data were used as reference data to evaluate and analyze the accuracy of the modified canopy height model and the GEDI_V27 canopy height model obtained by the proposed method.

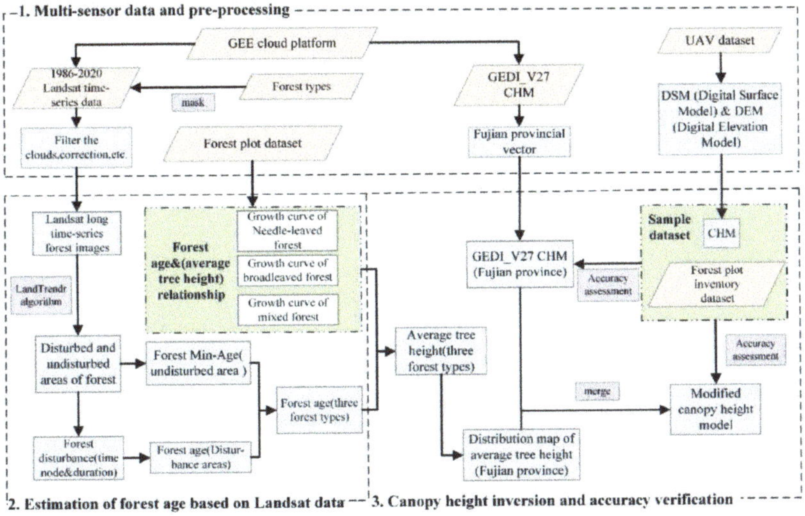

Figure 3. Overall technical flowchart for estimating forest canopy height combining time-series optical remote sensing and GEDI satellite-based LiDAR data.

3.1. Curve Fitting of Forest Age and Average Tree Height

In order to obtain growth curves for each forest type in Fujian, we collected a large amount of historical forest plot inventory samples distributed in Xiamen, Mingxi County, Jiangle County, and the Wanmulin reserve, as follows: 61,414 forest patches that contain forest structure information, such as average tree height, average diameter at breast height (DBH), forest age, and forest type. We statistically analyzed the samples and plotted the distribution of the average tree height data corresponding to each age in the forest patches (Figure 4a–c). We then used Origin software to determine the relationship between the forest age and the average tree height for the following three forest types: needleleaf, broadleaf, and mixed forests (Figure 5a–c). By comparing the accuracy of fitting results of various functions and analyzing the residual error, we found that the fitting effect of the exponential function was the most consistent with the relationship between the forest age and the average tree height. Therefore, the exponential function was selected as the optimal fitting function between the forest age and the average tree height in this study. The graph shows that the growth trends of the three forest types are basically the same, with faster growth before immaturity and a gradual flattening with maturity, at which point trees no longer increase in height. The negative values calculated for the coniferous forest equation are replaced by minimum positive values.

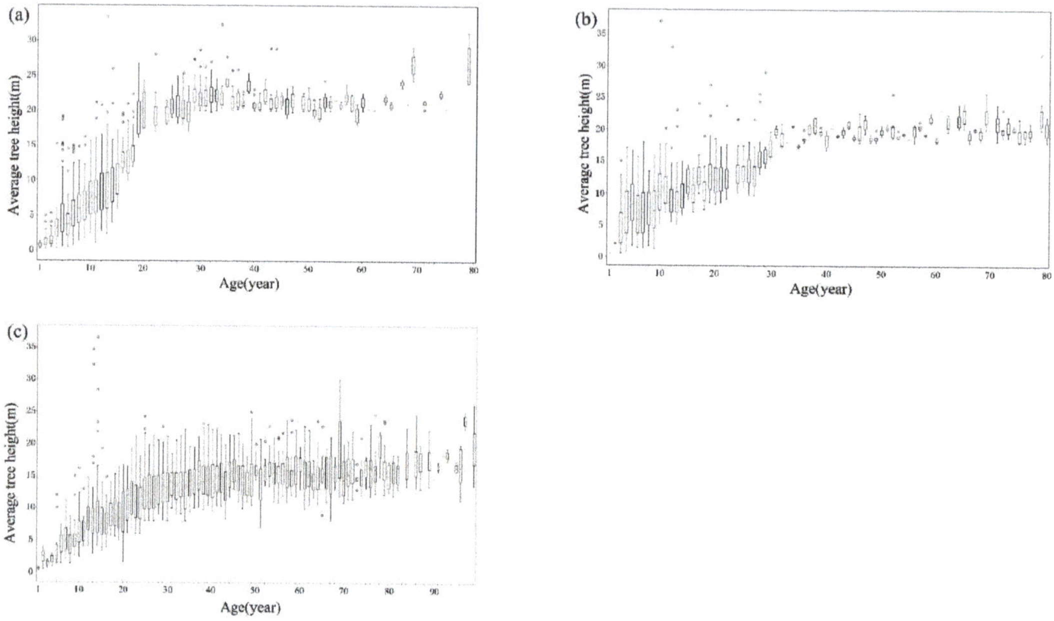

Figure 4. Box plot of the average tree height corresponding to each age. (**a**) Needleleaf forest. (**b**) Broadleaf forest. (**c**) Mixed forest.

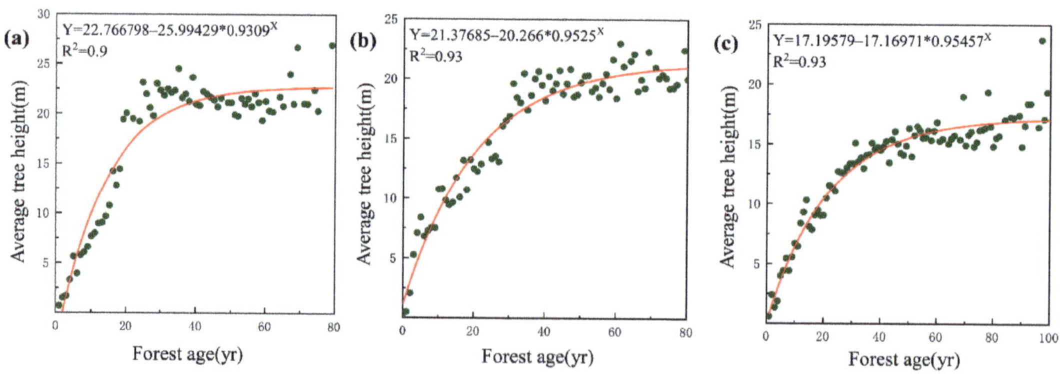

Figure 5. Growth curves of three forest types. (**a**) Needleleaf forest. (**b**) Broadleaf forest. (**c**) Mixed forest.

3.2. Forest Age Estimation

In order to estimate the forest age in 2020 in Fujian, the LandTrendr forest disturbance detection algorithm, which is based on long-time-series Landsat satellite data from 1986 to

2020, was adopted to distinguish the study area into forest disturbance and non-disturbance areas, and the forest age in 2020 was calculated based on the forest disturbance features extracted by the algorithm. A forest disturbance area refers to an area where the forest status is changed, and this paper takes the forest cover in 2020 as the research object, therefore the areas where disturbance occurs are restored to the forest, not other land types. A forest non-disturbance area is an area where no disturbance was detected in the Landsat time-series images during or after 1986. Since the tree height no longer changes significantly as the trees mature, the forest age in the undisturbed areas is based on the maximum forest age in the disturbed area as the minimum threshold.

3.2.1. The LandTrendr Algorithm

The forests in Fujian have the characteristics of short-term rotation and high productivity, and LandTrendr can identify both short-term drastic forest disturbances and determine long-term vegetation recovery times. Therefore, long-time-series Landsat data with a large amount of native Fujian data were used to invert the canopy height data, which can reduce the influence of complex topography and the problem of data saturation. The LandTrendr algorithm [39,42] is essentially a spectrum-time segmentation algorithm that focuses on the detection of changes in the time series of medium-resolution satellite images. The core of the algorithm lies in the segmented fitting of time series and in extracting spectral time trajectories image by image, by which both short-term drastic forest disturbances and long-term vegetation recoveries can be identified. The LandTrendr algorithm in general appears to only consider the trend component of the time-series data and fits a concise straight-line segment based on the dynamic trajectory of the spectrum (Figure 6).

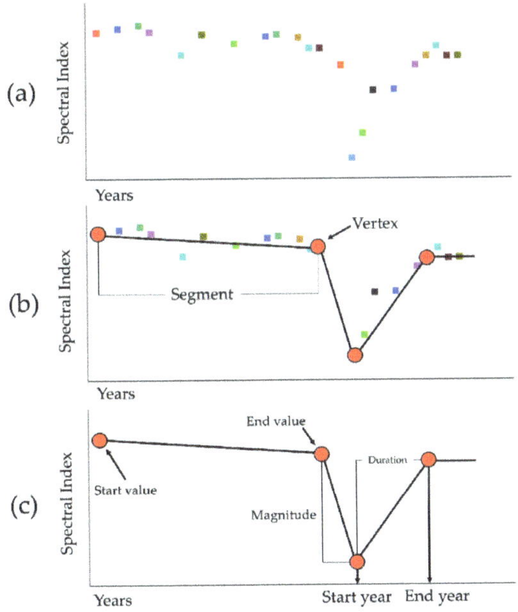

Figure 6. LandTrendr pixel time-series segmentation [42]. (**a**) Time series of individual bands or spectral indicators of image elements. (**b**) Segmentation into a series of linear segments by identifying vertices. (**c**) Significant features of each segment of the indicator trajectory.

For the selection of indices, scholars have investigated the differentiation of image reflectance and derived vegetation indices for forest disturbance types, such as deforestation and fire, showing that the shortwave infrared (SWIR) band reflectance has the highest differentiation among the disturbance types [43,44]. Kennedy et al. [39] showed that the corresponding normalized burn ratio (NBR) has the highest sensitivity for capturing the disturbance time of harvesting and fire. Therefore, in this study, NBR was chosen as the spectral index to construct the time series pixel by pixel, and the NBR was calculated as follows:

$$NBR = (NIR - SWIR)/(NIR + SWIR) \quad (1)$$

The spectral value is the ratio of the difference between the near-infrared (NIR) and the shortwave infrared (SWIR) values to the sum of those values. Healthy vegetation has a high NIR value and a low SWIR value, therefore, it will have a high NBR index, which will suddenly become smaller when healthy vegetation is disturbed, and forest status information is detected by such a change. The LandTrendr algorithm was implemented based on the GEE cloud platform, which can efficiently determine the forest disturbance and restoration in Fujian. After obtaining the forest disturbance time nodes and duration, while afforestation is generally within one year after harvesting, the afforestation time could be inferred, and the forest age of the disturbance area could be obtained. The forest age of the undetected disturbance area was estimated to be above the maximum forest age in the disturbance area, and this result was used to make the overall forest age distribution map at a 30 m resolution.

3.2.2. Accuracy Verification of Forest Age

The sample sets for verifying the accuracy of the forest age were the forest plot inventory data from Mingxi County in 2008, Xiamen in 2009, and Jiangle County in 2019. These data were used to verify the distribution of the forest age at different disturbance times to verify that the disturbance recovery time occurred in areas before 2009, and to compare the forest age of the data in 2009 with the estimated forest age results. Three hundred forest plots were selected from the data at three sites for the validation and accuracy evaluation of the forest age.

3.3. *Canopy Height Estimation*

The process of canopy height estimation has two steps (Figure 7). The first step is to combine the forest age factor and the forest type growth curve for tree height estimation for the average tree height distribution in Fujian. The second step is to combine that with the average tree height distribution with the GEDI_V27 to obtain the final canopy height model: the modified GEDI_V27 (MGEDI_V27). In terms of determining the average tree height, this modified model has the advantage of not being limited by environmental factors, but the tree height of the same stand age has a certain fluctuation range. The GEDI_V27 shows a large difference in the accuracy of the canopy height estimated by the model in different areas. Based on a large number of sample datasets, we found that the average tree height in Fujian at the same age fluctuates within a range of 2 m, and only a small fraction will have a relatively large error due to special tree species. Therefore, in this paper, 2 m is the threshold value for combining the average tree height with the GEDI_V27 canopy height products. In the area of Fujian where the slope is less than 15°, the canopy height of the GEDI_V27 is selected. For other areas, the following strategy was employed in the merging process: if the difference between the mean tree height and the height value of the GEDI_V27 was within the range of 2 m, then the GEDI_V27 was considered to be an accurate estimation, therefore, the canopy height of the MGEDI_V27 was used as the canopy height value of the GEDI_V27. Within a forest disturbance area, since the forest age and the average tree height were a one-to-one correspondence, when the difference between the average tree height and the height value of the GEDI_V27 was greater than 2 m, we decided that the GEDI_V27 deviated from the growing range of tree height and considered the average tree height as the canopy height of the area. In an undisturbed forest area, since the average tree height was the minimum threshold, when the height value of the GEDI_V27 was greater than this threshold, the value of the GEDI_V27 was selected as the canopy height of the MGEDI_V27, and when the GEDI_V27 value was lower than the average tree height and the difference was more than 2 m, the average tree height was used instead of the canopy height.

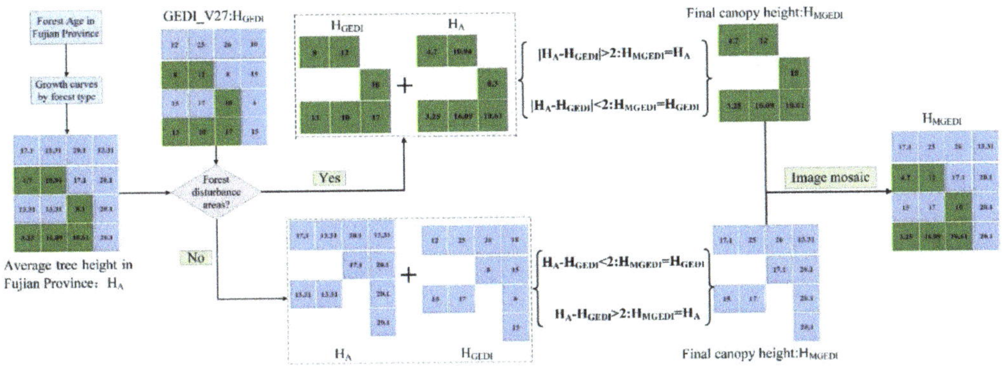

Figure 7. Flow chart for estimating canopy height. The average tree height is expressed as HA, the canopy height of the GEDI_V27 is expressed as HGEDI, and the canopy height of the MGEDI_V27 is expressed as HMGEDI. Green pixels indicate an interference area, and blue pixels indicate a non-interference area.

3.4. Accuracy Verification of Canopy Height

The canopy height accuracy validation process is divided into the following two steps: The first step is to use the sample set to statistically analyze the canopy height model product GEDI_V27 and assess its accuracy through statistical quantities. The second step is to use the sample set to evaluate the accuracy of the modified forest canopy height model and compare it with the accuracy of the GEDI_V27 to test the effectiveness of this study's approach.

The sample dataset was uniformly resampled to a 30 m resolution and was then compared with the canopy height model. The following three statistical parameters were calculated: the coefficient of determination (R^2), the root mean squared error (RMSE), and the mean absolute deviation (MAE), as follows:

$$R^2 = 1 - \frac{\sum_{i=1}^{n}(x_i - y_i)^2}{\sum_{i=1}^{n}(y_i - \bar{y})^2} \tag{2}$$

$$\text{RMSE} = \sqrt{\frac{1}{n} \times \sum_{i=1}^{n}(x_i - y_i)^2} \tag{3}$$

$$\text{MAE} = \frac{1}{n} \times \sum_{i=1}^{n}(|x_i - y_i|) \tag{4}$$

where x is the canopy height of the canopy height model; y is the canopy height in the sample data; n is the number of samples, and $\bar{y}$ is the mean of the canopy height of the sample set.

4. Results

4.1. Forest Age Estimation Results

4.1.1. Disturbed Area Feature Extraction Results

We extracted the forest disturbance area in Fujian based on the LandTrendr algorithm. From 1987 to 2020, strong forest disturbances, such as deforestation and forest fires, in Fujian accounted for 13.22% of the forest area, among which the western and central parts of Fujian were subject to more forest disturbance, and the distribution ratio of this area per square kilometer was mostly less than 30% (Figure 8). We also plotted the time and temporal trends of the forest disturbance in Fujian, from which we can see that the area of forest disturbance in Fujian did not fluctuate regularly, with a large area of forest disturbance occurring between 2005 and 2010 (Figure 9a,b).

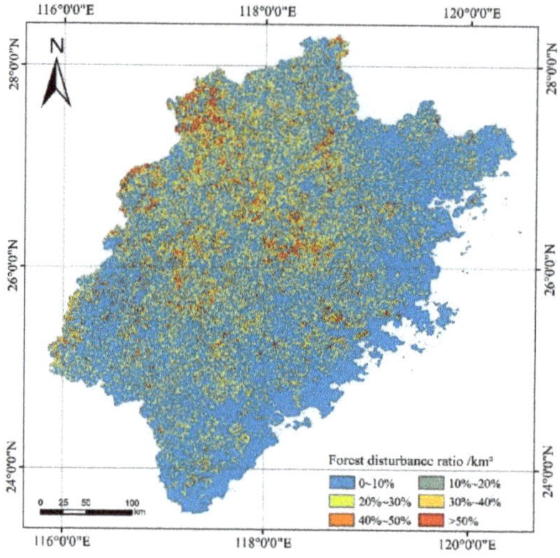

Figure 8. Proportion of the forest disturbance area in Fujian.

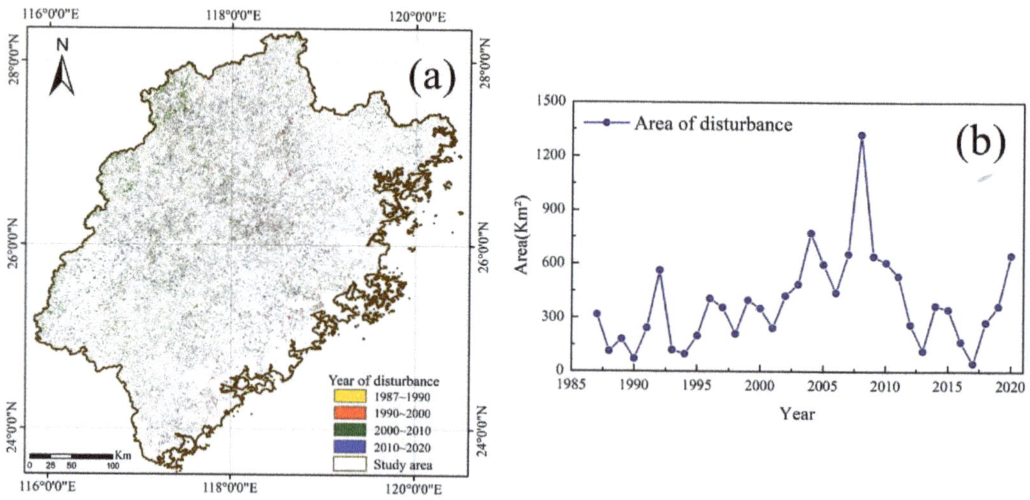

Figure 9. Time-series characteristics of forest disturbance in Fujian. (**a**) Disturbance years from 1987 to 2020. (**b**) Temporal characteristics of the forest disturbance area.

4.1.2. Accuracy Evaluation of Forest Age Results

Based on the characteristics of the disturbance area, the specific forest age of the disturbance area in Fujian can be calculated, while the age of the non-disturbance area takes the maximum forest age of the disturbance area as the minimum threshold, and the forest age distribution of Fujian in 2020 was mapped (Figure 10). Overall, most of the stands were older than 32 years, while within the disturbance area, the estimated forest age had a mean of 14.98 years, with a standard deviation of 7.5 years.

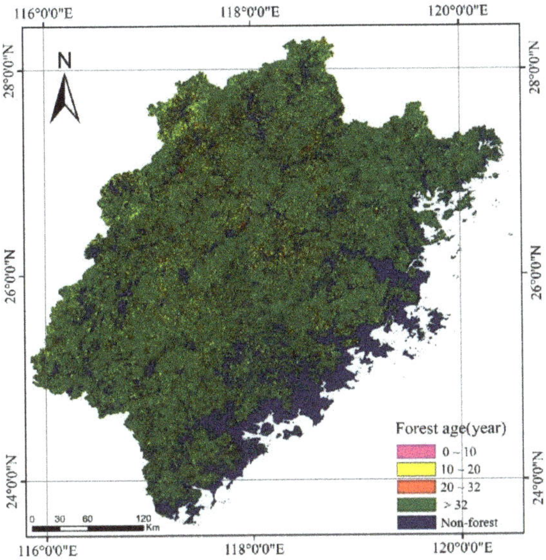

Figure 10. Forest age distribution map of Fujian in 2020.

In this study, forest sub-compartment datasets from three cities, Mingxi, Jiangle, and Xiamen, were selected as reference data, from which 300 samples were randomly selected in order to verify the accuracy of the forest age. The estimated stand age results showed good agreement with three validation datasets (Figure 11, R^2 = 0.91; RMSE = 1.98 years).

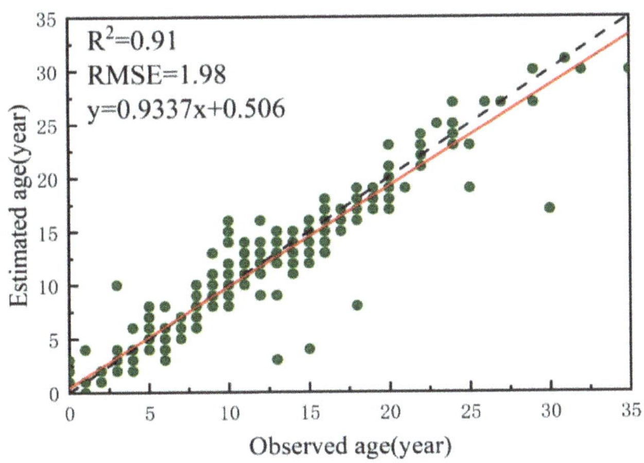

Figure 11. Accuracy verification of forest age.

4.2. Canopy Height Estimation and Accuracy Assessment

4.2.1. Canopy Height Estimation Results

The GEDI_V27 canopy height product, as well as the average tree height that was derived from the forest age (TH_A) in this paper, are shown in Figure 12a,b. The total average canopy heights of the TH_A and the GEDI_V27 in Fujian were 16.07 m and 11.92 m, with a standard deviation of 3.43 m and 4.68 m, respectively. According to the merging strategy in the method of this paper, the average tree height distribution was merged with the GEDI_V27 canopy height model in order to obtain the final canopy height distribution of Fujian in 2020 at a 30 m resolution (Figure 12c). Overall, the forest canopy heights in most of these areas ranged from 10 to 20 m. The mean of the MGEDI_V27 was 15.04 m, with a standard deviation of 4.98 m. Based on the spatial distribution, the forests in Northwestern and Central Fujian were relatively taller than the forests in Southeastern Fujian. The mean canopy height of all of the prefecture-level cities in Fujian ranged from 12.53 to 26.26 m. The differences between Xiamen and the other cities were larger than those between the other cities (Figure 12d).

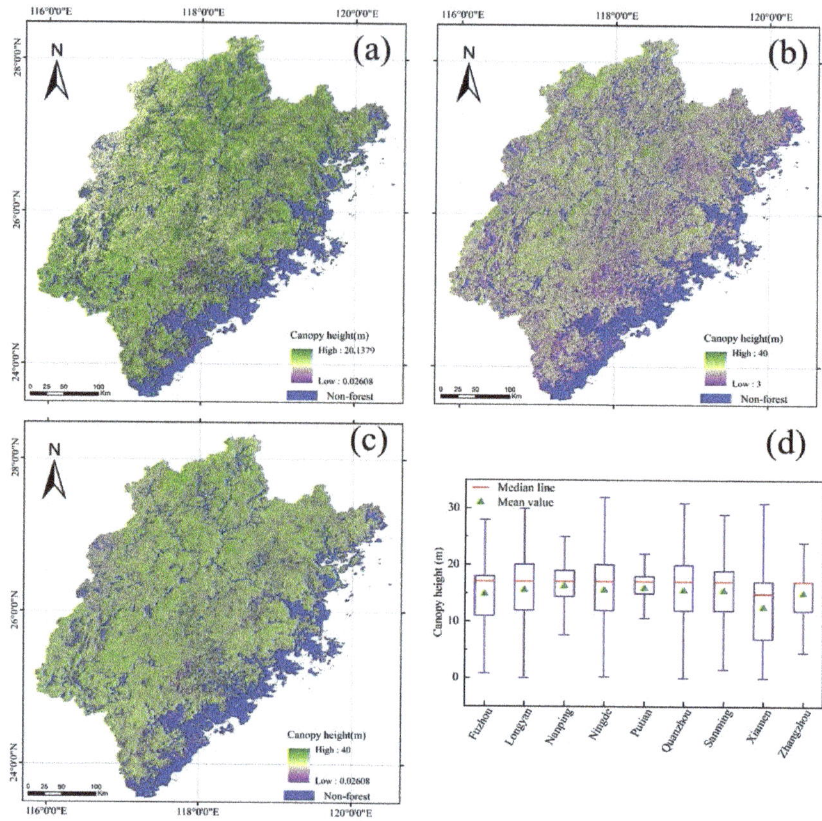

Figure 12. Canopy height estimation results. (**a**) Canopy height of Fujian estimated from time-series optical images. (**b**) The GEDI_V27 for the Fujian region. (**c**) Canopy height distribution of the MGEDI_V27 in Fujian obtained by merging time-series optical remote sensing and the GEDI LiDAR data. (**d**) Boxplots of the MGEDI_V27 forest canopy height of Fujian grouped by city.

4.2.2. Accuracy Evaluation

In order to evaluate the validity of the canopy height estimation method that was used in this study, the accuracy of the GEDI_V27 and the MGEDI_V27 canopy height model was evaluated using 5299 sample points. In the evaluation results, we observed that the GEDI_V27 canopy height model has a low accuracy and a poor agreement with the sample set (Figure 13a, $R^2 = 0.39$, RMSE = 3.15 m, MAE = 2.37 m), while the accuracy of the MGEDI_V27 canopy height model, which was obtained by combining the optical images with the GEDI data, is a significant improvement (Figure 13b, $R^2 = 0.67$, RMSE = 2.24 m, and MAE = 1.85 m). The results thus show that the MGEDI_V27 compensates for the defects of the GEDI satellite-based LiDAR canopy height model to a certain extent. In addition, in order to visualize the advantages of combining the time-series optical remote sensing and the satellite-based LiDAR methods to estimate forest canopy height in subtropical mountainous areas, we plotted the observed canopy height values of some of the sample points and the two canopy height models, as shown in Figure 14. The comparison chart shows that the MGEDI_V27 is more concentrated in the observed value range, which means that the interference that is caused by external factors is reduced. The accuracy of the forest canopy height estimation in Fujian is, therefore, effectively improved.

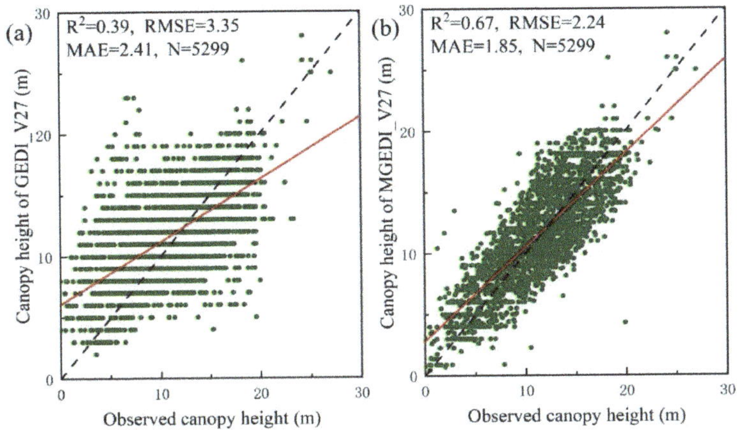

Figure 13. Accuracy assessment results of two canopy height models. (**a**) Linear regression of canopy height between the GEDI_V27 and the sample set. (**b**) Linear regression of canopy height between the MGEDI_V27 and the sample set. The dotted line represents a 1:1 match, and the red line represents the trend of the forest canopy height relationship between the field measurements and the two products.

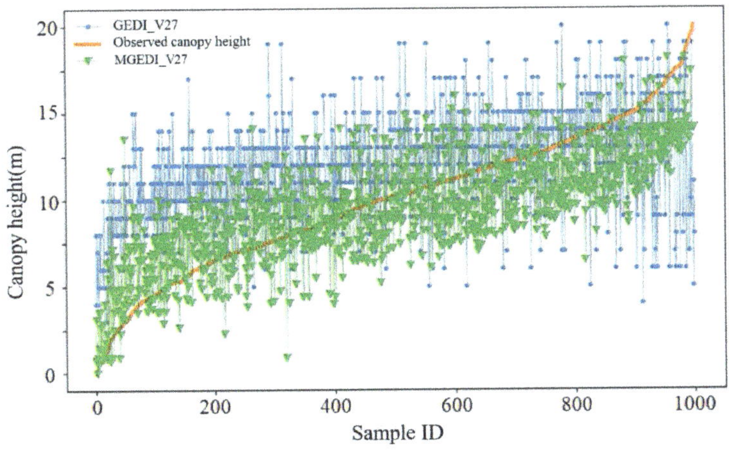

Figure 14. Comparison of canopy height of the GEDI_V27, the MGEDI_V27, and the observed canopy height.

5. Discussion

Fujian is extremely rich in forest resources, and the effective monitoring of forest resources is important for the stability of the Fujian ecosystem and the spatial analysis of forest carbon sinks [45]. Among the many forest parameters, canopy height is extremely important, but is difficult to obtain [46]. We attempted to use the methods from [14,15,47], however the results were poor, and the accuracy of the obtained results was very low. In addition, we evaluated the new canopy height product [19,20] using a sample dataset from Fujian and found that the accuracy was only about 0.4. The main reason for this is that the GEDI footprint data in Fujian are too sparse, while the environmental gradients in the mountainous region vary greatly, and the change process is highly complex and uncertain [48,49]. Therefore, it is important to address how these problems can be solved while preserving the advantages of satellite-based LiDAR for fast and effective measurements of forest canopy height. This paper combines long-time-series remote sensing images with satellite-based LiDAR canopy height in order to estimate the forest canopy height in a subtropical mountainous region, compensating for the fact that using long-time-series optical image features to estimate the forest tree height can easily reach saturation, while reducing the errors in satellite-based LiDAR canopy height acquisition [50–53]. In order to investigate the validity of the method, many local multi-source sample datasets were collected in this paper, such as the following: airborne LiDAR, UAV visible light data, tree height measurements in the field, and inventory plot data. We used these sample sets to compare the accuracy of the satellite-based LiDAR GEDI_V27 canopy height model and the MGEDI_V27 (R^2, RMSE, and MAE improved by 71.79%, 33.13%, and 22.53%, respectively), and the effectiveness of the method was confirmed. In addition, we also determined the distribution of forest disturbance in Fujian since 1986, the distribution of forest disturbance duration, the distribution of forest age in Fujian in 2020, and the growth curves of the major forest types in Fujian. The research results can provide important technical support in such tasks as the estimation of the stockpile volume, biomass estimation, and carbon sink analysis, and have important significance for forest ecological protection and management in Fujian.

In order to test the validity of this paper, we evaluated the accuracy of two canopy height models according to different forest age groups and slope classes. The forest age was divided into the following three age groups [54]: young forests (0–10 years), middle-aged forests (10–20 years), and mature forests (>20 years), and the slope class was divided into the following five classes: 0–10°, 10–20°, 20–30°, 30–40°, and >40°. The results suggest that the MGEDI_V27 shows an improved canopy height accuracy in all of the age groups and in the different slope classes. For the three age groups, we used 4981 forest plots from the forest plot survey data and included the forest age attributes as reference samples for accuracy verification. The accuracy of the GEDI_V27 was relatively low in all of the age groups (Figure 15), especially in the young forest plots (R^2 = 0.18, RMSE = 4.03 m, MAE = 3.14 m). The MGEDI_V27 concentrated more on the forest canopy height values from the field survey in all three of the age groups (Figure 16, R^2 = 0.33–0.55, RMSE = 2.15–2.27 m, MAE = 1.78–2.04 m). For the slope class, we used 5299 forest plotsas a reference sample for accuracy verification. The canopy height product GEDI_V27 had the highest accuracy in the slope range of 0–20°, but the accuracy decreased if the slope was greater than 20°. However, it is worth noting that it is not that the canopy height accuracy of the GEDI_V27 decreases with a higher slope, but that the stability of the overall accuracy of the product of the satellite-based LiDAR decreases if the slope is greater than 20° (Figure 17, R^2 = 0.32–0.45, RMSE = 2.47–3.38 m, MAE = 2.31–2.49 m). The MGEDI_V27 canopy height model is not affected by the slope (Figure 18, R^2 = 0.64–0.75, RMSE = 2.01–2.34 m, MAE = 1.71–1.99 m). The accuracy of the MGEDI_V27 had less variation under the different slope levels and was not significantly affected by topographic factors. There are two reasons for the lower accuracy of the GEDI_V27 (Figures 15 and 17). First, it uses the GEDI-derived RH95 index as the canopy height, which leads to a low forest canopy height in some areas. Second, the geographic positioning, especially in the mountainous areas where it is more complicated, suffers from a bias. Many other canopy height products have been obtained using satellite-based LiDAR [17,18,20,55]. In the future, we will further combine the advantages of various products in order to optimize the forest canopy height estimation approach.

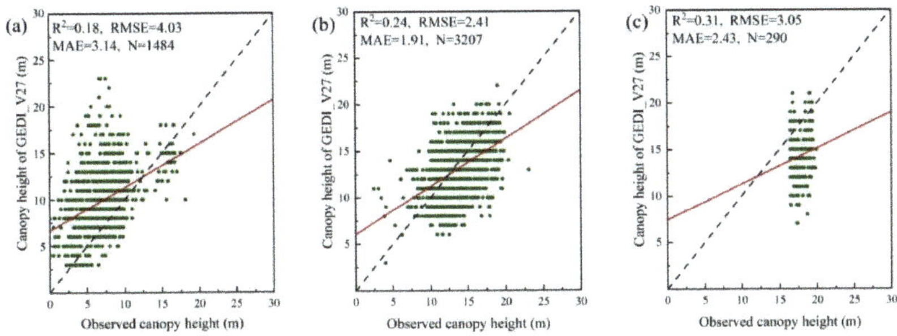

Figure 15. Canopy height accuracy for the three age groups of the GEDI_V27. (**a**) Young stands (0–10 years). (**b**) Middle-aged stands (10–20 years). (**c**) Mature stands (more than 20 years).

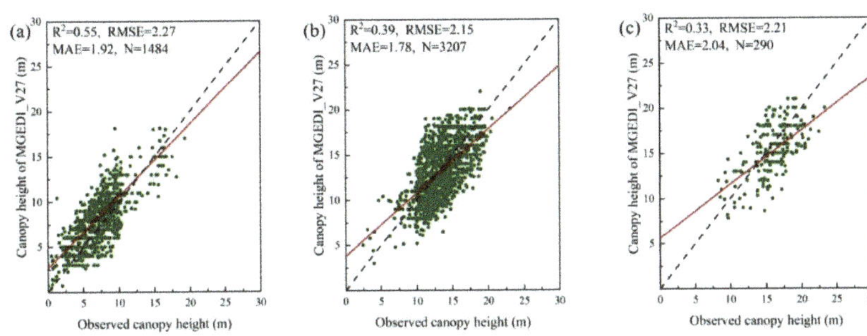

Figure 16. Canopy height accuracy for the three age groups of the MGEDI_V27. (**a**) Young stands (0–10 years). (**b**) Middle-aged stands (10–20 years). (**c**) Mature stands (more than 20 years).

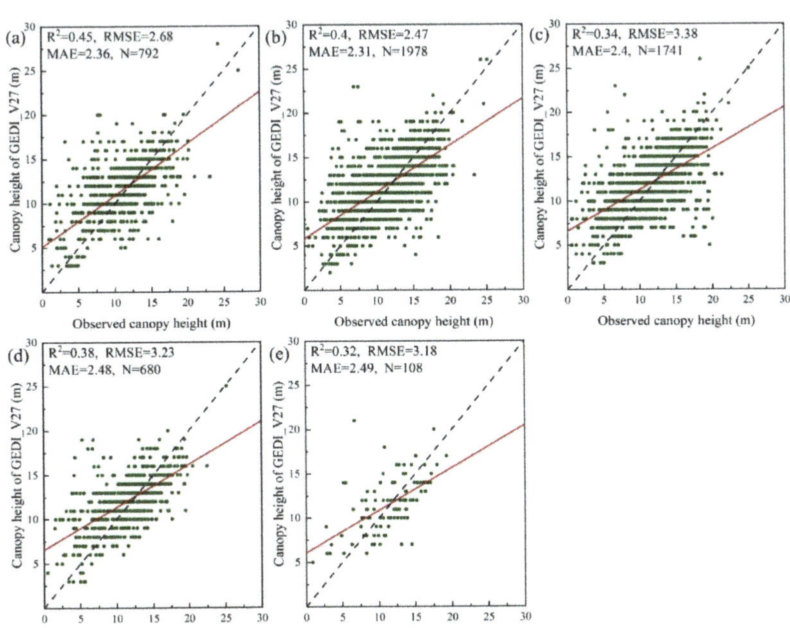

Figure 17. Canopy height accuracy of the GEDI_V27 at five slope levels. (**a**) 0–10°. (**b**) 10–20°. (**c**) 20–30°. (**d**) 30–40°. (**e**) >40°.

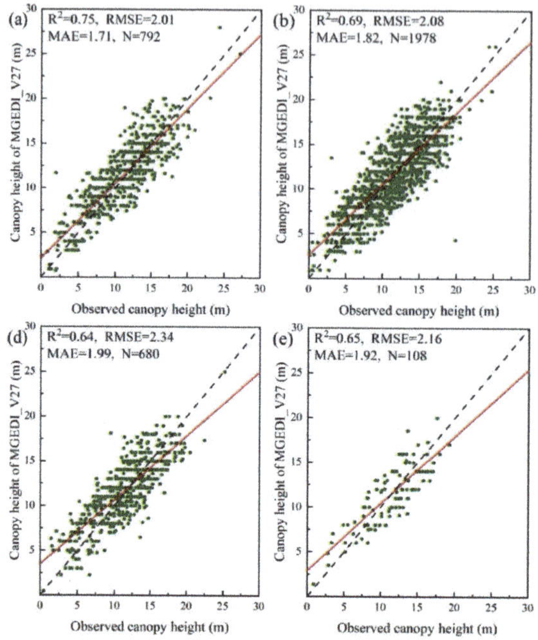

Figure 18. Canopy height accuracy of the MGEDI_V27 at five slope levels. (**a**) 0–10°. (**b**) 10–20°. (**c**) 20–30°. (**d**) 30–40°. (**e**) >40°.

Though the canopy height estimation method in this study shows great potential, there are still some limitations. First, the classification of the forest types is relatively rough, and some special tree species categories, such as eucalyptus and bamboo forests, have growth curves with unique characteristics. If more detailed tree species classification becomes available, then the method in this paper will more effectively improve the accuracy of the canopy height estimation. Second, the temporal resolution of the sample dataset that has been used in this paper is complex, and although it is as close as possible to the time of the GEDI data collection, there are still differences in the time of various data acquisitions, which may also cause some deviation in the canopy height in a small number of areas. Third, the growth curves of the forest types in this paper require a large amount of sample data, and, due to geographical limitations, the same forest types have different growth trajectories in different climates, therefore, the migration of the method in this paper needs to be enhanced. Fourth, in the undisturbed forest areas in Fujian, the forest sometimes does not grow normally, due to the influence of environmental factors. For example, the forests in Fujian that are above 1500 m in elevation do not exceed 10 m in height. Finally, although the GEDI footprint data are sparse in this study, the accuracy of the wall-to-wall satellite-based LiDAR data can be improved by using more advanced interpolation methods if the ICESat-2 data can be combined.

6. Conclusions

In this study, we demonstrate the utility of combining long time series Landsat satellite data with GEDI satellite-based LiDAR data to estimate forest tree height. In order to reduce the errors of satellite-based LiDAR data in complex areas, we produced a forest age distribution map of Fujian in 2020 that was based on long-time-series Landsat images and combined it with a large amount of historical forest plot inventory data in order to determine the relationship between the forest age and the average tree height of the coniferous forest, the broadleaf forest and the mixed forests in Fujian. We then combined this with the GEDI_V27 canopy height model in order to produce a modified forest canopy height in 2020. The main conclusions, which were verified by the accuracy of the estimation results, are as follows:

(1) The LandTrendr algorithm can effectively estimate forest age, and the error is about one to two years after sample verification. The results show that most of the forest disturbance lengths in Fujian can be recovered in a one-year time range, and there are few forests with disturbance lengths of more than two years.
(2) The canopy height products of the satellite-based LiDAR GEDI in Fujian are influenced by the topography and the climate, and the accuracy is lower than that of the global scale. Through the statistical analysis of the observed data, the results show that the R^2 is 0.39, the RMSE is 3.35 m, and the MAE is 2.41 m. Therefore, relying only on the satellite-based LiDAR canopy height model cannot meet the needs of subsequent studies on estimating the forest stock, estimating the biomass, and analyzing the forest carbon sink potential.
(3) The combination of long-time-series optical remote sensing images and native forest growth curves can improve the accuracy of satellite-based LiDAR canopy height, and, to a certain extent, compensate for the large errors of satellite-based LiDAR data acquisition in complex terrain areas. The accuracy evaluation results of the MGEDI_V27 canopy height that were obtained after the method of this paper were an R^2 of 0.67, an RMSE of 2.24 m, and an MAE of 1.85 m. This paper also compares the accuracy values of the GEDI_V27 and the MGEDI_V27 by age group and slope grade. In terms of the age groups, the accuracy of the canopy height that was obtained by the method in this paper was significantly higher (R^2, RMSE, and MAE increased by 91.5%, 27.33%, and 20.57%, on average), and the accuracy improvement of the young forests was the highest. The accuracy of the MGEDI_V27 was generally higher than that of the GEDI_V27 at different the slope levels, where the accuracy improvement was more obvious in the areas with a slope that was greater than 20° (R^2, RMSE, and MAE increased by 80.96%, 26.82%, and 22.53%, on average).

Author Contributions: Conceptualization: X.Z. and L.D.; methodology and software: Y.H. and Y.C.; data curation, writing—original draft preparation, and visualization: Y.H. and X.W.; writing—review and editing: Y.H. and X.Z.; revising: C.C., T.J. and G.N. All authors have read and agreed to the published version of the manuscript.

Funding: This research was supported by the International Cooperation Project of Fujian, China (No. 2022I0007), the Industry-University Cooperation Project of the Science and Technology Department of Fujian (No. 2022N5008), and the Science and Technology Project of Fujian Provincial Water Conservancy Department (No. MSK202214).

Data Availability Statement: The data presented in this study are available on request from the corresponding author.

Acknowledgments: The authors would like to thank Tan Fanglin, from th Fujian Academy of Forestry Sciences, for providing most of the verification data.

Conflicts of Interest: The authors declare that they have no conflict of interest.

References

1. Anaya, J.A.; Chuvieco, E.; Palacios-Orueta, A. Aboveground biomass assessment in Colombia: A remote sensing approach. *For. Ecol. Manag.* **2009**, *257*, 1237–1246. [CrossRef]
2. Peng, S.; Wen, D.; He, N.; Yu, G.; Ma, A.; Wang, Q. Carbon storage in China's forest ecosystems: Estimation by different integrative methods. *Ecol. Evol.* **2016**, *6*, 3129–3145. [CrossRef] [PubMed]
3. Bonan, G.B. Forests and climate change: Forcings, feedbacks, and the climate benefits of forests. *Science* **2008**, *320*, 1444–1449. [CrossRef] [PubMed]
4. Asner, G.P.; Mascaro, J.; Muller-Landau, H.C.; Vieilledent, G.; Vaudry, R.; Rasamoelina, M.; Hall, J.S.; Van Breugel, M. A universal airborne LiDAR approach for tropical forest carbon mapping. *Oecologia* **2012**, *168*, 1147–1160. [CrossRef]
5. Lin, X.; Xu, M.; Cao, C.; Dang, Y.; Bashir, B.; Xie, B.; Huang, Z. Estimates of forest canopy height using a combination of ICESat-2/ATLAS data and stereo-photogrammetry. *Remote Sens.* **2020**, *12*, 3649. [CrossRef]

6. Campbell, M.J.; Dennison, P.E.; Kerr, K.L.; Brewer, S.C.; Anderegg, W.R. Scaled biomass estimation in woodland ecosystems: Testing the individual and combined capacities of satellite multispectral and lidar data. *Remote Sens. Environ.* **2021**, *262*, 112511. [CrossRef]
7. Wang, M.; Sun, R.; Xiao, Z. Estimation of forest canopy height and aboveground biomass from spaceborne LiDAR and Landsat imageries in Maryland. *Remote Sens.* **2018**, *10*, 344. [CrossRef]
8. Pang, S.; Li, G.; Jiang, X.; Chen, Y.; Lu, Y.; Lu, D. Retrieval of forest canopy height in a mountainous region with ICESat-2 ATLAS. *For. Ecosyst.* **2022**, *9*, 100046. [CrossRef]
9. Davies, A.B.; Oram, F.; Ancrenaz, M.; Asner, G.P. Combining behavioural and LiDAR data to reveal relationships between canopy structure and orangutan nest site selection in disturbed forests. *Biol. Conserv.* **2019**, *232*, 97–107. [CrossRef]
10. Goetz, S.; Dubayah, R. Advances in remote sensing technology and implications for measuring and monitoring forest carbon stocks and change. *Carbon Manag.* **2011**, *2*, 231–244. [CrossRef]
11. Boudreau, J.; Nelson, R.F.; Margolis, H.A.; Beaudoin, A.; Guindon, L.; Kimes, D.S. Regional aboveground forest biomass using airborne and spaceborne LiDAR in Québec. *Remote Sens. Environ.* **2008**, *112*, 3876–3890. [CrossRef]
12. Fayad, I.; Baghdadi, N.; Riédi, J. Quality assessment of acquired gedi waveforms: Case study over france, tunisia and french guiana. *Remote Sens.* **2021**, *13*, 3144. [CrossRef]
13. Dubayah, R.; Blair, J.B.; Goetz, S.; Fatoyinbo, L.; Hansen, M.; Healey, S.; Hofton, M.; Hurtt, G.; Kellner, J.; Luthcke, S. The Global Ecosystem Dynamics Investigation: High-resolution laser ranging of the Earth's forests and topography. *Sci. Remote Sens.* **2020**, *1*, 100002. [CrossRef]
14. Fang, L.; Yang, J.; Zhang, W.; Zhang, W.; Yan, Q. Combining allometry and landsat-derived disturbance history to estimate tree biomass in subtropical planted forests. *Remote Sens. Environ.* **2019**, *235*, 111423. [CrossRef]
15. Helmer, E.H.; Ruzycki, T.S.; Wunderle, J.M., Jr.; Vogesser, S.; Ruefenacht, B.; Kwit, C.; Brandeis, T.J.; Ewert, D.N. Mapping tropical dry forest height, foliage height profiles and disturbance type and age with a time series of cloud-cleared Landsat and ALI image mosaics to characterize avian habitat. *Remote Sens. Environ.* **2010**, *114*, 2457–2473. [CrossRef]
16. Pflugmacher, D.; Cohen, W.B.; Kennedy, R.E.; Yang, Z. Using Landsat-derived disturbance and recovery history and lidar to map forest biomass dynamics. *Remote Sens. Environ.* **2014**, *151*, 124–137. [CrossRef]
17. Li, W.; Niu, Z.; Shang, R.; Qin, Y.; Wang, L.; Chen, H. High-resolution mapping of forest canopy height using machine learning by coupling ICESat-2 LiDAR with Sentinel-1, Sentinel-2 and Landsat-8 data. *Int. J. Appl. Earth Obs.* **2020**, *92*, 102163. [CrossRef]
18. Simard, M.; Pinto, N.; Fisher, J.B.; Baccini, A. Mapping forest canopy height globally with spaceborne lidar. *J. Geophys. Res. Biogeosciences* **2011**, *116*, G04021. [CrossRef]
19. Potapov, P.; Li, X.; Hernandez-Serna, A.; Tyukavina, A.; Hansen, M.C.; Kommareddy, A.; Pickens, A.; Turubanova, S.; Tang, H.; Silva, C.E.; et al. Mapping global forest canopy height through integration of GEDI and Landsat data. *Remote Sens. Environ.* **2021**, *253*, 112165. [CrossRef]
20. Adam, M.; Urbazaev, M.; Dubois, C.; Schmullius, C. Accuracy assessment of GEDI terrain elevation and canopy height estimates in European temperate forests: Influence of environmental and acquisition parameters. *Remote Sens.* **2020**, *12*, 3948. [CrossRef]
21. Liu, A.; Cheng, X.; Chen, Z. Performance evaluation of GEDI and ICESat-2 laser altimeter data for terrain and canopy height retrievals. *Remote Sens. Environ.* **2021**, *264*, 112571. [CrossRef]
22. Roy, D.P.; Kashongwe, H.B.; Armston, J. The impact of geolocation uncertainty on GEDI tropical forest canopy height estimation and change monitoring. *Sci. Remote Sens.* **2021**, *4*, 100024. [CrossRef]
23. Liu, X.; Su, Y.; Hu, T.; Yang, Q.; Liu, B.; Deng, Y.; Tang, H.; Tang, Z.; Fang, J.; Guo, Q. Neural network guided interpolation for mapping canopy height of China's forests by integrating GEDI and ICESat-2 data. *Remote Sens. Environ.* **2022**, *269*, 112844. [CrossRef]
24. Zhuang, F.; Qi, Z.; Duan, K.; Xi, D.; Zhu, Y.; Zhu, H.; Xiong, H.; He, Q. A comprehensive survey on transfer learning. *Proc. IEEE* **2020**, *109*, 43–76. [CrossRef]
25. Weiss, K.; Khoshgoftaar, T.M.; Wang, D. A survey of transfer learning. *J. Big Data* **2016**, *3*, 9. [CrossRef]
26. Maschler, B.; Weyrich, M. Deep transfer learning for industrial automation: A review and discussion of new techniques for data-driven machine learning. *IEEE Ind. Electron. Mag.* **2021**, *15*, 65–75. [CrossRef]
27. Rennolls, K. Forest height growth modelling. *Forest Ecol. Manag.* **1995**, *71*, 217–225. [CrossRef]
28. Zhang, C.; Ju, W.; Chen, J.M.; Li, D.; Wang, X.; Fan, W.; Li, M.; Zan, M. Mapping forest stand age in China using remotely sensed forest height and observation data. *J. Geophys. Res. Biogeosciences* **2014**, *119*, 1163–1179. [CrossRef]
29. Kayitakire, F.; Hamel, C.; Defourny, P. Retrieving forest structure variables based on image texture analysis and IKONOS-2 imagery. *Remote Sens. Environ.* **2006**, *102*, 390–401. [CrossRef]
30. Cutler, M.; Boyd, D.S.; Foody, G.M.; Vetrivel, A. Estimating tropical forest biomass with a combination of SAR image texture and Landsat TM data: An assessment of predictions between regions. *ISPRS J. Photogramm.* **2012**, *70*, 66–77. [CrossRef]
31. Yang, X.; Liu, Y.; Wu, Z.; Yu, Y.; Li, F.; Fan, W. Forest age mapping based on multiple-resource remote sensing data. *Environ. Monit. Assess.* **2020**, *192*, 734. [CrossRef] [PubMed]
32. Sun, C.; Cao, S.; Sanchez-Azofeifa, G.A. Mapping tropical dry forest age using airborne waveform LiDAR and hyperspectral metrics. *Int. J. Appl. Earth Obs.* **2019**, *83*, 101908. [CrossRef]

33. Hovi, A.; Lindberg, E.; Lang, M.; Arumäe, T.; Peuhkurinen, J.; Sirparanta, S.; Pyankov, S.; Rautiainen, M. Seasonal dynamics of albedo across European boreal forests: Analysis of MODIS albedo and structural metrics from airborne LiDAR. *Remote Sens. Environ.* **2019**, *224*, 365–381. [CrossRef]
34. Dong, J.; Xiao, X.; Chen, B.; Torbick, N.; Jin, C.; Zhang, G.; Biradar, C. Mapping deciduous rubber plantations through integration of PALSAR and multi-temporal Landsat imagery. *Remote Sens. Environ.* **2013**, *134*, 392–402. [CrossRef]
35. Hanna, K.D.; Schrader, D.L.; Cloutis, E.A.; Cody, G.D.; King, A.J.; McCoy, T.J.; Applin, D.M.; Mann, J.P.; Bowles, N.E.; Brucato, J.R. Spectral characterization of analog samples in anticipation of OSIRIS-REx's arrival at Bennu: A blind test study. *Icarus* **2019**, *319*, 701–723. [CrossRef]
36. Kuusinen, N.; Tomppo, E.; Shuai, Y.; Berninger, F. Effects of forest age on albedo in boreal forests estimated from MODIS and Landsat albedo retrievals. *Remote Sens. Environ.* **2014**, *145*, 145–153. [CrossRef]
37. Shen, J.; Chen, G.; Hua, J.; Huang, S.; Ma, J. Contrasting Forest Loss and Gain Patterns in Subtropical China Detected Using an Integrated LandTrendr and Machine-Learning Method. *Remote Sens.* **2022**, *14*, 3238. [CrossRef]
38. Peng, D.; Zhang, H.; Liu, L.; Huang, W.; Huete, A.R.; Zhang, X.; Wang, F.; Yu, L.; Xie, Q.; Wang, C. Estimating the aboveground biomass for planted forests based on stand age and environmental variables. *Remote Sens.* **2019**, *11*, 2270. [CrossRef]
39. Kennedy, R.E.; Yang, Z.; Cohen, W.B. Detecting trends in forest disturbance and recovery using yearly Landsat time series: 1. LandTrendr—Temporal segmentation algorithms. *Remote Sens. Environ.* **2010**, *114*, 2897–2910. [CrossRef]
40. Dubayah, R.; Hofton, M.; Blair, J.B.; Armston, J.; Tang, H.; Luthcke, S. GEDI L2A Elevation and Height Metrics Data Global Footprint Level V001. NASA EOSDIS Land Processes DAAC. Available online: https://doi.org/10.5067/GEDI/GEDI02_A.001 (accessed on 12 April 2021).
41. Serna, E.H.; Hernandez-Serna, A. pyGEDI: NASA's Global Ecosystem Dynamics Investigation (GEDI) Mission Data Extraction, Analysis, Processing and Visualization. Version 0.2, 5 April 2020. Available online: https://pypi.org/project/pyGEDI (accessed on 10 August 2021).
42. Kennedy, R.E.; Yang, Z.; Gorelick, N.; Braaten, J.; Cavalcante, L.; Cohen, W.B.; Healey, S. Implementation of the LandTrendr algorithm on google earth engine. *Remote Sens.* **2018**, *10*, 691. [CrossRef]
43. Ye, L.; Liu, M.; Liu, X.; Zhu, L. Developing a new disturbance index for tracking gradual change of forest ecosystems in the hilly red soil region of southern China using dense Landsat time series. *Ecol. Inform.* **2021**, *61*, 101221. [CrossRef]
44. Yu, Y.; Pan, J.; Xing, L.; Jiang, L.; Liu, S.; Yuan, Y.; Yu, H. Identification of high temperature targets in remote sensing imagery based on factor analysis. *J. Appl. Remote Sens.* **2014**, *8*, 83622. [CrossRef]
45. Ren, Y.; Wei, X.; Zhang, L.; Cui, S.; Chen, F.; Xiong, Y.; Xie, P. Potential for forest vegetation carbon storage in Fujian, China, determined from forest inventories. *Plant Soil* **2011**, *345*, 125–140. [CrossRef]
46. Guo, Q.; Su, Y.; Hu, T.; Guan, H.; Jin, S.; Zhang, J.; Zhao, X.; Xu, K.; Wei, D.; Kelly, M. Lidar boosts 3d ecological observations and modelings: A review and perspective. *IEEE Geosci. Remote Sens. Mag.* **2020**, *9*, 232–257. [CrossRef]
47. Ota, T.; Ahmed, O.S.; Franklin, S.E.; Wulder, M.A.; Kajisa, T.; Mizoue, N.; Yoshida, S.; Takao, G.; Hirata, Y.; Furuya, N. Estimation of airborne lidar-derived tropical forest canopy height using landsat time series in Cambodia. *Remote Sens.* **2014**, *6*, 10750–10772. [CrossRef]
48. Liu, M.; Cao, C.; Dang, Y.; Ni, X. Mapping forest canopy height in mountainous areas using ZiYuan-3 stereo images and Landsat data. *Forests* **2019**, *10*, 105. [CrossRef]
49. Zhang, J.; Nielsen, S.E.; Mao, L.; Chen, S.; Svenning, J.C. Regional and historical factors supplement current climate in shaping global forest canopy height. *J. Ecol.* **2016**, *104*, 469–478. [CrossRef]
50. Ahmed, O.S.; Franklin, S.E.; Wulder, M.A.; White, J.C. Characterizing stand-level forest canopy cover and height using Landsat time series, samples of airborne LiDAR, and the Random Forest algorithm. *ISPRS J. Photogramm.* **2015**, *101*, 89–101. [CrossRef]
51. Hansen, M.C.; Potapov, P.V.; Goetz, S.J.; Turubanova, S.; Tyukavina, A.; Krylov, A.; Kommareddy, A.; Egorov, A. Mapping tree height distributions in Sub-Saharan Africa using Landsat 7 and 8 data. *Remote Sens. Environ.* **2016**, *185*, 221–232. [CrossRef]
52. Shang, C.; Coops, N.C.; Wulder, M.A.; White, J.C.; Hermosilla, T. Update and spatial extension of strategic forest inventories using time series remote sensing and modeling. *Int. J. Appl. Earth Obs.* **2020**, *84*, 101956. [CrossRef]
53. Wang, C.; Elmore, A.J.; Numata, I.; Cochrane, M.A.; Shaogang, L.; Huang, J.; Zhao, Y.; Li, Y. Factors affecting relative height and ground elevation estimations of GEDI among forest types across the conterminous USA. *Gisci. Remote Sens.* **2022**, *59*, 975–999. [CrossRef]
54. Zhang, Y.; Yao, Y.; Wang, X.; Liu, Y.; Piao, S. Mapping spatial distribution of forest age in China. *Earth Space Sci.* **2017**, *4*, 108–116. [CrossRef]
55. Lang, N.; Kalischek, N.; Armston, J.; Schindler, K.; Dubayah, R.; Wegner, J.D. Global canopy height regression and uncertainty estimation from GEDI LIDAR waveforms with deep ensembles. *Remote Sens. Environ.* **2022**, *268*, 112760. [CrossRef]

remote sensing

MDPI

Article

Very High Resolution Images and Superpixel-Enhanced Deep Neural Forest Promote Urban Tree Canopy Detection

Yang Liu [1,2,3], Huaiqing Zhang [1,2,3,*], Zeyu Cui [1,2,3], Kexin Lei [1,2,3], Yuanqing Zuo [1,2,3], Jiansen Wang [1,2,3], Xingtao Hu [1,2,3,4] and Hanqing Qiu [1,2,3]

1 Institute of Forest Resource Information Techniques, Chinese Academy of Forestry, Beijing 100091, China
2 Key Laboratory of Forestry Remote Sensing and Information System, The National Forestry and Grassland Administration, Beijing 100091, China
3 Dongting Lake Remote Sensing Product Validation Station, Beijing 100091, China
4 School of Geography and Environmental Sciences, Guizhou Normal University, Guiyang 550025, China
* Correspondence: zhang@ifrit.ac.cn

Abstract: Urban tree canopy (UTC) area is an important index for evaluating the urban ecological environment; the very high resolution (VHR) images are essential for improving urban tree canopy survey efficiency. However, the traditional image classification methods often show low robustness when extracting complex objects from VHR images, with insufficient feature learning, object edge blur and noise. Our objective was to develop a repeatable method—superpixel-enhanced deep neural forests (SDNF)—to detect the UTC distribution from VHR images. Eight data expansion methods was used to construct the UTC training sample sets, four sample size gradients were set to test the optimal sample size selection of SDNF method, and the best training times with the shortest model convergence and time-consumption was selected. The accuracy performance of SDNF was tested by three indexes: F1 score (F1), intersection over union (IoU) and overall accuracy (OA). To compare the detection accuracy of SDNF, the random forest (RF) was used to conduct a control experiment with synchronization. Compared with the RF model, SDNF always performed better in OA under the same training sample size. SDNF had more epoch times than RF, converged at the 200 and 160 epoch, respectively. When SDNF and RF are kept in a convergence state, the training accuracy is 95.16% and 83.16%, and the verification accuracy is 94.87% and 87.73%, respectively. The OA of SDNF improved 10.00%, reaching 89.00% compared with the RF model. This study proves the effectiveness of SDNF in UTC detection based on VHR images. It can provide a more accurate solution for UTC detection in urban environmental monitoring, urban forest resource survey, and national forest city assessment.

Keywords: VHR; urban tree canopy; superpixel-enhanced deep neural forest; remote sensing

1. Introduction

Urban forest is an important material basis for urban economic development and can assure the physical and mental health of urban residents [1]. Canopy cover in urban regions is one of the most important indicators for monitoring planning and evaluating urban forests [2]. In 2007, China issued The National Forest City Assessment Indicators [3], it set tree canopy cover as an important indicator for evaluating the forest city. Urban tree canopy (UTC) is generally estimated as the area covered by tree leaf layers, branches and trunks when viewed vertically above the trees [4]. UTC is the most relevant stability optimization factor with the urban forest health assessment [5]. In this study, the term "UTC" refers to all trees in the urban region, including individual street trees and clusters of park trees, while peri-urban woods extend to the outskirts of the metropolitan area. City parks include urban forests larger than 0.5 ha, pocket parks and gardens with trees, trees on streets or in public squares, any other green areas with trees, and riparian corridors and roofs [1].

In recent years, the analysis of very high resolution (VHR) images has proven very useful for extracting UTC [2–7]. With the unique spatial location (streetside, roadside, waterside, and building-side) [8], there have been great challenges for the segmentation of the accurate and consistent boundary based on VHR image classification [9]. The traditional remote sensing classification methods for medium- and low-resolution images are not suitable for VHR images, which contain abundant texture information [10]. Superpixel clusters image regions through grouping pixels; comparing the pixel, it provides image data in a more natural representation [11–13]. Learning-based superpixel segmentation method can effectively alleviate edge blur and classification noise [14–16].

Traditional machine learning methods heavily depend on the feature extraction process, such as Markov random fields (MRFs), random forest (RF) and conditional random fields (CRFs) [13]. However, deep convolutional neural networks (DCNNs) have better performance in image classification and learning feature representation simultaneously [17,18]. Although the powerful representation capability of DCNNs has promoted the development of image semantic segmentation, there are still many challenges: (1) The complexity of the surface feature spectrum, the features of high intraclass variance and low interclass variance are widely distributed, which easily confused in VHR images. DCNNs still exhibit a significant imbalance between the training of classifiers and representations for VHR images [19,20]. Simple categorization criteria are still insufficient to differentiate between ground objects. For instance, trees and submerged vegetation are frequently confused by spectral similarity [21–24]. (2) Ground objects in VHR images have strong requirements for consistency and boundary accuracy [25]. DCNNs have made remarkable achievements in semantic segmentation. Expanding the receptive field also increases edge blur and noise, since it employs a broad receptive field to obtain context information for classification [26]. In order to address these two challenges, we introduced a new VHR image classification framework—superpixel-enhanced deep neural forests (SDNF) [13].

The purpose of this study is to improve SDNF semantic segmentation based on Gaofen 2 (GF-2) VHR images, end-to-end integration of deep neural network and decision tree, and add a superpixel-enhanced region module (SRM) to conduct the framework increase the consistency and boundary precision for objects and to improve the sparse and small object detection ability for VHR images.

2. Materials and Methods

2.1. Study Area

The study area is Hui'an county, which is located on the southeast coast of Fujian province (Figure 1A,B), between Quanzhou Bay and Meizhou Bay. The total area of the study area is 793.04 km^2 (Figure 1C,D). The climate of Hui'an has four basic characteristics: high temperature, rich light and heat, abundant precipitation and monsoon climate. The precipitation is mainly concentrated in spring and summer, annual average precipitation approximately 1000.00 mm.

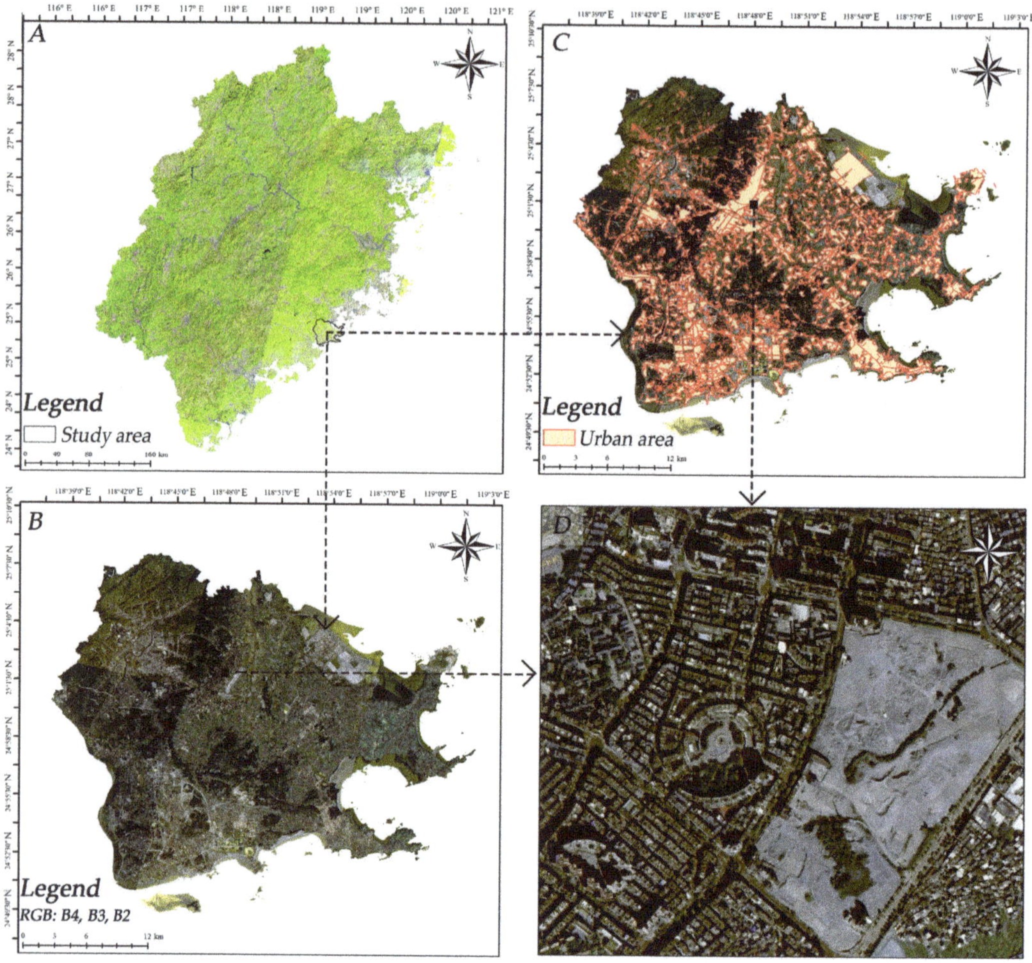

Figure 1. Study area: (**A**) the location of study area in Fujian province; (**B**) the GF-2 images of study area; (**C**) the distribution of urban areas of study area; (**D**) the images after superpixel-enhancement.

2.2. Materials

2.2.1. Data Source

In order to accurately detect the distribution of tree crowns in the urban area, seven VHR Gaofen-2 (GF-2) images with no cloud coverage were collected in 2021. GF-2 includes two PMS sensors that can obtain multispectral image with four bands, the spectral range is 0.45–0.89 μm and the spatial resolution is 4 m. In addition, PMS sensor can also obtain panchromatic images with higher spatial resolution, for which spectral range is 0.45~0.89 μm with one band, and the spatial resolution is 1 m. See Table 1 for details. The auxiliary data include digital elevation model (DEM) data of 30 m, the forest resource planning and design survey data at the study area in 2020 (FRPDSD2020) and a Google Earth visual selection of 40,000 images of sample data.

Table 1. GF-2 image band information.

ID	Acquisition Date	Center Longitude and Latitude	Scene Identifier	Senor	Spatial Resolution (Multispectral and Panchromatic)
1	2021/11/25	E118.7_N25.1	9600277	PMS 1	4 m and 1 m
2	2021/11/25	E118.6_N24.9	9600278	PMS 1	4 m and 1 m
3	2021/11/30	E119.1_N25.1	9617729	PMS 1	4 m and 1 m
4	2021/11/30	E119.1_N24.9	9617730	PMS 1	4 m and 1 m
5	2021/11/30	E119.0_N24.8	9617731	PMS 1	4 m and 1 m
6	2021/11/25	E118.9_N25.1	9600565	PMS 2	4 m and 1 m
7	2021/11/25	E118.8_N24.9	9600566	PMS 2	4 m and 1 m

2.2.2. Data Preprocessing

(1) The GF-2 absolute radiometric calibration coefficient provided by China Centre for Resources Satellite Data and Application (http://www.cresda.com/CN/ (accessed on 17 November 2021) to perform radiometric calibration calibrated panchromatic data as surface reflectance (SR). (2) The FLAASH module was used to calibrate the atmospheric correction on SR images. (3) The DEM and FRPDSD2020 were used to process orthorectification and geo-spatial registration for GF-2 images, registration error < 1 pixel. (4) Fused the corrected multispectral and panchromatic images. (5) Clipped the fused images and used the study area boundary to obtain the region of interest (ROI). The above operations were implemented in ENVI 5.3 and ArcGIS 10.2. Finally, the ROI images with 1m spatial resolution and four multispectral bands, and the root mean square error is 0.039 m. The image size of the study area was 34,620 rows and 41,948 columns, with a total area of 793.04 km^2, as shown in Figure 1C.

2.3. *Method*

In this study, based on the GF-2 VHR images, the SDNF framework was used [13] to solve the problems in UTC extraction where detection difficulty and boundary blur always occur.

2.3.1. SDNF

SDNF combined the advantages of DCNNs and tree-based model in encoder-decoder framework, to enhance decision tree (DT) and deep neural forest (DNF) image classification ability. There are two main parts: DNF and SRM. In order to improve the classification ability of DCNNs, the encoder was based on atrous spatial pyramid pooling (ASPP) and the residual neural network 101 (ResNet101), and closely combined with deep neural forest (DNF). The image features encoded in ResNet101's first four modules were delivered to the ASPP. As the following part, we focused on the interpretation of ASPP to represent different proportions of local and global data at various scales. Then the differentiable forest decision nodes were predicted through random paths using the fully differentiable forest as the decoder. In the SRM module, a learning-based superpixel segmentation task was introduced to emphasize the homogeneity inner the superpixel and the heterogeneity between superpixels to reduce classification noise and edge blur. Superpixel constraints were added in an end-to-end manner with designing superpixel enhanced segmentation loss. The SDNF framework is shown in Figure 2.

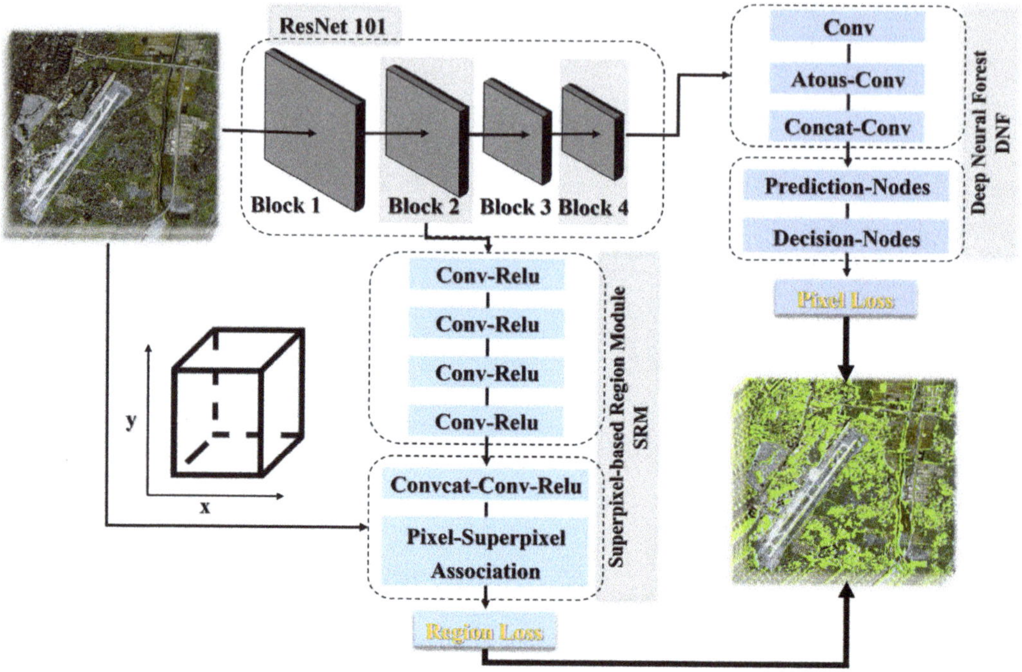

Figure 2. The structure of SDNF model.

(1) Deep neural forest (DNF). Based on ASPP, the feature representation of image classification considered more context information from the image scenario. However, if the number of convolution kernels was increased, the repeated combination of maximum pool and step size would lead to the number of parameters increased while the size of feature mapping would decrease. The former made network training more difficult, while the latter made difficult application of the decoder for upsampling. Atrous-Conv can avoid these problems (Figure 3). For each location (i, j) on mapping of input feature x_{conv} and the filter "f", the formulation of atrous convolution be expressed as following:

$$y_{conv}[i\,,\,j] \equiv \sum_{k=1}^{|K|} x_{conv}[i+rk\,,\,j+rk]\; f[k\,,\,k], \tag{1}$$

where K was the kernel size of filter f, y_{conv} was the output convolution result and r was the corresponding atrous rate.

In Figure 4, when rate = 6, the receptive field of the Atrous Conv convolution nucleus was 23 × 23; when rate = 12, the receptive field of Atrous Conv was 47 × 47; when rate = 18, the receptive field of Atrous Conv was 71 × 71; when rate = 24, the receptive field of Atrous Conv was 99 × 99, respectively.

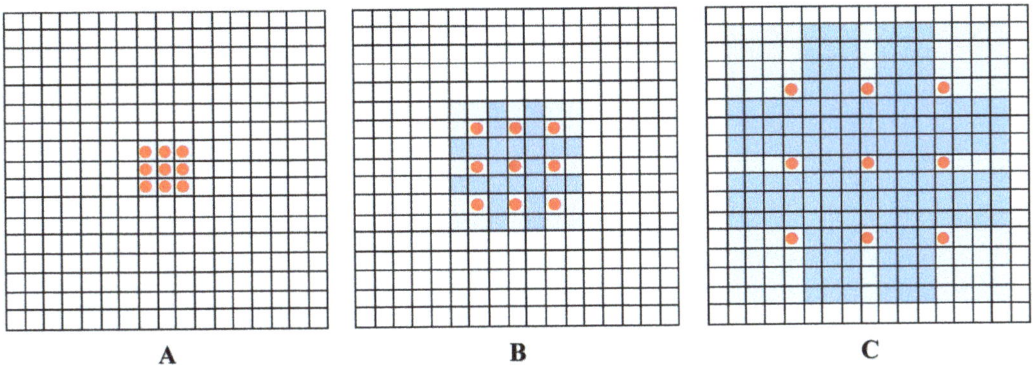

Figure 3. Conceptual graph of atrous convolution. (**A**) The 1-dilated conv of 3 × 3; (**B**) the 2-dilated conv of 3 × 3; (**C**) the 4-dilated conv of 3 × 3.

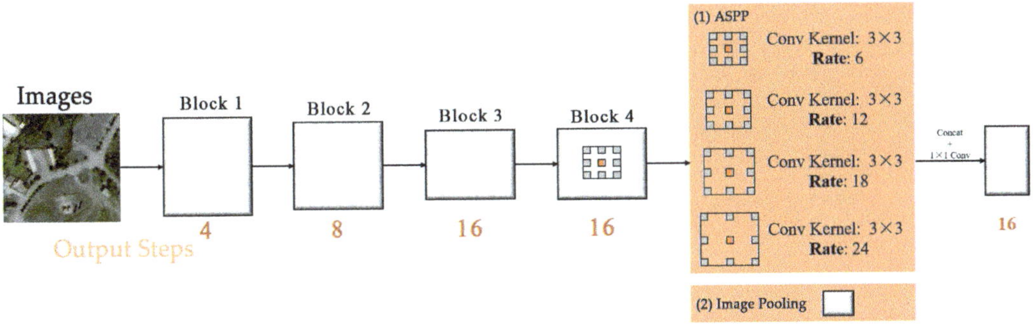

Figure 4. The structure of ASPP.

A differentiable forest was introduced as a decoder to conduct the representation learning of the DCNN. The decision tree was a tree-structured classifier consisting of split nodes and leaf nodes. The differentiable forest can be implemented to the overall framework of DCNNs through stochastic routing, which was different from traditional decision forest with deterministic routing rules. Decision nodes indexed by $\mathcal{N}$ were internal nodes of the tree, while prediction nodes indexed by $\mathcal{L}$ were the final nodes of the tree. Each prediction node $l \in \mathcal{L}$ held a probability distribution π_l over y. Each decision node $n \in \mathcal{N}$ was instead assigned a decision function $d_n(\,;\Theta) : x \to [0,1]$ parametrized by Θ, which was responsible for routing samples along the tree. When a sample $x \in \mathcal{X}$ reached a decision node n it was sent to the left or right subtree based on the output of $d_n(\,;\Theta)$. In standard decision forests, d_n was binary and the routing is deterministic. Rather, in this paper, we consider a probabilistic route. Once a sample ends in a leaf node l, the related tree prediction is provided by the class label distribution π_l. In the case of stochastic routings, the leaf predictions are averaged by the probability of reaching the leaf [27]. The final prediction for sample x from tree $\mathcal{T}$ with decision nodes parametrized by Θ as the following:

$$P_T(y|x,\Theta,\pi) = \sum_{l \in \mathcal{L}} \pi_{ly}\mu_l(x|\Theta), \tag{2}$$

where $\mu_l(x|\Theta)$ is regarded as the routing function providing the probability that sample x will reach leaf l. $\sum_l \mu_l(x|\Theta) = 1$ for all $x \in \mathcal{X}$, $\pi = (\pi_l)_{l\in\mathcal{L}}$ and π_{ly} denotes the probability of a sample reaching leaf l to take on y.

(2) Superpixel-based region module (SRM). A superpixel is a small area composed of adjacent pixels with similar color, brightness, texture and other features. Most of these regions have effective information for image segmentation and have generally not destroyed the boundary information of objects in the image. A superpixel is the abstraction of basic information elements, dividing a pixel level image into a district level image. To enhance the target boundary and ensure the regional continuity, SDNF introduced superpixel segmentation as an auxiliary module. The block 2 mapping features of residual network (ResNet) was sent to a various of convolution layers to activate was rectified linear unit (ReLU) and batch normalization (BN). The maximum pool's window size was set to 2, which was used to expand the receptive field after secondary convolution. The convolution filter output channel was set to 64, and the convolutional neural network (CNN) output channel was set to m. Concatenation was conducted on the original image, the input of ResNet block 2, and two pooled features in XYLab. The outcome after combining XYLab channels was m+5 dimensional pixel features, m+5 dimensional feature was transferred to two modules of simple linear iterative cluster (SLIC) [28], which have updated the correlation between the v iteration pixel and the super pixel center. The complexity of SLIC was linear in the number of pixels in the image O (N), while the conventional k-means algorithm was O (kNI), where I was the number of iterations. This provided a search space for each cluster center in the allocation step. Figure 5A shows that in the conventional k-means algorithm, the distance from each cluster center to each pixel in the image was calculated. Figure 5B shows that SLIC only calculates data from each cluster center to 2S × Distance of pixels in 2S area. The expected superpixel size here was only S × S, represented by a smaller square. This method not only reduces the distance calculation, but also makes the complexity of SLIC independent of the number of superpixels.

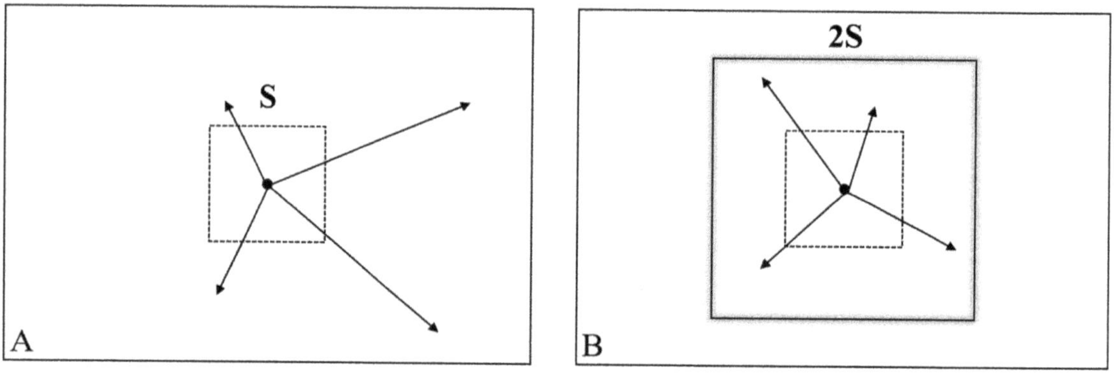

Figure 5. Superpixel reduction search region. (**A**) The standard k-means search region; (**B**) SLIC model search region.

For pixel (P) and super pixel (SP) in iteration t, the correlation between pixel and superpixel can be expressed as following:

$$Q^{t}_{(p,sp)} = e^{-\|I_p - C^{t-1}_{sp}\|}, \quad (3)$$

where I represented feature mapping from convolution layer, super pixel center C^t_i was modified to the weighted sum of pixel characteristics: $C^t_i = \frac{1}{M^t_i}\sum^n_{p=1} Q^t_{pi} I_p$ and $M^t_i = \sum_p Q^t_{pi}$ were a general term.

2.3.2. Expand Training Samples

Remote sensing images provide a wealth of spectral information that is simple to operate, such as color transformation, band combination and band computation. Due to the limitation of time and space in the study area, in order to obtain enough sample data to train the depth network model, this study expanded the sample on the basis of the original image data. For model training, the image of the whole experimental area was cut into 256 × 256 pixel blocks, the number of pixel blocks at the entire study area was 14,077 blocks, and eight-image transformation processing was designed on each block. The transformation results are shown in Figure 6, where Figure 6A is the original image of RGB band combination B1, B2, B3; Figure 6B is the original image of B1, B2 and B3 after high contrast retention processing; Figure 6C is the maximum processed image of the original image based on RGB band combination B1, B2, B3; Figure 6D is the minimum processed image of the original image based on RGB band combination B1, B2, B3; Figure 6E is the uniform color processed image of the original image based on RGB band combination B1, B2, B3; Figure 6F is the processed image with Gaussian noise added to the original image based on RGB band combination B1, B2, B3; Figure 6G is RGB band combination B3, B2, B1—false color image; Figure 6H is the RGB band combination B3, B4, B2—false color image; Figure 6I is the RGB band combination B4, B3, B2—false color image. To evaluate the effect of sample size on the model, using random sampling, 5000, 10,000, 20,000 and 40,000 different gradient sample size gradients were set, respectively, in which 2/3 of the samples in each gradient were used for training the model and 1/3 of the samples were used for model accuracy verification.

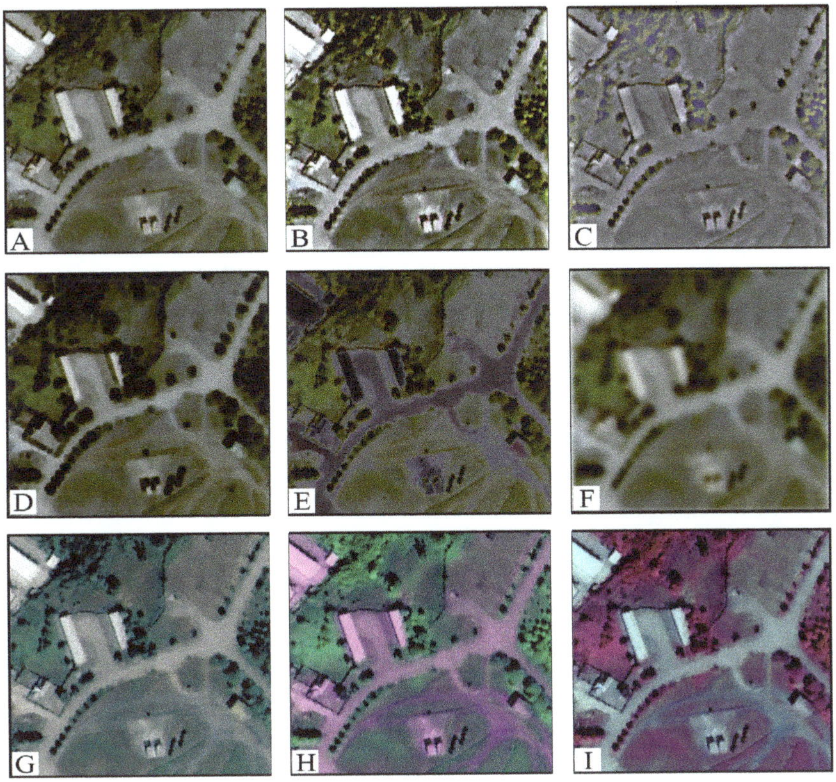

Figure 6. Sample expanded images: (**A**) original picture (band 1, band 2, band 3 combination); (**B**) high contrast retention; (**C**) maximum value; (**D**) minimum value; (**E**) overexposure; (**F**) Gaussian noise; (**G**) band 3, band 2, band 1 combination; (**H**) band 3, band 4, band 2 combination; (**I**) band 4, band 3, band 2 combination.

2.3.3. Model Parameter Setting and Training

After expanding samples, L2 regularization was used, the single input sample size of the training model was 64. The initial learning rate of Adam Optimizer was set a 0.000001. In the BN layer, ε was set as 0.000001, and attenuation was set as 0.9997. In ASPP, the multi-grids were set as [2,4,8], the atrous rate was [1,6,12,18], and the output step size was 16. For DNF module, the number of trees was set as 3 and the depth of trees was 9. For the SRM module, the superpixel value was set as 64 and the number of iterations was 10. The memory of graphics processing unit (GPU) was 24 G, which was used to speed up the training process. The details of parameters were shown in Figure 7.

Module	Name	Value
Global	Batch Size	64
	Starting learning rate	0.00001
ASPP	Multi-grids	[2, 4, 8]
	Atrous rate	[1, 6, 12, 18]
	Output stride	16
DNF	Number of trees	3
	Depth of tree	9
SRM	Number of superpixels	64
	Iteration	10

Figure 7. Model parameters.

2.3.4. Comparison Method

In this study, VHR image semantic segmentation methods have three significant characteristics: (1) The segmentation process in one stage: SDNF combined differentiable forest and deep neural network in an end-to-end manner to improve model performance. (2) Without post-processing. (3) Without other information. Therefore, RF classifier was selected as the comparison method, which used the training samples in Section 2.2.2 to extract UTC synchronously. RF classifier is the mainstream method of remote sensing image supervised classification at present. Based on decision tree learner, which was constructed with Bagging integration, introduced random attribute selection.

2.3.5. Accuracy Assessment

In order to evaluate the performance of SDNF in UTC extraction, three overall benchmark indicators were used, such as F1 score (F1), intersection over union (IoU) and overall accuracy (OA). The formula of indicators are as follows:

$$F_1 = 2 \times \frac{P \times R}{P + R},\ P = \frac{TP}{TP + FP},\ R = \frac{TP}{TP + FN}, \tag{4}$$

$$IoU = \frac{TP}{TP + FP + FN}, \tag{5}$$

$$OA = \frac{TP + FN}{N}, \tag{6}$$

where TP, FP and FN are the number of true positives, false positives and false negatives, respectively. P represents precision, R represents recall rate and N represents the total number of pixels.

3. Results

3.1. Sample Size

In order to evaluate the impact of the sample size on the SDNF model, four gradients: 5000, 10,000, 20,000 and 40,000 were input. The model fitting accuracy, classification accuracy and running time of SNDF models were tested, respectively. The final classification accuracy OA of the SNDF model and RF classifier are shown in Figure 8A. The performance of SNDF and RF in training time Figure 8B. The performance of the SNDF model in F1, IoU and OA are shown in Figure 8C. The overall results shown that when the sample size was 5000, the overall classification of the two models was lower, and the stitching line of the extracted results was obvious. With the increase in sample size, the accuracy of both models improved, but the RF model trend of improvement was small. After the sample size was increased eightfold, the accuracy had only increased by 0.13, compared with SNDF model, which increased by 0.21. In terms of training time, with the increase of sample size, the training time increased significantly. But in all four sample gradients, the RF model took more time than the SNDF model. For the SNDF model, when 40,000 samples were selected, the OA had reached a higher level, up to 0.89.

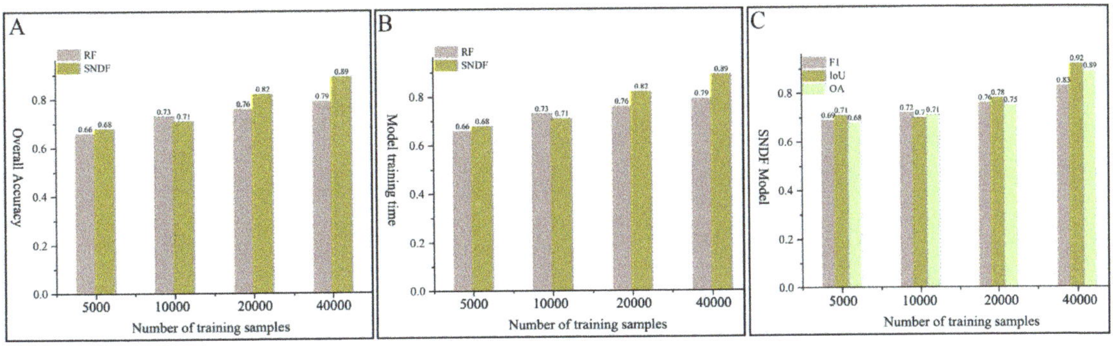

Figure 8. The influence of sample size on model accuracy and efficiency. (**A**) The final classification accuracy OA of the SNDF model and RF classifier; (**B**) the performance of SNDF and RF in training time; (**C**) the performance of SNDF model in F1, IoU and OA.

3.2. Training Loss and Elapsed Time

In the SNDF, the loss function was used to assess the model training procedure, the model convergence, and if the model was overfitted, it may also have reflected the error between the target detection model's final prediction result and the actual true value [29]. Figure 9 shows the change of the loss function in the model training process with different epoch times. With the model training, the loss function started to converge gradually, and after 160 training times, it was at the stable point of 0.14. Figure 10 shows the pictures of pixel reconstruction with different training data. Line 1 to line 4 shows the low-resolution image, after model-reconstructed images and the original images. When epoch = 30, 80, 160 and 220, with the increase in iteration times, the better the image restoration effect. The image restored after the 80th epoch had reduced the impact of the noise block greatly, and the images were more smooth and clear.

Regarding the operation environment of this study, the training time of SDNF was longer than that of RF. On a single GPU, it took 2 h and 45 min to the test study area based on RF, while the process of SDNF only took 1 h and 37 min. There was no significant difference between the SDNF and the RF in the test time of each sample. RF took an average of 0.012 s to predict the semantic segmentation results of samples, while SDNF took 0.014 s at the same time.

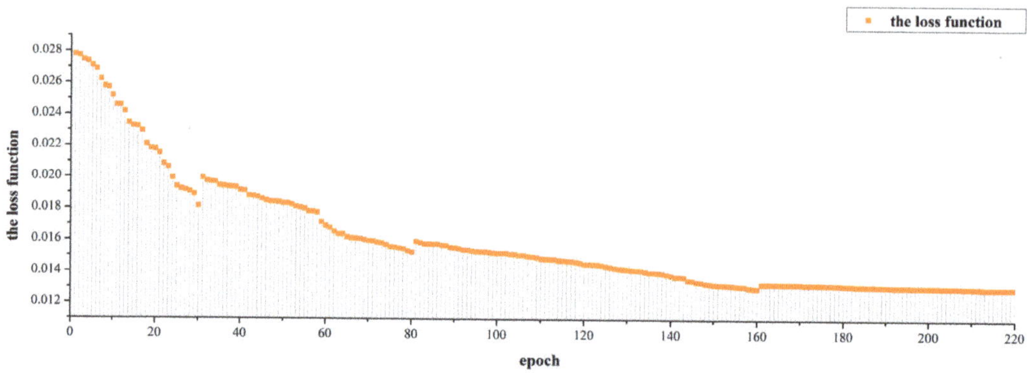

Figure 9. The variation curve of loss function value with the number of training rounds.

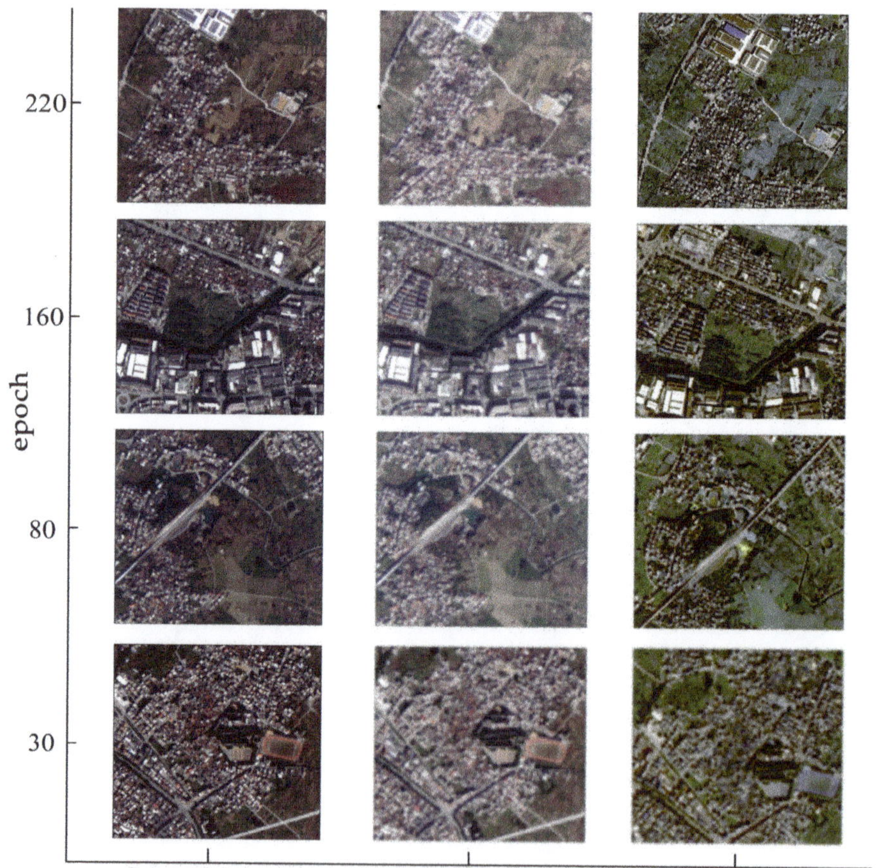

Figure 10. Superpixel reconstruction of training sample change.

3.3. Superpixel Segmentation

In this study, the superpixel segmentation results did not involve the training and testing process of the SDNF model. To present the effectiveness of the SRM module more visually, and better analyze the contribution of SLIC in this study, Figure 11 shows the comparison of the superpixel segmentation results before and after simple noniterative clustering (SNIC), SLIC and modified SLIC (MSLIC) were put into the SRM module. Figure 11A1,A2 show the segmentation results before and after SNIC, respectively; Figure 11B1,B2 show the segmentation results before and after SLIC, respectively; Figure 11C1,C2 show the segmentation results before and after MSLIC. The experimental results show that the segmentation results of SLIC-based superpixel segmentation in the SRM module were more accurate and suitable for VHR images, especially for the edge confusion areas of buildings and their shadows.

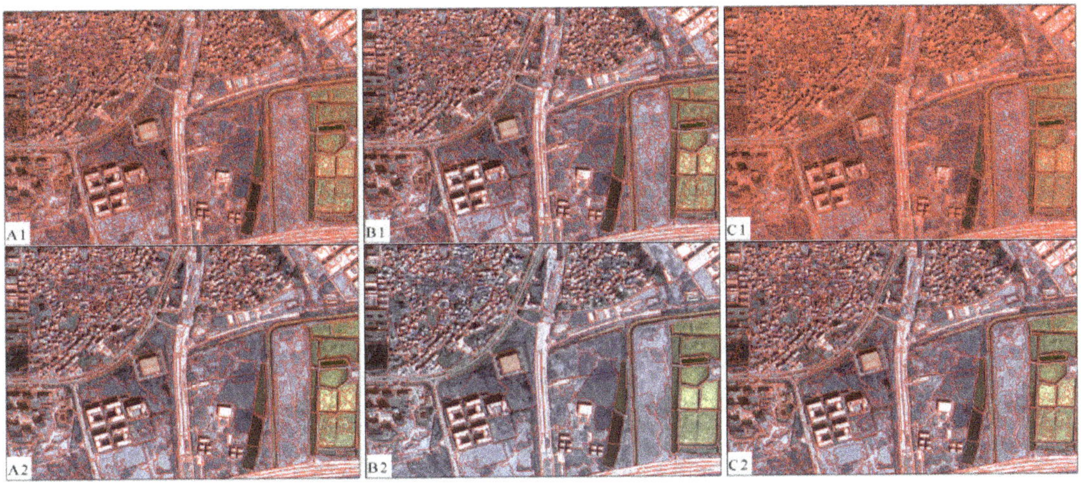

Figure 11. Comparison of SRM module segmentation effect. (**A1**,**A2**) The segmentation results before and after SNIC, respectively; (**B1**,**B2**) The segmentation results before and after SLIC, respectively, and (**C1**,**C2**) are the segmentation results before and after MSLIC.

3.4. Model Accuracy

The SDNF model was trained with the training sample, which was designed in Section 2.2, and the accuracy of the trained model was verified with the verification sample. After iterative training, the SDNF model converged when the epoch was 200, with training accuracy of 95.16% and verification accuracy of 94.87%; the RF model converged in the 160th epoch, with the training accuracy being 83.16% and verification accuracy being 87.73%.

Compared with the training accuracy, (1) the SDNF model has three deeper network layers, and the training accuracy improved significantly; (2) combining the semantic segmentation with semantic information edge detection can improve the prediction accuracy of object boundaries. The SDNF model enhanced object boundaries using superpixel segmentation; (3) the SDNF model end-to-end self-cascade network, and the sequential context aggregation from global to local improved the consistency of sample labels, the strategy from coarse to thinning improved the accuracy of the labels. Compared with the verification accuracy, the results show that the convergence effect of the SDNF model was better when it was at deeper training depths.

3.5. Classification Accuracy

As shown in Table 2, compared with RF model, the OA of SDNF improved significantly from 0.79 to 0.89, achieving higher detection accuracy, particularly for low and sparse trees (Figure 11). However, SDNF still made some error pixel in impervious surfaces and UTCs. Figure 12 shows the classification results, where column A is the superpixel-enhanced images; column B is the training sample images; column C is the classification result based on SDNF; D is the classification result of RF. The results show that compared with RF, SDNF improved its performance through end-to-end training, and combined the region loss with superpixels enhanced into the whole framework, emphasized the effective of object boundary zonation through edge detection. The performance of SDNF model varies with specific objects, and the confusion of low trees, sparse trees and building shadows improved significantly. For SNDF, the F1 improved 5%, IoU improved 12% and OA improved 10%. The results (Figure 13) show that, through the representation learning of DCNN under the guidance of differentiable forests (DF), the classification accuracy of easily confused objects was improved. This indicates that the SDNF made significant progress in UTC extraction.

Table 2. Comparison of classification accuracy between SDNF and RF Model.

Model	Accuracy Assessment		
	F1	**IoU**	**OA**
RF	0.78	0.80	0.79
SDNF	0.83	0.92	0.89

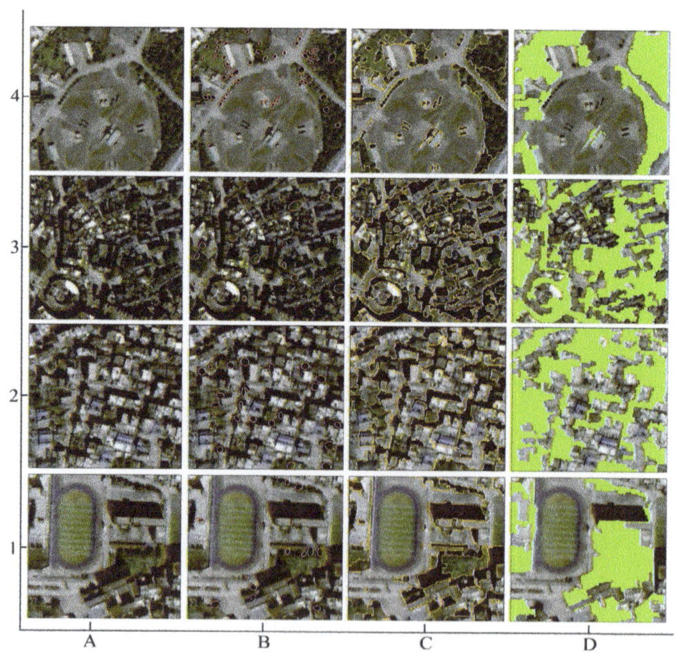

Figure 12. Detailed display of classification results of SDNF and RF methods. (**A**) The superpixel-enhanced images; (**B**) training sample images; (**C**) classification result based on SDNF; (**D**) classification result of RF.

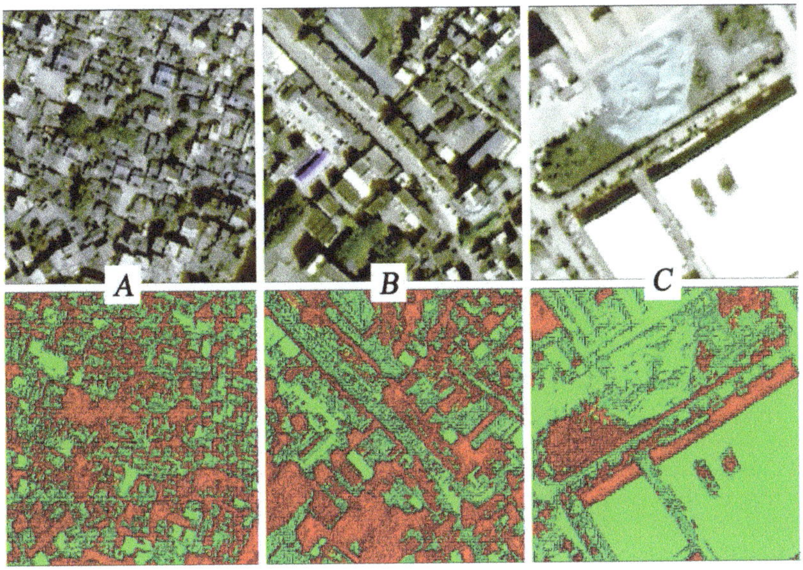

Figure 13. Detailed display of classification results of SDNF methods. Red represents the canopy coverage area; green represents the other land coverage area. (**A**) Urban residential region; (**B**) urban road region; (**C**) urban large building region.

4. Discussion

According to data published by pertinent authorities, nearly 70% of the world's population will reside in urban settings by 2050 [30]. The rapid growth of the global urban population drives the dynamic development of cities [31]; urban areas provide a home to 10 billion trees [1], which will play an increasingly essential role in preserving livable cities in the face of population and climatic pressures. Assessment methods are crucial for preserving and enhancing the ecosystem services offered by urban forests as well as determining the efficacy of policies intended to improve this important resource. Here, a new VHR image semantic segmentation framework—SDNF—was introduced to improve the sparse and small objects detection ability for VHR images.

In this study, satellite submicron VHR images were used to detect UTC. The traditional methods for VHR image classification, representation learning and classifier training were unbalanced, and the edge blur and noise problems of sparse and low objects were present (Figure 14). The SDNF framework can be used in any urban area for UTC detection. The framework is designed in two modules: DNF and SRM, which are responsible for pixel loss and region loss, respectively. SDNF is emphasized for improving DCNNs classification ability, combining DNF in the coding and decoding architecture, and reducing classification edge blur and noise. The introduced SRM module is used for learning-based superpixel segmentation tasks to dissolve the homogeneity within the superpixel and the heterogeneity between superpixels. The SDNF method goes beyond the capability of traditional remote sensing monitoring technology in UTC detection.

The availability and quality of training data are likely to be the most important impediments to applying this method more generally. Here, the sample data were expanded in the limited study region to produce a dense training dataset, which contains derived metrics in a limited spatial range (i.e., not directly measured), provides direct but limited observations of canopy structure throughout the study area. Other training data sources include current government tree datasets and Google Street View [32]. When substituting training data sets, the understanding of expected changes in model uncertainty and variance should be attained [33].

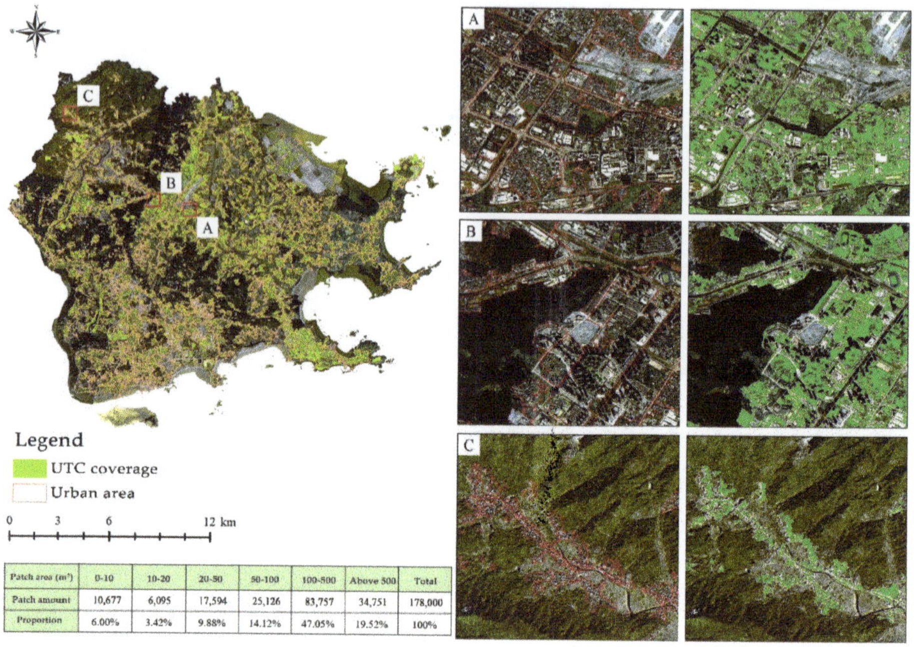

Patch area (m^2)	0–10	10–20	20–50	50–100	100–500	Above 500	Total
Patch amount	10,677	6,095	17,594	25,126	83,757	34,751	178,000
Proportion	6.00%	3.42%	9.88%	14.12%	47.05%	19.52%	100%

Figure 14. Detailed results of UTC coverage in the study area. (**A**) The center of city; (**B**) The suburban region of the city; (**C**) Outermost region of the city.

Owing to the special location of urban forest distribution, unclear boundaries, and ease of obfuscation by the shadows of buildings, the regions with high canopy density of single tree segmentation in the process of target detection using the SDNF method cannot be addressed. How to detect single tree target of urban forest through optimization model or combining with other auxiliary data will be the content of the next step.

5. Conclusions

In this study, a method based on semantic segmentation object extraction—superpixel-enhanced deep neural forest (SDNF)—was used to solve the representation learning and classifier training unbalanced problem and the edge blur and noise of sparse and low objects in VHR images, such as UTC. The SDNF model balanced the generalization ability of the DCNNs model in classification and deep learning, introducing a completely differentiable forest to control the learning ability of deep convolution layer. At the same time, a superpixel-enhanced region module (SRM) was used to reduce the classification noise and enhance the edge details of objects in VHR images. The experimental results show that the training accuracy of the superpixel-enhanced deep neural forest (SDNF) was 95.16%, the verification accuracy was 94.87%, and the OA of UTC extraction was 89.17%, evidencing that SDNF reached a new level in UTC extraction. It has obvious advantages compared with the existing unified detection methods using the RF model: it can quickly, accurately and automatically detect the canopy cover in urban regions. Therefore, the SDNF method will provide technical support for improving the location extraction, spatial distribution planning and protection of canopy cover distribution in urban areas.

Author Contributions: Conceptualization, Y.L. and H.Z.; methodology, Y.L.; software, Y.L.; validation, Z.C., K.L. and Y.Z.; formal analysis, K.L. and Y.Z.; investigation, X.H. and J.W.; resources, H.Q. and Y.Z.; data curation, J.W.; writing—original draft preparation, Y.L.; writing—review and editing, Y.L.; visualization, X.H., Y.L. and Z.C.; supervision, H.Z.; project administration, H.Z.; funding acquisition, H.Z. All authors have read and agreed to the published version of the manuscript.

Funding: This research was funded by the Foundation Research Funds of Institute of Forest Resource Information Techniques (IFRIT), grant number CAFYBB2019SZ004 and the National Natural Science Foundation of China, grant number 32071681.

Conflicts of Interest: The authors declare no conflict of interest.

References

1. Endreny, T.A. Strategically Growing the Urban Forest Will Improve Our World. *Nat. Commun.* **2018**, *9*, 10–12. [CrossRef] [PubMed]
2. Ucar, Z.; Bettinger, P.; Merry, K.; Siry, J.; Bowker, J.M.; Akbulut, R. A Comparison of Two Sampling Approaches for Assessing the Urban Forest Canopy Cover from Aerial Photography. *Urban For. Urban Green.* **2016**, *16*, 221–230. [CrossRef]
3. The State Forestry Administration of the People's Republic of China. *National Forest City Evaluation Indicators*; Urban Forestry in China: Beijing, China, 2007; Volume 5, p. 2.
4. Jia, B.Q.; Liu, X.P. *Canopy Coverage Characteristics and Landscape Ecological Changes in the First Green Isolation Area in Beijing*; Scientia Silvae Sinicae: Beijing, China, 2017; Volume 53, p. 10.
5. Krajter Ostoić, S.; Salbitano, F.; Borelli, S.; Verlič, A. Urban Forest Research in the Mediterranean: A Systematic Review. *Urban For. Urban Green.* **2018**, *31*, 185–196. [CrossRef]
6. Erker, T.; Wang, L.; Lorentz, L.; Stoltman, A.; Townsend, P.A. A Statewide Urban Tree Canopy Mapping Method. *Remote Sens. Environ.* **2019**, *229*, 148–158. [CrossRef]
7. Alonzo, M.; Bookhagen, B.; Roberts, D.A. Urban Tree Species Mapping Using Hyperspectral and Lidar Data Fusion. *Remote Sens. Environ.* **2014**, *148*, 70–83. [CrossRef]
8. Mi, L.; Chen, Z. Superpixel-Enhanced Deep Neural Forest for Remote Sensing Image Semantic Segmentation. *ISPRS J. Photogramm. Remote Sens.* **2020**, *159*, 140–152. [CrossRef]
9. Kuffer, M.; Barros, J.; Sliuzas, R.V. The Development of a Morphological Unplanned Settlement Index Using Very-High-Resolution (VHR) Imagery. *Comput. Environ. Urban Syst.* **2014**, *48*, 138–152. [CrossRef]
10. Achanta, R.; Shaji, A.; Smith, K.; Lucchi, A.; Fua, P.; Süsstrunk, S. SLIC superpixels compared to state-of-the-art superpixel methods. *IEEE Trans. Pattern Anal. Mach. Intell.* **2012**, *34*, 2274–2282. [CrossRef]
11. Niu, S.F.; Qiu, E.F.; Zhang, Z.Y.; Xi, L. Gradient change of population diversity of woody plants in the urban riverbank forest of Beijing. *Sciatica Silvia Sin.* **2020**, *56*, 198–206.
12. Liu, B.; Hu, H.; Wang, H.; Wang, K.; Liu, X.; Yu, W. Superpixel-Based Classification with an Adaptive Number of Classes for Polarimetric SAR Images. *IEEE Trans. Geosci. Remote Sens.* **2013**, *51*, 907–924. [CrossRef]
13. Ren, C.Y.; Reid, I. *gSLIC: A Real-Time Implementation of SLIC Superpixel Segmentation*; University of Oxford, Department of Engineering: Oxford, UK, 2011; pp. 1–6.
14. Wei, J.; Huang, W.; Li, Z.; Sun, L.; Zhu, X.; Yuan, Q.; Liu, L.; Cribb, M. Cloud Detection for Landsat Imagery by Combining the Random Forest and Superpixels Extracted via Energy-Driven Sampling Segmentation Approaches. *Remote Sens. Environ.* **2020**, *248*, 112005. [CrossRef]
15. Huang, X.; Cao, Y.; Li, J. An Automatic Change Detection Method for Monitoring Newly Constructed Building Areas Using Time-Series Multi-View High-Resolution Optical Satellite Images. *Remote Sens. Environ.* **2020**, *244*, 111802. [CrossRef]
16. Wang, S.; Guan, K.; Zhang, C.; Zhou, Q.; Wang, S. Remote Sensing of Environment Cross-Scale Sensing of Field-Level Crop Residue Cover: Integrating Field Photos, Airborne Hyperspectral Imaging, and Satellite Data. *Remote Sens. Environ.* **2023**, *285*, 113366. [CrossRef]
17. Aloysius, N.; Geetha, M. A Review on Deep Convolutional Neural Networks. In Proceedings of the 2017 International Conference on Communication and Signal Processing (ICCSP 2017), Chennai, India, 6–8 April 2017; pp. 588–592. [CrossRef]
18. Liu, M.; Shi, J.; Li, Z.; Li, C.; Zhu, J.; Liu, S. Towards Better Analysis of Deep Convolutional Neural Networks. *IEEE Trans. Vis. Comput. Graph.* **2017**, *23*, 91–100. [CrossRef]
19. Waldner, F.; Diakogiannis, F.I. Deep Learning on Edge: Extracting Field Boundaries from Satellite Images with a Convolutional Neural Network. *Remote Sens. Environ.* **2020**, *245*, 111741. [CrossRef]
20. Gallwey, J.; Robiati, C.; Coggan, J.; Vogt, D.; Eyre, M. A Sentinel-2 Based Multispectral Convolutional Neural Network for Detecting Artisanal Small-Scale Mining in Ghana: Applying Deep Learning to Shallow Mining. *Remote Sens. Environ.* **2020**, *248*, 111970. [CrossRef]
21. Huang, B.; Zhao, B.; Song, Y. Urban Land-Use Mapping Using a Deep Convolutional Neural Network with High Spatial Resolution Multispectral Remote Sensing Imagery. *Remote Sens. Environ.* **2018**, *214*, 73–86. [CrossRef]
22. Zhang, D.; Pan, Y.; Zhang, J.; Hu, T.; Zhao, J.; Li, N.; Chen, Q. A Generalized Approach Based on Convolutional Neural Networks for Large Area Cropland Mapping at Very High Resolution. *Remote Sens. Environ.* **2020**, *247*, 111912. [CrossRef]

23. Zhang, C.; Sargent, I.; Pan, X.; Li, H.; Gardiner, A.; Hare, J.; Atkinson, P.M. An Object-Based Convolutional Neural Network (OCNN) for Urban Land Use Classification. *Remote Sens. Environ.* **2018**, *216*, 57–70. [CrossRef]
24. Persello, C.; Tolpekin, V.A.; Bergado, J.R.; de By, R.A. Delineation of Agricultural Fields in Smallholder Farms from Satellite Images Using Fully Convolutional Networks and Combinatorial Grouping. *Remote Sens. Environ.* **2019**, *231*, 111253. [CrossRef]
25. Wang, J.; Ma, A.; Zhong, Y.; Zheng, Z.; Zhang, L. Cross-Sensor Domain Adaptation for High Spatial Resolution Urban Land-Cover Mapping: From Airborne to Spaceborne Imagery. *Remote Sens. Environ.* **2022**, *277*, 113058. [CrossRef]
26. Cao, Y.; Huang, X. A Deep Learning Method for Building Height Estimation Using High-Resolution Multi-View Imagery over Urban Areas: A Case Study of 42 Chinese Cities. *Remote Sens. Environ.* **2021**, *264*, 112590. [CrossRef]
27. Kontschieder, P.; Fiterau, M.; Criminisi, A.; Bulo, S.R. Deep neural decision forests. In Proceedings of the IEEE International Conference on Computer Vision, Santiago, Chile, 7–13 December 2015; pp. 1467–1475. [CrossRef]
28. Chen, L.C.; Papandreou, G.; Kokkinos, I.; Murphy, K.; Yuille, A.L. DeepLab: Semantic Image Segmentation with Deep Convolutional Nets, Atrous Convolution, and Fully Connected CRFs. *IEEE Trans. Pattern Anal. Mach. Intell.* **2018**, *40*, 834–848. [CrossRef] [PubMed]
29. Ma, Y.K.; Liu, H.; Lin, C.X.; Zhao, F.; Jiang, X.; Zhang, Y.T. Study on target detection of mangrove single tree based on improved YOLOv5. *Adv. Lasers Optoelectron.* **2022**, *29*, 1828003.
30. Al-Surf, M.; Balabel, A.; Alwetaishi, M.; Abdelhafiz, A.; Issa, U.; Sharaky, I.; Shamseldin, A.; Al-Harthi, M. Stakeholder's Perspective on Green Building Rating Systems in Saudi Arabia: The Case of LEED, Mostadam, and the SDGS. *Sustainability* **2021**, *13*, 8463. [CrossRef]
31. Parnell, S.; Walawege, R. Sub-Saharan African Urbanisation and Global Environmental Change. *Glob. Environ. Chang.* **2011**, *21*, S12–S20. [CrossRef]
32. Li, X.; Zhang, C.; Li, W.; Ricard, R.; Meng, Q.; Zhang, W. Assessing Street-Level Urban Greenery Using Google Street View and a Modified Green View Index. *Urban For. Urban Green.* **2015**, *14*, 675–685. [CrossRef]
33. Baines, O.; Wilkes, P.; Disney, M. Quantifying Urban Forest Structure with Open-Access Remote Sensing Data Sets. *Urban For. Urban Green.* **2020**, *50*, 126653. [CrossRef]

forests

MDPI

Article

Comprehensive Decision Index of Logging (CDIL) and Visual Simulation Based on Horizontal and Vertical Structure Parameters

Kexin Lei [1,2,3], Huaiqing Zhang [1,2,3,*], Hanqing Qiu [1,2], Yang Liu [1,2,3], Xingtao Hu [1,2,3,4], Jiansen Wang [1,2,3], Zeyu Cui [1,2,3] and Yuanqing Zuo [1,2,3]

1 Institute of Forest Resource Information Techniques, Chinese Academy of Forestry, Beijing 100091, China
2 Key Laboratory of Forest Management and Growth Modelling, NFGA, Beijing 100091, China
3 National Long Term Scientific Research Base of Huangfengqiao Forest Monitoring and Simulation in Hunan Province, Beijing 100091, China
4 School of Geography and Environmental Sciences, Guizhou Normal University, Guiyang 550025, China
* Correspondence: zhang@ifrit.ac.cn; Tel.: +86-1-365-116-6672

Abstract: The comprehensive indexes approach based on stand structure parameters is mainly used to select trees for harvest. However, these indexes do not consider the comprehensive impact of horizontal and vertical structures, leading to an incomplete analysis of the forest structure and an inaccurate selection of trees for harvest. To solve this problem, we constructed a comprehensive decision index of logging (CDIL), integrating horizontal and vertical structure parameters which can identify harvest trees more scientifically. In this study, we took the Shanxia Forest Farm in the Jiangxi Province of China as the experimental area and used mixed broadleaf/conifer forests at different ages as our experimental sample. We selected eight horizontal and vertical spatial structure parameters to establish an efficient, objective, and accurate comprehensive decision index of logging. We combined 3D visualization technology to realize the dynamic visualization simulation of the index at different intensities of tending and felling management. The results indicated that the proposed CDIL-index could effectively optimize the forest spatial structure. From the perspective of stand structure adjustment, the optimal thinning intensity was 20%. The average CDIL in each plot decreased by more than 80% after logging, while the change range of each plot was between 30% and 70% after the F index was applied to implement tending and logging. The CDIL was 11.4% more accurate in selecting trees for harvesting than the F index. In this study, the main conclusion is that the CDIL would enable forest managers to more accurately choose trees for harvesting, leading to forest adjustment that would reduce the competition pressure among trees and improve the distribution and health of trees, possibly making the forest structure more stable.

Keywords: stand spatial structure; comprehensive decision index of logging (CDIL); stand structure adjustment; visual simulation

1. Introduction

As the most basic and important feature of forest ecosystems, the forest structure has spatial and nonspatial characteristics [1]. An important part of forest management is forest structure analysis since it facilitates our understanding of forest communities [2]. The forest spatial structure is more important than the nonspatial structure since the spatial structure largely determines the distribution and competition of trees, and reflects the health of the trees, the diversity of tree species, and the stability of the stand [3]. In recent years, to transform large areas of coniferous pure forests with low ecological and economic benefits into broad-leaved mixed forests with high biodiversity and stability, European forestry-developed countries, such as Germany and the United Kingdom, have carried out research on forest spatial structure adjustment with selective harvest trees as the main measure [4–8]. North American countries have paid attention to analyzing the forest spatial structure, and providing the basis for forest growth and stand dynamic simulation [6,9,10].

These studies have shown that the analysis and optimization of the forest spatial structure is the most active research field [7].

An important issue in forest management is forest thinning, which can directly affect the stand structure. By regularly cutting some trees in the forest, we can improve the competitive environment of the trees, enhance the living space of the trees and the forest structure, and adjust the total amount and distribution mode of forest resources [11–13]. Trees of different species, in different health conditions, and of different diameter classes can respond differently to the distribution of natural resources and the improved competitive environment [14,15]. Forest thinning may change the understory light, temperature, and humidity conditions, thereby improving the physical and chemical properties of the soil, promoting the regeneration of undergrowth vegetation, maintaining soil fertility, and providing higher ecological and environmental benefits as well as social and economic benefits [16]. However, unreasonable thinning may have negative consequences, such as damaging vegetation regeneration, interfering with nutrient cycles, hindering animal adaptation, increasing forest pests, and even causing forest fires [17–19]. Therefore, how to effectively optimize the stand spatial structure, scientifically select harvest trees, and practice reasonable tending and felling is the premise for maintaining stand stability for the sake of sustainable development [20,21].

At the core of thinning is the selection of trees to be harvested [22]. In recent decades, a large number of scholars have proposed some forest management methods, including the comprehensive indexes approach [23], the stem number guide curve [24], and target tree-oriented management [25]. Although each of these methods include several parameters, they lack information about the spatial structure. To evaluate characteristics of the stand spatial structure, recently, Ye et al. [21], and Pastorella et al. [26] developed comprehensive indexes to evaluate the characteristics of the stand spatial structure based on several common parameters, such as the U-index, the M-index, and the D-index. However, none of these comprehensive indexes considers the impact of the stand vertical structure. For example, Li et al. used the composite index to select harvest trees according to the bivariate distributions of the three spatial parameters (i.e., the U, the M, and the D index) [22]. Though the three spatial structure indicators are used extensively, they are not comprehensive enough. Courbaud et al. simulated the influence of different selective cutting methods, including cluster selective cutting and single tree selective cutting, on the spatial structure and growth of Norwegian spruce uneven-aged forests in the French Alps Mountains [27]. Cao et al. used multiplication and division to carry out multiobjective planning for the selected spatial structure parameters and proposed the objective function for optimizing the stand spatial structure and exploring the stand spatial structure of trees of different ages and different thinning intensities [28]. Lv et al. used multiplication and division to develop a comprehensive cutting index to verify the effectiveness of adjusting the spatial structure of the stand. The index put forward by Lv et al. covered eight parameters, including horizontal spatial structure parameters, which produced good results. Therefore, this paper used the index constructed by Lv et al. for comparison [29]. In most of these studies, stand spatial and nonspatial structure indicators or horizontal structure parameters were selected to construct the cutting index [20,22,28,29]. They did not consider the comprehensive effects of vertical spatial structural parameters and other structural parameters. However, the vertical spatial structure parameters of the stand play an important role in improving the forest growth space and accurately and efficiently selecting trees for harvest.

In a study on the setting of sample plots for forest structure optimization, Dong selected four permanent plots, each measuring 100 m × 100 m [20]. Ahmad et al. selected 82 sample plots, some measuring 30 m × 30 m and some of 20 m × 20 m [30]. Bhandari et al. selected 30 sample plots measuring 40 m × 40 m each [31]. These studies had different sample plot sizes due to different topographic factors. In our study, on the basis of the influence of topographic and human factors in the study area, we selected five sample plots of 50 m × 50 m each.

The purpose of our study was to put forward a new comprehensive decision index of logging (CDIL) based on horizontal and vertical spatial parameters of the stand. We then employed the comprehensive index to determine the harvest trees for the five plots of the Shanxia Forest Farm in Jiangxi Province of China, determined the number of harvest trees in the plot according to the optimal thinning intensity, and then conducted a thinning simulation, The development of the CDIL involved four steps: (1) comprehensively analyzing and judging harvest trees in the stand, (2) identifying any correlation between vertical spatial structure parameters and the CDIL, and quantifying the advantages of the CDIL in adjusting the stand spatial structure, (3) comparing the CDIL with the logging index by Lv et al., and calculating the accuracy of the CDIL in selecting harvest trees [21], and (4) carrying out the dynamic visual simulation of stand management, and evaluating the simulation results, based on Unreal Engine 4 [32].

The CDIL helps to adjust the spatial structure of the stand and transform the unreasonable spatial structure of the forest into an optimal state. Moreover, visual simulation enables managers to respond in a timely manner to the changes and distribution of the spatial structure characteristics of the forest stand and the problems that may occur in the management process so as to modify the forest management plan when necessary.

2. Materials and Methods

2.1. Study Areas

The study area was located in Shanxia Experimental Forest Farm in the southwest of Fenyi County, Xinyu City, Jiangxi Province, China (Figure 1). The region has 1920 kinds of plants, and the forest coverage rate is 64.3%. The zonal vegetation is the midsubtropical evergreen broad-leaved forest. The dominant tree species are Cunninghamia lanceolate, Pinus massoniana, and Schima Superba. The area belongs to the subtropical monsoon climate with sunshine. The frost-free period is of 270 days. It has abundant rainfall, with an annual precipitation of 1600 mm and a mean annual temperature of 15~17 °C. The soil in the forest farm is fertile, and the soil type is mainly loam soil (Figure 1).

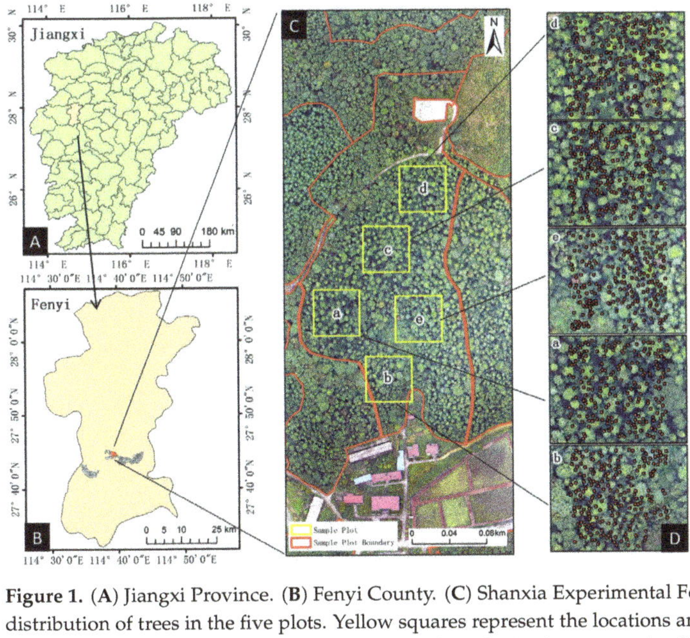

Figure 1. (**A**) Jiangxi Province. (**B**) Fenyi County. (**C**) Shanxia Experimental Forest Farm. (**D**) The distribution of trees in the five plots. Yellow squares represent the locations and sizes of the plots; the red lines represent the subplot boundaries; and graphs (**a–e**) represent the distribution of trees in each plot respectively.

2.2. Data Collection

In May 2022, five fixed plots, each measuring 50 m × 50 m, with artificial uneven-aged coniferous and broad-leaved mixed forests were set up on the forest farm. In each plot, the relative coordinates (x, y, z) of the root of each tree were measured using the total station and prism. The tree species were recorded and provided by forest farm staff. The diameter at breast height (DBH) of each tree, was measured using the DBH tape measure. A laser altimeter was used to measure the tree height(H) and the height under the living branches (UBH) of trees. The crown widths (CWs) of the trees were measured using a box staff. Table 1 lists the basic stand characteristics of the five plots. To avoid inaccurate calculation results of the single tree competition index, we set up 5 m buffer zones among the five plots. To build 3D models of the trees, we used a digital camera to photograph the bark and leaf texture of trees, shrubs, and herbs.

Table 1. The basic stand characteristics of the five plots.

Plots	Plot Area (hm^2)	Stand Density (Trees/ha)	Average DBH (cm)	Average Height (m)	Average Crown Width (m)	Average UBH (m)	Number of Tree Species
a	0.25	215	17.9	12.4	3.7	4.8	8
b	0.25	216	16.9	11.4	3.2	4.1	16
c	0.25	221	17.7	14.6	3.4	4.5	6
d	0.25	203	18.1	12.4	3.8	4.3	9
e	0.25	225	16.9	12.9	3.6	4.5	7

2.3. Data Analysis

2.3.1. Construction of the Comprehensive Decision Index of Logging (CDIL)

The CDIL consists of eight parameters: neighborhood Pattern (W), neighborhood comparison (U), mingling degree (M), openness (B), health index (H), spatial density index (D), Heygi competition index (C), and vertical spatial structure parameter (PV). They, respectively, reflect the following eight horizontal and vertical structural parameters: the spatial distribution pattern of trees, the diversity of tree species, the degree of crowding, the competitiveness of trees, the spatial advantage degree, openness, the health of trees, and the number of trees covered by adjacent trees. Table 2 describes the eight horizontal and vertical structure parameters. Figure 2 presents the development process of the CDIL.

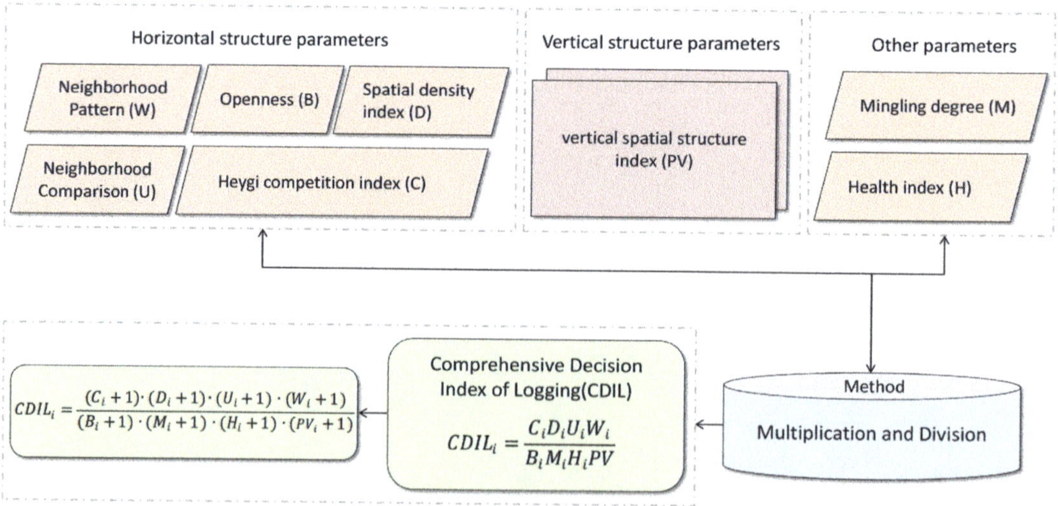

Figure 2. The construction process of the CDIL.

Table 2. Descriptions of the eight horizontal and vertical structure parameters.

Type	Indexes	Abbreviation	Equation	Describe/Rule	References
Horizontal structure parameters	Neighborhood pattern	W	$W_i = \frac{1}{n}\sum_{j=1}^{n} Z_{ij}$	It is used to describe the evenness of adjacent trees around the target tree. $Z_{ij} = \begin{cases} 1, \text{ When the angle j was smaller than the standard angle } \alpha_0 \\ 0, \text{ otherwise} \end{cases}$	[26,33,34]
	Neighborhood comparison	U	$U_i = \frac{1}{n}\sum_{j=1}^{n} k_{ij}$	It reflects the dominance of trees in the growth process. $k_{ij} = \begin{cases} 1, \text{ If the adjacent tree i was larger than the reference tree} \\ 0, \text{ otherwise} \end{cases}$	[35]
	Openness	B	$B_i = \frac{1}{n}\sum_{j=1}^{n} \frac{D_{ij}}{H_j}$	It is used to describe the growth of trees in the plot space.	[36]
	Spatial density index	D	$D_i = 1 - \frac{r_i}{r_{max}}$	It refers to the degree of tree crowding in the spatial structure unit.	[37]
	The Heygi competition index	C	$C_i = \sum_{j=1}^{n} \frac{D_j}{D_i \cdot L_{ij}}$	It is defined as the distance between the target tree and the the competitive tree and the ratio of the diameter of the competitive tree to the target tree.	[32]
Vertical structure parameters	Vertical spatial structure parameter	PV	$PV_i = \sum_{j=1}^{n} \begin{cases} \tan^{-1}\left(\frac{(H_j - H_i)}{dist_{ij}}\right) \\ -arctg\left(\frac{(H_j - H_i)}{dist_{ij}}\right) \end{cases}$	It indicates the degree to which the target tree is covered by the competitive tree.	[38]
Other parameters	Mixing degree	M	$M_i = \frac{1}{n}\sum_{j=1}^{n} v_{ij}$	It describes the isolation degree of tree species in the mixed forest. $v_{ij} = \begin{cases} 1, \text{ The target tree i and the adjacent tree j are different species} \\ 0, \text{ otherwise} \end{cases}$	[26,36]
	Health index	H	$H_i = \frac{1}{n}\sum_{j=1}^{n} x_{ij}$	It indicates that when the health of the target tree i is better than the that of adjacent tree j, $x_{ij} = 1$, otherwise $x_{ij} = 0$.	[39]

All the quantitative horizontal and vertical indexes described above are significant for stand spatial structure adjustment. To select the harvest trees, the relationship between each target tree and its four nearest trees should be evaluated. Therefore, coupled with the above eight stand structure parameters, we used multiplication and division to construct a comprehensive decision index of logging to comprehensively analyze the trees. The formula is:

$$CDIL_i = \frac{C_i D_i U_i W_i}{B_i M_i H_i PV}, \tag{1}$$

When Formula (1) is applied, the value of factors such as M_i and W_i may be zero, resulting in large errors in the results. Therefore, according to the method of multiplication and division, 1 is added to all parameters in Formula (1) to get Formula (2):

$$CDIL_i = \frac{(C_i+1)\cdot(D_i+1)\cdot(U_i+1)\cdot(W_i+1)}{(B_i+1)\cdot(M_i+1)\cdot(H_i+1)\cdot(PV_i+1)}, \tag{2}$$

The formula for the average comprehensive decision index of logging the stand is stated as:

$$CDIL = \frac{1}{N}\sum_{i=1}^{N} CDIL_i, \tag{3}$$

where $CDIL_i$ represents the comprehensive decision index of logging; $CDIL$ represents the average comprehensive decision index of logging of the stand; and C_i, D_i, U_i, W_i, B_i, M_i, H_i, and PV_i represent, respectively, the Heygi competition index, the spatial density index, neighborhood comparison, neighborhood pattern, openness, the mingling degree, the health index, and the vertical spatial structure parameter of the target tree.

2.3.2. Dynamic Visual Simulation of Management

Using Unreal Engine 4 (https://www.unrealengine.com/zh-CN, accessed on 10 December 2023), which is an efficient and versatile game development engine launched by Epic Games, we can directly preview the development effect. We connect to the MySQL database with UE4 and read the stand data table. The three-dimensional model of a tree is divided into two parts, the trunk and the crown. The original point of the coordinates of the trunk and the crown of the tree classification model is assumed to be at the bottom. The initial DBH of the trees in the forest is D_0, the height of the trees is H_0, the crown width is CW_0, the height under branches is UBH_0, and the crown height is CH_0. The trunk diameter at breast height scaling factor DBH_XY_scale = D_0/D (D is the diameter at breast height of the initial model), the height under the branch scaling coefficient UBH_Z_scale = UBH_0/UBH (UBH is the height under the branch of the initial model), the scaling coefficient of the crown width CW_ XY_scale = CW_0/CW (CW is the crown size of the initial model), and the crown height scaling coefficient CH_Z_scale = CH_0/CH (CH is the crown height of the initial model). Figure 3 presents the virtual reality scene of the real initial forest scene of the five sample plots.

On the basis of the index calculation formula listed in Table 2 and the UE4 blueprint function script, which is a visual script that provides an intuitive, node-based interface and realizes code programming by dragging and dropping nodes, a function for calculating the eight stand structural parameters and the CDIL is written. As per the calculation results, we determined the trees for harvest in each plot according to thinning intensities of 10%, 20%, and 30% of the number of trees and established the cutting data table. We then removed the harvest trees determined by each thinning intensity value and replaced them with stump models. The stand structure changed after cutting. We recalculated the stand structure parameters and the CDIL and analyzed the accuracy of harvest tree selection and the results of stand structure adjustment.

Figure 3. (**A**) represents the texture of *Cunninghamia lanceolata* bark. (**B**) represents the texture of *Cunninghamia lanceolata* leaves. (**C**) is the initial three-dimensional model and stump model of *Cunninghamia lanceolata* created by using (**A**,**B**), and real measurement data.

According to the blueprint function script, the timeline animation event, and the forest growth model library (the forest growth model library used the growth model that conforms to the growth of the dominant species provided by the forest farm locally), the tree growth functions were written and the growth of stands of different ages was dynamically simulated, in which the DBH growth model and the UBH growth model were used for the tree trunk, while the CW growth model and the CH growth model were used for the tree crown. We assume that the DBH of trees of different ages is D_i, the tree height is H_i, the crown width is CW_i, the height under branches is UBH_i, and the crown height is CH_i. The DBH scaling coefficient of the trunk is xiongjing_XYscale = D_i/D_{i-1} $(i > 1)$, and the UBH scaling coefficient zhixiagao_Zscale = UBH_i/UBH_{i-1} $(i > 1)$. The crown width scaling coefficient guanfu_XYscale = CW_i/CW_{i-1} $(i > 1)$, and the crown height scaling coefficient guangao_Zscale = $CH_i/CH_{i-1}(i > 1)$.

3. Results

3.1. Determination of Logging

To identify the accuracy of the CDIL in selecting trees for harvest, we selected the logging index with comprehensive indicators by Lv et al. [29] for comparison.

The thinning intensities of 10%, 20%, and 30% were simulated according to the growth and quality of the stand in the plot. The harvest trees were selected according to the calculation results of the CDIL and the F index [29]. The average index when 10% of the trees were identified for harvest in the five plots is illustrated in Table 3. First, trees numbered 13, 17, 19, 23, and 37, marked as unhealthy, broken, and withered in the plot survey, had to be cut down. According to the value of the CDIL, it was concluded that the comprehensive decision index of the tree numbered 177 was the largest. Its H-index was 0.25, close to zero, and thus it could be regarded as an unhealthy tree. The spatial density index was 0.99, which meant the area around it was extremely crowded and not conducive to its growth. Competitive tree numbered 94 was simultaneously squeezed by trees numbered 178 and 93. Therefore, it was removed. The CDIL was recalculated after removing the trees, and this operation was repeated until the forest spatial structure after the trees were cut reached the ideal state of management objectives.

Table 3. The average index when 10% of the trees were identified for harvest in the five plots.

Plots	Harvest Tree Number		W-Index		U-Index		M-Index		B-Index		H-Index		D-Index		C-Index		PV-Index
	CDIL	F Index	CDIL	F Index	CDIL	F Index	CDIL	F Index	CDIL	F Index	CDIL	F Index	CDIL	F Index	CDIL	F Index	CDIL
a	48, 197, 103, 47, 49, 198, 70, 102, 56, 25, 41, 114, 7, 16, 2, 55, 121, 108, 36, 180, 69	25, 149, 1, 213, 55, 206, 106, 118, 132, 35, 161, 119, 40, 8, 169, 215, 9, 66, 81, 188, 184	0.59	0.58	0.66	0.65	0.52	0.61	0.2	0.23	0.29	0.39	0.94	0.81	6.40	9.48	1.33
b	210, 188, 204, 9, 181, 167, 165, 205, 182, 192, 137, 50, 209, 121, 202, 89, 88, 189, 142, 166, 197	82, 180, 157, 108, 117, 20, 153, 53, 211, 11, 58, 126, 3, 40, 176, 8, 54, 56, 13, 107, 139	0.61	0.49	0.69	0.81	0.28	0.59	0.31	0.25	0.32	0.51	0.83	0.7	3.92	3.47	1.16
c	149, 177, 108, 13, 107, 178, 127, 82, 183, 33, 193, 55, 35, 58, 42, 122, 20, 78, 95, 18, 126, 83	177, 107, 43, 215, 57, 126, 106, 148, 129, 185, 191, 45, 71, 145, 189, 11, 214, 82, 9, 186, 209, 199	0.56	0.55	0.75	0.68	0.45	0.59	0.2	0.23	0.24	0.47	0.86	0.82	8.88	3.84	2.06
d	51, 118, 97, 99, 179, 164, 184, 107, 7, 91, 50, 82, 158, 182, 53, 189, 49, 18, 77, 136	7, 80, 175, 164, 57, 67, 38, 101, 117, 126, 12, 193, 100, 55, 181, 28, 4, 33, 169, 180, 42	0.56	0.43	0.81	0.79	0.39	0.62	0.22	0.26	0.14	0.35	0.78	0.73	8.14	3.38	2.76
e	195, 190, 161, 167, 171, 165, 216, 181, 222, 40, 208, 197, 193, 203, 225, 2, 196, 215, 185, 186, 192, 223	12, 43, 134, 32, 88, 18, 110, 197, 221, 145, 102, 106, 223, 154, 108, 98, 91, 195, 24, 60, 136, 169, 51	0.59	0.58	0.63	0.43	0.15	0.53	0.24	0.25	0.29	0.56	0.85	0.74	4.17	2.48	1.02

For a target tree to be in an advantageous position, its competition neighborhood comparison should not have been greater than 0.25 [22,24]. In the five plots, the average W-index values of the harvest trees were all above 0.5 and the C-index was large. Therefore, identifying them as harvest trees reflected the rule of cutting small trees while retaining large ones. The average H-index was close to zero, indicating that most of the selected trees were unhealthy and this reflected the necessity of cutting the inferior trees and retaining the superior ones. In terms of the average D-index, the values of the five plots were all higher than 0.7. The D-index of Plot a was close to one. Plot a was in an extremely crowded state, indicating the need to reduce the density and retain sparsity. The PV-index was small, implying that the trees identified for harvest were considerably covered by the surrounding trees and should be harvested.

The harvest trees selected by the CDIL were different from those selected using the F index. A W-index close to 0.5 indicates that the distribution pattern of the stand was random and the stand tended to be stable [40]. The harvest trees selected by the CDIL in Plots a-e had average W-indexes of 0.59, 0.61, 0.56, 0.56, and 0.59, respectively, which belonged to aggregated distribution. However, the harvest trees selected by the F index had average W-indexes of 0.58, 0.49, 0.55, 0.43, and 0.58, respectively. Compared with the CDIL, the F index produced a result closer to 0.5, signifying that Plot b had random distribution and no adjustment was required. This indicates that the trees identified for harvest by the CDIL did not conform to the random distribution and the adjustment range was larger than the F index. The larger the U-index, the less is the growth advantage [41]. The average U-indexes of the harvest trees selected by the CDIL in the five plots were larger than the harvest trees selected by the F index, with the exception being Plot b. This indicated that the harvest trees selected by the CDIL were more unfavorable to growth than those selected by the F index. The harvest trees selected by the CDIL in the five plots had average M-indexes of 0.09, 0.31, 0.14, 0.23, and 0.38, which were smaller than those of the harvest trees selected by the F index. To improve the diversity of tree species, trees with a small M-index should be harvested. However, the harvest trees selected by the F index in the five plots had M-indexes greater than 0.5, which was inconsistent with the research of other scholars [3,36,42]. The average H-index of the harvest trees selected by the CDIL in each of the five plots was close to zero, indicating that the trees identified for harvest were unhealthy. However, the harvest trees selected by the F index had an H-index close to 0.5, so it was difficult to judge the health status of the trees. The harvest trees selected by the CDIL in Plots a, c, d, and e had an average B-index smaller than that of the harvest trees selected by the F index. The trees selected for harvest by the CDIL in Plots a-e had average D-indexes of 0.13, 0.13, 0.04, 0.05, and 0.11, respectively, and all values were greater than the D-indexes of the harvest trees selected by the F index. The B-index and the D-index indicated that the harvest trees selected by the CDIL were distributed in the more clustered areas. The average C-index of the harvest trees selected by the CDIL in each of Plots b, c, d, and e was larger than that of the harvest trees selected by the F index, indicating that the harvest trees selected by the CDIL have a higher competition pressure, which was not conducive to the growth of trees.

These results indicate that the CDIL constructed by comprehensively considering the horizontal spatial structure parameters and the vertical spatial structure parameters can help in a more reasonable and accurate selection of harvest trees.

3.2. Results of Vertical Spatial Structure Parameters

We analyzed the calculation results of the eight spatial structure parameters and the CDIL of stands in the five plots. The relative strength of the CDIL affected by the stand spatial structure index was calculated by using the grey correlation analysis method in the multifactor statistical analysis [32]. Table 4 presents the coefficient of correlation between the spatial structure parameters of stands and the CDIL. The influence of the spatial structure parameters of stands on the CDIL varied in different plots. In Plots a and b, the index with the largest correlation with the CDIL was C_i, followed by PV_i. The correlation coefficient of PV_i was the second, better than other selected indexes. In Plot c, the correlation coefficient ranking first was still C_i. The PV-index ranked second from the bottom. The correlation between the CDIL and the PV-index was not strong, possibly related to the spatial pattern and site conditions of the stand in Plot c. In Plot d, the strongest correlation index was B_i, and the correlation coefficient of PV_i was the third. In Plot e, the strongest index of correlation coefficient index was C_i, and the correlation coefficient of PV_i was ranked fourth. The correlation between the eight stand structure parameters and the CDIL was ranked in the following order: $C_i > B_i > M_i > PV_i > H_i > U_i > D_i > W_i$. In this study, the vertical spatial structure parameters in this study were ranked fourth in the comprehensive list, before H_i, U_i, D_i, and W_i. In general, the average correlation between the eight structural parameters and the CDIL was greater than 0.4, and the results showed that there was a significant correlation between them [43].

Table 4. The coefficient of correlation between stand spatial structure parameters and the CDIL.

Plots	M-Index	U-Index	W-Index	H-Index	D-Index	C-Index	B-Index	PV-Index
a	0.356	0.353	0.351	0.352	0.353	0.842	0.345	0.369
b	0.555	0.531	0.544	0.549	0.544	0.747	0.528	0.576
c	0.672	0.726	0.688	0.735	0.706	0.914	0.761	0.696
d	0.954	0.828	0.795	0.834	0.789	0.609	0.983	0.862
e	0.653	0.575	0.563	0.547	0.551	0.734	0.605	0.601
average	0.638	0.602	0.588	0.603	0.589	0.769	0.644	0.621

3.3. Adjustment of the Stand Spatial Structure after Logging

We used ArcMap software to obtain weighted Delaunay and Voronoi diagrams. The distance and the angle between the target tree and its nearest four trees were calculated in the attribute table. We used relevant formulas to calculate the average U-index, PV-index, M-index, and W-index of the five plots under the nonharvesting scenario and obtained the 10%, 20%, and 30% thinning intensities in the Unreal Engine 4 blueprint function. Table 5 shows the spatial structure parameters of different thinning intensities in different plots.

After logging, the average U-index of the five plots was close to 0.5, indicating that there was no significant difference in the size of the trees after management. The stands were in the middle state as per the quantitative index of the stand distribution state proposed by Hui Gangying [2]. The results showed that the average W-index of the stands in each plot was between 0.475 and 0.517 when the thinning intensity was 20%. The distribution of each plot gradually changed from aggregated distribution to random distribution. The variables of the average M-index of the five plots increased their thinning intensity. The average M-index of the stands in each plot was between 0.5 and 0.75, which belonged to medium-intensity mixing. The average PV-indexes of Plots a, c, and d gradually decreased with increasing thinning intensity. This was because the larger the PV-index, the more densely were the target trees covered, and the smaller the PV value, the better the stand adjustment effect. When the thinning intensity of Plot b was 30%, the value of the PV-index was lower than the original value. When the thinning intensities of Plot e were 20% and 30%, the value of the PV-index decreased compared with the original value, indicating that the PV-index reduced the degree to which the target tree was covered. The spatial structure parameters of all plots reached the optimal value when the thinning intensity was 20%, the exception being Plot e. Therefore, the optimal thinning intensity of the five plots was 20%, which was more conducive to adjusting the stand structure.

3.4. Interactive Visual Simulation of Forest Management

According to the measured data of each tree and the morphological structure data, the 3D models of 19 kinds of trees, such as Cunninghamia lanceolate, Schima Superba, and Machilus pauhoi Kaneh were built using by SpeedTree software. The 3D models of trees were added to the 3D scene as a class, and we read the coordinates in the stands data table of the plot through the inaccurate transform longitude latitude height to the unreal node under the CesiumGeoreference reference in the UE4 3D rendering engine. According to the scaling ratio calculated by the trunk and crown grading model, we loaded the trees in each plot according to the location. Figure 4 shows the simulation results of the initial stand scenarios of the five plots.

The 2D point data representing the distribution of trees could reflect the characteristics of single trees, such as the crown, the diameter at breast height, the tree height, and the crowding degree (Figure 5). However, these data were incomplete. On the other hand, the trees displayed in the 3D scene using 3D visual simulation technology helped not only directly observe the compression between tree crowns, the size difference, the spatial distribution pattern of trees, and the diversity of tree species, it also dynamically expressed the whole cutting and tending process (Figure 4), offering obvious advantages in analyzing, adjusting, and harvesting the stand structure.

Table 5. Comparison of spatial structure parameters of different thinning intensities in different plots.

Plots	Parameters	Intensity/0%	Intensity/10%	Intensity/20%	Intensity/30%
a	$\overline{M}$	0.663	0.692	0.718	0.733
	$\overline{W}$	0.521	0.489	0.480	0.472
	$\overline{U}$	0.511	0.497	0.496	0.5
	PV_i	1.919	1.855	1.908	1.903
b	$\overline{M}$	0.635	0.679	0.702	0.725
	$\overline{W}$	0.519	0.494	0.479	0.477
	$\overline{U}$	0.521	0.51	0.503	0.501
	PV_i	1.607	1.647	1.619	1.605
c	$\overline{M}$	0.694	0.728	0.729	0.739
	$\overline{W}$	0.547	0.503	0.488	0.469
	$\overline{U}$	0.502	0.484	0.49	0.489
	PV_i	1.805	1.791	1.746	1.718
d	$\overline{M}$	0.719	0.735	0.724	0.726
	$\overline{W}$	0.522	0.475	0.478	0.447
	$\overline{U}$	0.506	0.482	0.492	0.484
	PV_i	1.797	1.706	1.675	1.612
e	$\overline{M}$	0.532	0.576	0.617	0.661
	$\overline{W}$	0.518	0.519	0.496	0.475
	$\overline{U}$	0.488	0.476	0.491	0.484
	PV_i	1.571	1.567	1.563	1.604

Figure 4. (**A**) The result of visual simulation of the five sample plots in the virtual scene. (**B**) The distribution result of the five sample plots; graphs (**a**–**e**) represent the real position distribution of all the trees in each plot in the virtual scene.

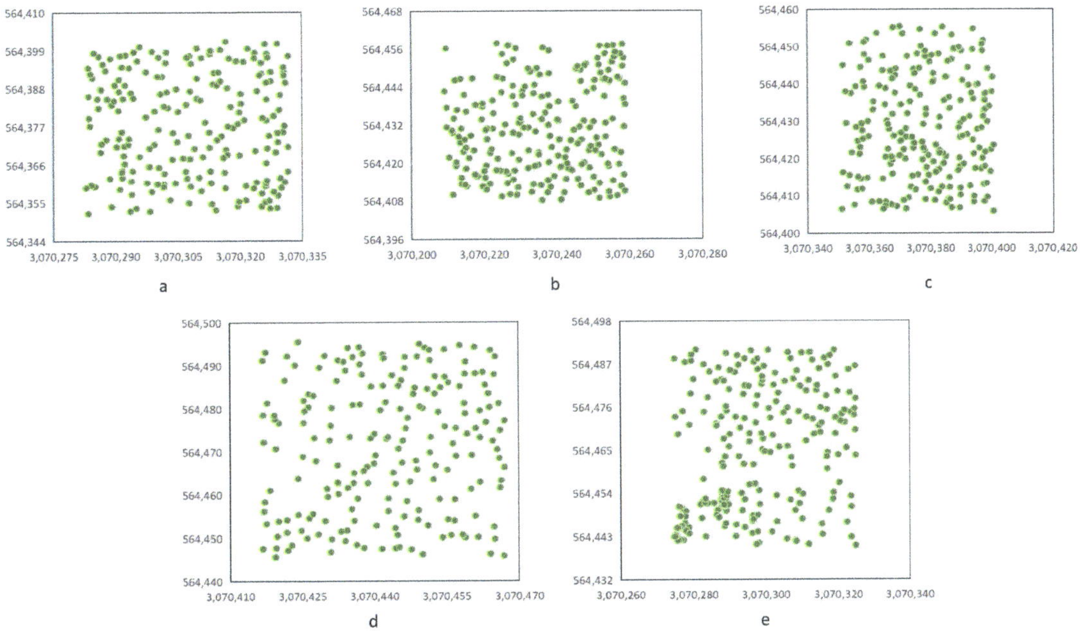

Figure 5. Two-dimensional distribution of tree position. The green dots represent the location of trees; graphs (**a–e**) represent the distribution of trees in each plot respectively.

To calculate the spatial structure parameters of each tree and each parameter index of the tree by the UE4 blueprint function script, a set of methods were developed. On the basis of the eight spatial structure parameters, the CDIL was calculated in real time in UE4 and a table providing information on the trees for harvest was generated. The system automatically logged according to the information table and the implemented growth simulation. Considering that the spatial structure of the stand would change in real time during the process of thinning and growth, the whole management simulation process would change dynamically in real time according to the management operation. Taking Plot c as an example, panels 1–3 in Figure 6 show that most of the trees selected for harvest by the CDIL were distributed in densely forested areas. With the naked eye, we could directly see from the top that the harvest trees were mainly trees with a small crown and poor growth, following the principle of cutting trees in dense areas and retaining trees in sparse areas. However, panels 4–6 in Figure 6 show that most of the trees selected for harvest by the F index were distributed in the areas with sparse trees, and the crowns of the trees identified for harvest tree were relatively large, which did not conform to the principle of cutting the small trees and retaining the large one. Studies by, for example, Li et al. [22], Pérezdelis et al. [44], and Lv et al. [29] have achieved good results in the research on the adjustment of the stand spatial structure, however, the accuracy of the results was reflected only by the parameter calculation, they lacked the visual expression of the simulation results from a 3D perspective (Figure 6).

Furthermore, panels 1–3 in Figure 7 represent the stand state after management by the CDIL, indicating that many trees with a small crown covered by adjacent trees were cut. The spatial distribution pattern of Plot c was changed from aggregate distribution to random distribution, which improved the diversity of tree species and the stability of the stand. Additionally, panels 4–6 in Figure 7 represented the stand state after management by the F index, from which we can intuitively see that many trees with large crowns were cut. The number of tree species in the stand decreased significantly after cutting, which violated the adjustment principle of improving the diversity of tree species. These results show that the CDIL gave a more reasonable selection of trees for harvest than the F index.

Figure 6. Visual simulation of Plot c by the CDIL and the F index. (**a–c**) The distribution of harvest trees identified by the CDIL. (**d–f**) The distribution of harvest trees identified by the F index. The red dots show logged trees.

Figure 7. Visual simulation of Plot c by the CDIL and the F index. (**a–c**) The adjusted stand state by the CDIL. (**d–f**) The adjusted stand state by the F index.

According to the growth model of each tree species, the growth of the stand was simulated. No.1 in Figure 8 represents the growth with a 20% thinning intensity by the CDIL, and No.2 in Figure 8 represents the growth after a 20% thinning intensity by the F index. The stand growth state after adjustment of the stand spatial structure can be seen.

Figure 8. Growth simulation of Plot c after thinning. with (**a**) representing the growth with a 20% thinning intensity by the CDIL and (**b**) representing the growth with a 20% thinning intensity by the F index.

3.5. Evaluation of the Stand Adjustment Effect

Forest management generally focuses on the analysis of the stand status, generally reflected in space utilization, species diversity, competition of constructive species, and stand composition. Therefore, this study analyzed and evaluated the effect of stand structure adjustment before and after stand management from seven aspects: the C-index, the H-index, the D-index, the B-index, the M-index, the PV-index, and the CDIL. The logging index constructed by Lv et al. was calculated and applied to this study for comparison. Table 6 presents the results of changes in stand structure parameters and the value of the F index constructed by Lv et al. before and after thinning in each plot.

After the application of the CDIL to the tending and cutting management of the five plots, the change range of the stand H-index was more than 10%, and the health status of the stand was better than that before the management. The PV-index of Plots a, c, d, and e decreased after management and varied from 1.919 to 1.855, 1.815 to 1.781, 1.797 to 1.706, and 1.571 to 1.547, respectively, indicating that the overall coverage of the stand was also decreasing. The PV-index of Plot b increased slightly, by 0.04. The M-indexes of the five plots after the management increased by 10.6%, 14.2%, 6.3%, 1%, and 24.2%, respectively, maintaining the diversity of the tree species in stands. The change in the C-index was also obvious after the management. From the results, the C-index of Plot a had the most obvious change, with a decline of 61.2%, followed by that of Plot d, with a decline of 29%, indicating that this management reduced the competitive pressure on trees and improved the competitiveness of trees. The B-indexes of all plots increased. The D-index decreased after the management, which improved the living space of trees and protected the stability of the forest structure.

The average CDIL in each plot changed by 92.6%, 88.7%, 86.7%, 84.9%, and 83.4%, respectively, after management, with an increase of more than 80%. The change in Plot a was the greatest, indicating that the management effect was the best. However, the values of the F index by Lv et al. of Plots a-e had changed by 45.4%, 36.4%, 67.8%, 54.2%, and 63.1% respectively. Different from the study by Lv et al., the CDIL adds the vertical spatial structure parameter based on the horizontal spatial structure parameters, which takes into consideration the phenomenon of adjacent trees covering the target trees. Therefore, after logging, the change range of the average CDIL of the stand was greater than that of the F index.

Table 6. Changes in stand spatial structure parameters before and after 20% thinning.

Plots	Management [1]	H-Index	C-Index	D-Index	B-Index	W-Index	M-Index	PV-Index	CDIL	F_i [2]
a	Before management	0.525	4.775	0.695	0.291	0.521	0.663	1.919	34.925	2.408
	After management	0.628	1.855	0.648	0.317	0.489	0.733	1.855	2.59	1.313
	Change range/%	19.6	61.2	6.8	8.9	6.1	10.6	3.3	92.6	45.4
b	Before management	0.517	2.193	0.686	0.315	0.519	0.635	1.607	23.583	2.054
	After management	0.6	1.929	0.652	0.331	0.494	0.725	1.647	2.671	1.306
	Change range/%	16.1	12	5	5.1	4.8	14.2	2.5	88.7	36.4
c	Before management	0.516	2.775	0.705	0.285	0.547	0.694	1.815	17.301	4.207
	After management	0.632	2.467	0.671	0.303	0.503	0.738	1.781	2.315	1.354
	Change range/%	22.5	11.1	4.8	6.3	8	6.3	1.9	86.7	67.8
d	Before management	0.524	2.369	0.654	0.35	0.522	0.719	1.797	9.357	3.094
	After management	0.65	1.682	0.617	0.367	0.475	0.726	1.706	1.414	1.416
	Change range/%	24	29	5.7	4.9	9	1	5.1	84.9	54.2
e	Before management	0.536	2.389	0.699	0.269	0.518	0.532	1.571	26.248	3.41
	After management	0.608	2.172	0.669	0.288	0.496	0.661	1.547	4.355	1.26
	Change range/%	13.4	9.1	4.3	7.1	4.2	24.2	1.5	83.4	63.1

Note: [1] The relative changes in stand spatial structure parameters for all plots before and after 10% thinning. [2] The composite spatial structure index proposed by Lv et al. [30].

We normalized the average W-indexes, U-indexes, M-indexes, B-indexes, H-indexes, D-indexes, and C-indexes in the five plots of trees selected for harvest by the CDIL and the F index in Table 2. Then, we obtained the average of the normalized values of each indicator of the five plots. The result showed that the value of the trees selected for harvest by the CDIL was 0.703 and the value of the trees selected for harvest by the F index was 0.528, indicating that the CDIL method was 11.4% more accurate than the F index. These results show that under the same horizontal spatial structure parameters, considering the influence of vertical spatial structure parameters on trees made the identification of trees for harvest more reasonable and accurate.

4. Discussion

In the forest ecosystem, the interaction between trees is essential to controlling the competition status, health, living space, and crown formation of trees. Analyzing and adjusting the complex forest structure can improve the biodiversity and stability of the forest. Therefore, reasonable thinning is the key factor in improving the stand structure and adjusting the stand density. We should consider the influence of horizontal and vertical parameters on the stand spatial structure. In this study, we selected eight spatial structure parameters and successfully constructed a comprehensive decision index of logging. On the basis of the UE4 3D rendering engine, according to the calculation results of the CDIL, the data table of harvest trees in each plot was established and management logging of 10%, 20%, and 30% intensities was simulated. The changes in stand structure indexes before and after management under different thinning intensities were compared and analyzed. The results indicated that the CDIL could effectively and reasonably select trees in the stand for harvest, making the adjusted forest structure stable and healthy.

4.1. Dynamic Changes in Stand Structure Parameters

To express dynamic changes in stand structure characters, we used eight parameters to describe subtler features of each tree: W-index, U-index, PV-index, H-index, B-Index, D-index, M-index, and C-index.

The W-index was used to analyze the individual distribution pattern of the trees in a forest and can be used to reconstruct complex forest structures [40,45]. To date, many scholars studying neighborhood patterns have shown that the distribution pattern of most trees is random [41,46]. In this study, the average W-index for all of the plots before logging was higher than 0.517, indicating aggregate distribution. However, after the implementation of the 20% thinning intensity measure, the average W-index of the stand in each plot was between

0.475 and 0.517, indicating a gradual change from aggregate to random distribution [2,22]. This is similar to the results of previous research [40,45]. Our study showed that this management can increase the proportion of the random distribution of trees.

The U-index represents the spatial dominance of tree species [2,42]. In this study, the average U-index for all plots was closer to 0.5 after cutting than before cutting, indicating that the size of trees was uniform after management and the stand was in the middle state. These results are the same as the findings of other studies [44,47].

The average M-indexes for all plots increased after logging. The greater the M-index, the higher the diversity of tree species [3,44,48]. Therefore, the M-index was beneficial to the growth and diversity of trees.

PV indicates the degree of influence of tree cover in the vertical direction. The smaller the PV-index, the less the trees are covered and the better the trees grow [32]. In this study, The PV-index for all plots decreased with the increase in the thinning intensity and the best effect was achieved when the thinning intensity was 20%. This result shows that the effect of the PV-index on the stand structure was effective, which was not found in other studies [17,25,29].

The H-index and the B-index represent the health status of trees and the openness of stands. The H-indexes and B-indexes for all plots increased after the trees were cut, indicating that the H-index and the B-index increased the growth and development space of the trees. The health status of the stands improved.

The C-index and the D-index represent the competitiveness of the trees and the degree of crowding of stands [37]. In this study, the values of the C-index and the D-index for all plots changed greatly after logging, especially the values of the C-index, which showed that management reduced the competitive pressure on trees and improved their competitiveness.

The above eight indicators considered the characteristics of trees in the horizontal and vertical structural directions respectively. As per our results, the optimal thinning intensity of the five plots was 20%, which was more conducive to adjusting the stand structure. The thinning intensity mainly concentrated on the trees with a high aggregation, poor health, low mingling, and a high competition status. However, different optimization objectives led to different optimal intensities [44,49]. Therefore, more research is needed to explore the optimal thinning intensity that affects all aspects of the ecosystem.

4.2. Evaluation of the CDIL

In this study, the index that had the largest impact on the CDIL was the C-index. The correlation between the PV-index and the CDIL was 0.61, ranking fourth in terms of correlation. The CDIL in the five plots decreased by 92.6%, 88.7%, 86.7%, 84.9%, and 83.4%, respectively. Different from the cutting index proposed by Gadow et al. [24], Li et al. [22], and Lv et al. [29], which only considered some stand structure characteristics of trees, the comprehensive decision index of logging proposed in this paper selected eight indexes that influence the horizontal and vertical structure of trees. The simulation results showed that with an increase in the thinning intensity, the CDIL values for all plots were significantly reduced. The change range of the eight stand structure parameters before and after management showed that thinning could increase the dominance, openness, stability, and mixing degree of trees and reduce the competitive pressure and density of the target trees. These results are consistent with those of other research [11,22,24]. Taking Plot c as an example, the PV-index of the stand in Plot c decreased from 1.815 before management to 1.781 afterward, indicating that the degree of coverage above the trees was greatly reduced after management. It not only considered the vertical distribution of trees, it also improved the growth space of trees. In addition, in Plot c, the CDIL index was 17.301 before management and 2.315 after management, with a change of 14.986, and a decrease of about 7.5 times. These results show that the CDIL could reasonably and accurately select trees for harvest and improve the stability of the stand structure.

This form of management greatly reduced the competitive pressure on the top tree species and improved the competitive ability and tree species advantage of the top tree species. The degree of crowding in the forest was reduced, the living space of the forest was improved, and the stability of the forest structure was maintained. Compared with the harvesting index constructed by other scholars, our index was advantageous in that it comprehensively considered the impact of horizontal and vertical directions on spatial structure and used eight horizontal and vertical spatial structure parameters. A comparison with the index constructed by Lv et al. shows that the CDIL was 11.4% more accurate than the F index in selecting harvest trees. The real process of cutting management and growth simulation was simulated in the UE4 engine by using visualization techniques. Compared with the 2D results of other scholars' studies [22,29,42], the simulation in this paper enabled managers to observe the change process of the stand spatial structure more accurately.

The CDIL has proven to be accurate and reasonable for comprehensively considering horizontal and vertical spatial structure parameters to make stand structure adjustment more effective. It is easy to use and less dependent on experience. It can greatly improve the growth and development space of trees. However, the application of CDIL to other types of forests and landscapes is still restricted due to numerous reasons, such as the limited plot number and tree number, and the fact that climatic conditions and topographic factors are not considered. Moreover, the CDIL considers the spatial structure characteristics of trees and does not consider the collision and extrusion between tree crowns or reflect them in the virtual scene. In the next research, the collision and extrusion between tree crowns will be added to the index as a factor and will be presented in a visual way.

5. Conclusions

In this study, we constructed a highly efficient, objective, and accurate comprehensive decision index of logging (CDIL) by combining the quantitative indicators representing the health status, density, distribution pattern, and competition status of the stands and used them to select trees for harvest. The results of the stand spatial structure adjustment of the index under different thinning intensities show that the optimum thinning intensity was 20%. The numbers of trees to be cut in the five sample plots were 43, 43, 44, 40, and 45 and these were mainly distributed in areas with an uneven forest size and dense forest size differentiation. After stand adjustment, the average CDIL in each plot decreased by more than 90%. The average W-index of the stand improved from 0.475 to 0.517, showing a gradual change from aggregate distribution to random distribution. This showed that the CDIL application effectively adjusts the spatial structure of the stand and makes the stand structure more stable. The CDIL index was compared with the F index constructed by Lv et al., and the results show that the CDIL was 11.4% more accurate than the F index in identifying trees for harvest. The visual simulation technology was closely combined with tending and felling and displayed the entire process of actual forest tending and felling, which will enable forest managers to observe the structural characteristics of forests more intuitively and formulate forest management plans in time. However, to make the selection of logging more accurate, we will consider the impact of crown extrusion in the future.

Author Contributions: Conceptualization, H.Z. and K.L.; methodology, K.L.; software, K.L. and J.W.; formal analysis, Y.L., H.Q. and Z.C.; resources, X.H., Y.Z. and J.W.; writing—original draft preparation, K.L.; writing—review and editing, K.L., H.Z. and H.Q.; visualization, H.Z.; supervision, H.Z. and Y.L.; All authors have read and agreed to the published version of the manuscript.

Funding: This research was funded by the National Natural Science Foundation of China, grant number 32071681, 32271877 and the Foundation Research Funds of IFRIT, grant number CAFYBB2021ZE005, CAFYBB2019SZ004.

Data Availability Statement: Not applicable.

Conflicts of Interest: The authors declare that there are no conflict of interest.

References

1. Zhang, L.; Hui, G.; Hu, Y.; Zhao, Z. Spatial structural characteristics of forests dominated by *Pinus tabulaeformis* Carr. *PLoS ONE* **2018**, *13*, e0194710. [CrossRef]
2. Hui, G.; Zhang, G.; Zhao, Z.; Yang, A. Methods of forest structure research: A review. *Curr. For. Rep.* **2019**, *5*, 142–154. [CrossRef]
3. Wan, P.; Zhang, G.; Wang, H.; Zhao, Z.; Hu, Y.; Zhang, G.; Hui, G.; Liu, W. Impacts of different forest management methods on the stand spatial structure of a natural *Quercus aliena* var. acuteserrata forest in Xiaolongshan, China. *Ecol. Inform.* **2019**, *50*, 86–94. [CrossRef]
4. Kerr, G. The use of silvicultural systems to enhance the biological diversity of plantation forests in Britain. *For. An Int. J. For. Res.* **1999**, *72*, 191–205. [CrossRef]
5. Hanewinkel, M.; Pretzsch, H. Modelling the conversion from even-aged to uneven-aged stands of *Norway spruce* (*Picea abies* L. Karst.) with a distance-dependent growth simulator. *For. Ecol. Manage.* **2000**, *134*, 55–70. [CrossRef]
6. Mickaël, H.; Michaël, A.; Fabrice, B.; Pierre, M.; Thibaud, D. Soil detritivore macro-invertebrate assemblages throughout a managed beech rotation. *Ann. For. Sci.* **2007**, *64*, 219–228. [CrossRef]
7. Aguirre, O.; Hui, G.Y.; von Gadow, K.; Javier, J. An analysis of spatial forest structure using neighbourhood-based variables. *For. Ecol. Manag.* **2003**, *183*, 137–145. [CrossRef]
8. Kint, V.; Meirvenne, M.; Nachtergale, L.; Geudens, G.; Lust, N. Spatial methods for quantifying forest stand structure development: A comparison between nearest-neighbor indices and variogram analysis. *For. Sci.* **2003**, *49*, 36–49. [CrossRef]
9. Antos, J.A.; Parish, R. Dynamics of an old-growth, fire-initiated, subalpine forest in southern interior British Columbia: Tree size, age, and spatial structure. *Can. J. For. Res.* **2002**, *32*, 1935–1946. [CrossRef]
10. North, M.; Chen, J.; Oakley, B.; Song, B.; Rudnicki, M.; Gray, A.; Innes, J. Forest stand structure and pattern of old-growth western hemlock/Douglas-fir and mixed-conifer forests. *For. Sci.* **2004**, *50*, 299–311. [CrossRef]
11. Li, Y.; Hui, G.; Ye, S.; Hui, G.; Hu, Y.; Zhao, Z. Spatial structure of timber harvested according to structure-based forest management. *For. Ecol. Manag.* **2014**, *322*, 106–116. [CrossRef]
12. Adams, D.M.; Latta, G.S. Effects of a forest health thinning program on land and timber values in eastern Oregon. *J. For.* **2004**, *102*, 9–13. [CrossRef]
13. Paletto, A.; Meo, I.D.; Grilli, G.; Nikodinoska, N. Effects of different thinning systems on the economic value of ecosystem services: A case-study in a black pine peri-urban forest in Central Italy. *Ann. For. Res.* **2017**, *60*, 313–326.
14. Forrester, D.I.; Elms, S.R.; Baker, T.G. Tree growth-competition relationships in thinned *Eucalyptus plantations* vary with stand structure and site quality. *Eur. J. For. Res.* **2013**, *132*, 241–252. [CrossRef]
15. Forrester, D.I. Linking forest growth with stand structure: Tree size inequality, tree growth or resource partitioning and the asymmetry of competition. *For. Ecol. Manage.* **2019**, *447*, 139–157. [CrossRef]
16. Pommerening, A. Evaluating structural indices by reversing forest structural analysis. *For. Ecol. Manag.* **2006**, *224*, 266–277. [CrossRef]
17. Darenova, E.; Crabbe, R.A.; Knott, R.; Uherková, B.; Kadavý, J. Effect of coppicing, thinning and throughfall reduction on soil water content and soil CO_2 efflux in a sessile oak forest. *Silva Fenn.* **2018**, *52*, 9927. [CrossRef]
18. Hill, M.; Pospíšil, M. On the relation between the secretion of the perivascular mast cells and the serum level of mucoproteins. *Experientia* **1959**, *15*, 267–269. [CrossRef]
19. Johnson, D.W.; Murphy, J.D.; Walker, R.F.; Miller, W.W.; Glass, D.W.; Todd, D.E. The combined effects of thinning and prescribed fire on carbon and nutrient budgets in a *Jeffrey pine* forest. *Ann. For. Sci.* **2008**, *65*, 601. [CrossRef]
20. Dong, L.; Wei, H.; Liu, Z. Optimizing forest spatial structure with neighborhood-based indices: Four case studies from northeast China. *Forests* **2020**, *11*, 413. [CrossRef]
21. Ye, S.X.; Zheng, Z.R.; Diao, Z.Y.; Ding, G.D.; Bao, Y.F.; Liu, Y.D.; Gao, G.L. Effects of thinning on the spatial structure of *Larix principis-rupprechtii* plantation. *Sustainability* **2018**, *10*, 1250. [CrossRef]
22. Li, Y.; Hui, G.Y.; Wang, H.X.; Zhang, L.J.; Ye, S.X. Selection priority for harvested trees according to stand structural indices. *iForest* **2017**, *10*, 561–566. [CrossRef]
23. Von Gadow, K.; Hui, G.Y. Modelling forest development. *For. Sci.* **1999**, *57*, 46–58.
24. Von Gadow, K.; Zhang, C.; Wehenkel, C.; Pommerening, A.; Corral-Rivas, J.; Korol, M.; Myklush, S.; Hui, G.Y.; Kiviste, A.; Zhao, X.H. *Forest Structure and Diversity*; Springer: Berlin, Germany, 2012; pp. 30–62.
25. Song, Y.F. Individual Tree Growth Models and Competitors Harvesting Simulation for Target Tree-Oriented Management. Ph.D. Thesis, Chinese Academy of Forestry, Beijing, China, 2015. (In Chinese)
26. Pastorella, F.; Paletto, A. Stand Structure Indices as Tools to Support Forest Management: An Application in Trentino Forests (Italy). *J. For. Sci.* **2013**, *59*, 159–168. [CrossRef]
27. Courbaud, B.; Goreaud, F.; Dreyfus, P.; Bonnet, F.R. Evaluating thinning strategies using a tree distance dependent growth model: Some examples based on the CAPSIS software "uneven-aged spruce forests" module. *For. Ecol. Manage.* **2001**, *145*, 15–28. [CrossRef]
28. Cao, X.Y.; Li, J.P.; Hu, Y.J.; Yang, J. Spatial structure optimizing model of stand thinning of *Cunninghamia lanceolata* ecological forest. *J. Chin. J. Ecol.* **2017**, *36*, 1134–1141. (In Chinese)
29. Lv, Z.S.; Liu, Z.G.; Dong, L.B.; Zhang, L.Y.; Sun, Y.X. Simulation of Mixed Forest Structure Optimization with Comprehensive Cutting Index in Maoer Mountain. *J. Northeast. For. Univ.* **2018**, *46*, 12–17. (In Chinese)

30. Ahmad, B.; Wang, Y.; Hao, J.; Liu, Y.; Bohnett, E.; Zhang, K. Optimizing Stand Structure for Tradeoffs between Overstory and Understory Vegetation Biomass in a Larch Plantation of Liupan Mountains, Northwest China. *For. Ecol. Manage.* **2019**, *443*, 43–50. [CrossRef]
31. Bhandari, S.K.; Veneklaas, E.J.; McCaw, L.; Mazanec, R.; Whitford, K.; Renton, M. Individual Tree Growth in Jarrah (*Eucalyptus marginata*) Forest Is Explained by Size and Distance of Neighbouring Trees in Thinned and Non-Thinned Plots. *For. Ecol. Manage.* **2021**, *494*, 119364. [CrossRef]
32. Li, S.J.; Zhang, H.Q.; Li, Y.L.; Yang, T.D.; He, J.P.; Ma, Z.Y.; Seng, K. Dynamic Visual Simulation of Chinese Fir Stand Growth Based on Sample Library. *J. For. Res.* **2019**, *32*, 21–30. (In Chinese)
33. Pommerening, A. Approaches to Quantifying Forest Structures. *For. An Int. J. For. Res.* **2002**, *75*, 305–324. [CrossRef]
34. Hui, G.Y.; Gadow, K.V.; Albert, M. The neighborhood pattern-a new structure parameter for describing distribution of forest tree position. *Sci. Silvae Sin.* **1999**, *35*, 37–42. (In Chinese)
35. Hui, G.Y.; Gadow, K.V.; ALbert, M. A New Parameter for Stand Spatial Structure—Neighbourhood Comparison. *J. For. Res.* **1999**, *12*, 1–6.
36. Tanaka, K.; Tokuda, M. Negative correlation between dispersal investment and canopy openness among populations of the ant-dispersed sedge, *Carex lanceolata*. *Plant Ecol.* **2020**, *221*, 1105–1115. [CrossRef]
37. Hegyi, F. A simulation model for managing jack-pine stands. In *Growth Models for Tree and Stand Simulation*; Fries, J., Ed.; IUFRO: Vienna, Austria, 1974; pp. 74–90.
38. Hui, G.; Zhao, X.; Zhao, Z.; Gadow, K. Evaluating tree species spatial diversity based on neighborhood relationships. *For. Sci.* **2011**, *57*, 292–300. [CrossRef]
39. Zeng, Q.Y.; Zhou, Y.M.; Li, J.P.; Liu, S.Q. Decision-making Methodology of Forest Ecosystem Management Based on Spatial Structure Factors of Forest. *J. Northeast. For. Univ.* **2010**, *38*, 31–35. (In Chinese)
40. Li, Y.; Hui, G.; Zhao, Z.; Hu, Y. The bivariate distribution characteristics of spatial structure in natural Korean pine broad-leaved forest. *J. Veg. Sci.* **2012**, *23*, 1180–1190. [CrossRef]
41. Li, Y.; He, J.; Yu, S.; Wang, H.; Ye, S. Spatial structures of different-sized tree species in a secondary forest in the early succession stage. *Eur. J. For. Res.* **2020**, *139*, 709–719. [CrossRef]
42. Cao, X.; Li, J.; Feng, Y.; Hu, Y.; Zhang, C.; Fang, X.; Deng, C. Analysis and evaluation of the stand spatial structure of *Cunninghamia lanceolata* ecological forest. *Sci. Silvae Sin.* **2015**, *51*, 37–48.
43. Azzeh, M.; Neagu, D.; Cowling, P.I. *Fuzzy Grey Relational Analysis for Software Effort Estimation*; Kluwer Academic Publishers: Alphen am Rhein, The Netherlands, 2010.
44. Pérezdelis, G.; Garcíagonzález, I.; Rozas, V.; Arevalo, J.R. Effects of thinning intensity on radial growth patterns and temperature sensitivity in *Pinus canariensis* afforestations on Tenerife Island, Spain. *Ann. For. Sci.* **2011**, *68*, 1093–1104. [CrossRef]
45. Zhao, Z.; Hui, G.; Hu, Y.; Wang, H.; Zhang, G.; Gadow, K. Testing the significance of different tree spatial distribution patterns based on the uniform angle index. *Can. J. For. Res.* **2014**, *44*, 1417–1425. [CrossRef]
46. Zhang, G.; Hui, G.; Zhao, Z.; Hu, Y.; Wang, H.; Liu, W.; Zang, R. Composition of basal area in natural forests based on the uniform angle index. *Ecol. Inf.* **2018**, *45*, 1–8. [CrossRef]
47. Fang, X.; Tan, W.; Gao, X.; Chai, Z. Close-to-nature management positively improves the spatial structure of *Masson pine* forest stands. *Web. Ecol.* **2021**, *21*, 45–54. [CrossRef]
48. Zhao, Z.; Hui, G.; Hu, Y.; Li, Y.; Wang, H. Method and application of stand spatial advantage degree based on the neighborhood comparison. *J. Beijing For. Univ.* **2014**, *36*, 78–82.
49. Settineri, G.; Mallamaci, C.; Mitrović, M.; Sidari, M.; Muscolo, A. Effects of different thinning intensities on soil carbon storage in Pinus laricio forest of Apennine South Italy. *Eur. J. For. Res.* **2018**, *137*, 131–141. [CrossRef]

forests

MDPI

Article

A New Tree-Level Multi-Objective Forest Harvest Model (MO-PSO): Integrating Neighborhood Indices and PSO Algorithm to Improve the Optimization Effect of Spatial Structure

Hanqing Qiu [1,2,3], Huaiqing Zhang [1,2,3,*], Kexin Lei [1,2,3], Xingtao Hu [4], Tingdong Yang [1,2,3] and Xian Jiang [1,2,3]

[1] Institute of Forest Resource Information Techniques, Chinese Academy of Forestry, Beijing 100091, China
[2] Key Laboratory of Forest Management and Growth Modelling, NFGA, Beijing 100091, China
[3] National Long term Scientific Research Base of Huangfengqiao Forest Monitoring and Simulation in Hunan Province, Beijing 100091, China
[4] School of Geography and Environmental Sciences, Guizhou Normal University, Guiyang 550025, China
* Correspondence: zhang@ifrit.ac.cn; Tel.: +1-365-116-6672

Abstract: Accurate, efficient, impersonal harvesting models play a very important role in optimizing stand spatial structural and guiding forest harvest practices. However, existing studies mainly focus on the single-objective optimization and evaluation of forest at the stand- or landscape-level, lacking considerations of tree-level neighborhood interactions. Therefore, the study explored the combination of the PSO algorithm and neighborhood indices to construct a tree-level multi-objective forest harvest model (MO-PSO) covering multi-dimensional spatial characteristics of stands. Taking five natural secondary forest plots and thirty simulated plots as the study area, the MO-PSO was used to simulate and evaluate the process of thinning operations. The results showed that the MO-PSO model was superior to the basic PSO model (PSO) and random thinning model Monte Carlo-based (RD-TH), DBH dominance (DOMI), uniform angle (ANGL), and species mingling (MING) were better than those before thinning. The multi-dimensional stand spatial structure index (L-index) increased by 1.0%~11.3%, indicating that the forest planning model (MO-PSO) could significantly improve the spatial distribution pattern, increase the tree species mixing, and reduce the degree of stand competition. In addition, under the four thinning intensities of 0% (T1), 15% (T2), 30% (T3), and 45% (T4), L-index increased and T2 was the optimal thinning intensity from the perspective of stand spatial structure overall optimization. The study explored the effect of thinning on forest spatial structure by constructing a multi-objective harvesting model, which can help to make reasonable and scientific forest management decisions under the concept of multi-objective forest management.

Keywords: forest planning; stand spatial structure; selection thinning priority; PSO algorithm; neighborhood-based indices

1. Introduction

Forest planning mainly consists of three stages: assessment, planning, and monitoring [1,2]. It is a complex task highly dependent on mathematical planning and information technology [3], which provides a decision-making framework for forest managers: when, where, how, and how much to harvest [4]. Modern forest planning involves the coordination of multi-function relationships. The trade-offs between forest ecosystem services supply and forest management costs may especially complicate the decision-making process [5,6]. Therefore, combinatorial optimization is the key to optimizing stand structure, maintaining species diversity, improving forest productivity, and increasing habitat heterogeneity [7–9].

Characterization of stand spatial structure is an essential part of sustainable forest management [10–12]. Among them, neighborhood indexes that consider the location of individual trees and interactions between neighboring trees have received widespread attention. For instance, DBH (diameter at breast height) dominance, species mingling, and uniform angle are distance-dependent, neighborhood-based indices. DBH dominance is widely used to describe tree competition, Species mingling is used to evaluate the degree of tree species mixing, and uniform angle represents tree spatial distribution pattern [11]. Some studies evaluated forest spatial structure using neighborhood indices and developed corresponding strategies to increase the diversity of tree species, optimize stand spatial distribution pattern, and reduce the degree of competition, thus enhancing forest quality [11]. Unfortunately, few studies use the multi-dimensions neighborhood indices to develop forest planning strategies.

Thinning, as a silvicultural treatment in forest planning, can promote forest growth, improve stand structure, maintain biodiversity, and prevent pests and fires [12]. Trees should be removed, and strategies should be taken in tree-level forest planning. These are the key elements, as the thinning process is limited by multiple factors. Previous studies focused on constructing thinning models using stem quality attributes [13], tree species diversity [11], adjacencies between trees [14], and economic criteria [15,16] as the criteria for selecting harvested trees, which have certain limitations, including the ignoration of the potential impact of the spatial distribution characteristics [17,18]. Multi-dimension optimization objectives often conflict with each other, showing nonlinearity and discontinuity characteristics. Therefore, multi-objective optimization algorithms are needed in tree-level forest planning.

Recently, heuristics are increasingly being used to solve forest planning problems [19,20]. For instance, Fotakis et al. [21] proposed the constrained NSGA-II, which is used for multi-objective forest planning problems, including maximization of timber volume, minimization of division levels, and even the flows and minimum timber yield. Bettinger and Tang [11] combined the threshold accepting heuristic and the species mingling index to achieve tree-level operational harvest scheduling optimization. Compared with other algorithms, the particle swarm optimization (PSO) algorithm has significant advantages in solving large-scale mathematical optimization problems by using memory and feedback mechanisms: (1) fast calculation speed and strong global search ability; (2) faster convergence than evolutionary algorithms and genetic algorithms; (3) simple parameter setting, strong local search ability, and low possibility of falling into local optimum [22,23]. Multi-objective forest planning is also a similar optimization process: calculate the changes of comprehensive indices before- and after-harvesting under the constraints of multiple management objectives and in a given solution space (e.g., trees in the plot) to determine the optimal harvesting intensity.

However, the application of PSO in tree-level forest planning has not been explored in depth. Therefore, from the perspective of tree-level multi-objective forest planning, it is a novel concept and method to combine particle swarm optimization (PSO) with multi-dimensional stand structure indices to construct a harvest model. It contributes to the expansion and extension of sustainable forest management. Based on five natural secondary forest plots in South Dongting Lake and thirty simulated plots, the following studies were conducted: (1) develop a tree-level multi-objective forest planning model by integrating PSO algorithm and neighborhood indexes (DOMI, ANGL, and MING). (2) explore the response of species diversity, competition, and forest spatial distribution pattern to forest thinning, as well as the optimal thinning intensity.

2. Materials and Methods

2.1. Data

2.1.1. Natural Plots

The natural plots were located in the Dongting Lake region in Hunan province, China. Dongting Lake is the second largest freshwater lake in China, located in the middle of the Yangtze River (28°39′ N–30°14′ N, 111°42′ E–113°93′ E). The zonal vegetation in the area was the subtropical evergreen broad-leaved forest in the middle subtropics.

Cyclobalanopsis glauca and Schima superb were dominant tree species. The frost-free period was 263~277 d. The mean annual rainfall was 1400~1500 mm. The mean annual temperature was 16.6~17.0 °C, the lowest temperature in January was 4.1~4.5 °C, and the highest temperature in July was 28.7~29.2 °C.

We prepared five natural plots, which were located in Lutou Forest Farm (LT), Daweishan Nature Reserve (DW), Longhushan Forest Farm (LH), Dashanchong Forest Farm (DS), and Wuyunjie Nature Reserve (WT). The area of each plot was 20 m × 30 m. The tree height, DBH (≥5 cm), tree position, tree crown width, and tree species of all surveyed trees were recorded. The tree species were recorded and provided by forest farm staff. The area of each plot was 20 m × 30 m. The tree height, DBH (≥5 cm), tree position, tree crown width, and tree species of all surveyed trees were recorded. The DBH was measured using a ruler. Tree height was measured using a laser altimeter. The crown width was measured with a tape measure and the spatial position (x,y) of each tree was recorded using RTK. The spatial distributions of five natural plots and the statistical characteristics of trees are illustrated in Figure 1 and Table 1.

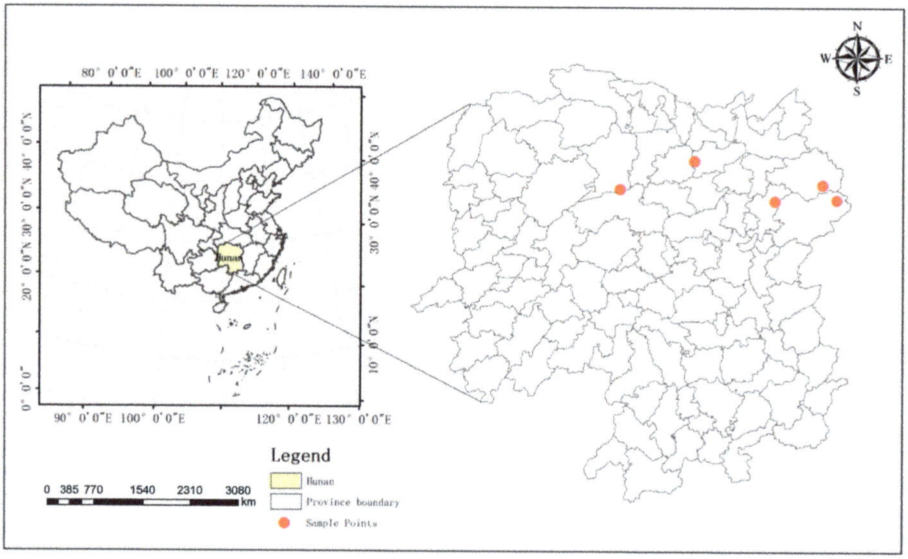

Figure 1. Location of the natural plots.

Table 1. Basic statistical characteristics of the five studied plots.

Site	DW	LH	DS	LT	WT
Altitude (m)	1300	68	217	335	713
Slope (°)	47	35	27	15	38
Aspect	E	W	ES	EN	S
Canopy density	0.75	0.65	0.70	0.80	0.60
Mean DBH (cm)	12.7	10.3	14.7	11.3	9.4
Mean Height (m)	11.3	9.1	13.8	9.7	7.1
Mean Crown (m)	2.4	2.7	2.0	3.1	2.5
Number of species	8	7	10	6	8

2.1.2. Simulated Plots

The study sample plots included two parts: simulated plot and natural plot. The spatial distribution pattern of simulated plots was divided into three types: uniform distribution (A), random distribution (B), and aggregated distribution (C). Each type contained ten plots, for a total of 30 plots (Figure 2). In our study, Python (version 3.6.0) was used to generate 30 plots for each type, numbered 1–30, with 30–70 trees per plot, and a plot area was 20 m × 30 m, the dominant species were *Cyclobalanopsis glauca and Schima superb*. The standard deviation of tree size (DBH, tree height, crown) in each group increased as the number of groups increased: the minimum standard deviation of tree size of group 1 was 0.014, and for group 30, the standard deviation of tree size was at 0.613, which was the largest.

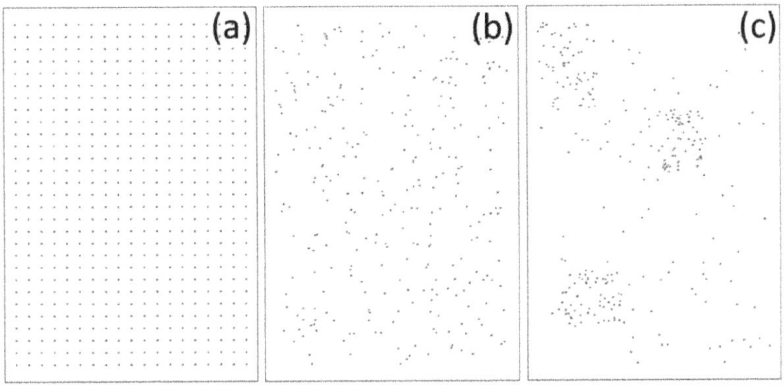

Figure 2. Sample plots with three distributions: (**a**) uniform distribution, (**b**) random distribution, and (**c**) aggregated distribution.

2.2. Neighborhood Indices

In order to accurately demonstrate the relationship of neighboring trees, such as the degree of tree species mixing, tree competition, and tree spatial distribution pattern, we compiled the widely used and representative stand spatial structure indices (Table 2). Hui et al. proposed the "Structure-based Forest Management" theory. Among them, the division of forest spatial structure units by "1 + 4" had been recognized by the international public and widely used. According to the recommendation and the "1 + 4" criterion, the number of neighboring trees n selected in this paper was 4 in our study [24,25]. The index distribution of five natural plots was shown in Figure 3.

Table 2. Stand spatial structure index.

Index	Calculation Formula	Variable Definition
Uniform angle (ANGL)	$W_i = \frac{1}{n}\sum_{j=1}^{n} Z_{ij}$	W_i is the uniform angle of central tree i, n is the number of neighboring trees, Z_{ij} is the variable of uniform angle. When the angle between central tree i and neighboring tree j is less than the standard angle, $Z_{ij} = 1$, otherwise, $Z_{ij} = 0$.
DBH Dominance (DOMI)	$U_i = \frac{1}{n}\sum_{j=1}^{n} k_{ij}$	U_i is the neighborhood comparison of central tree i, k_{ij} is the value variable of neighborhood comparison. When the DBH of neighboring tree j is smaller than that of central tree i, $k_{ij} = 0$, otherwise, $k_{ij} = 1$.
Species mingling (MING)	$M_i = \frac{1}{n}\sum_{j=1}^{n} v_{ij}$	M_i is the species mingling of central tree i, v_{ij} is the value variable of the mingling degree. When the central tree i and neighboring tree j are the same trees, $v_{ij} = 0$, otherwise $v_{ij} = 1$.

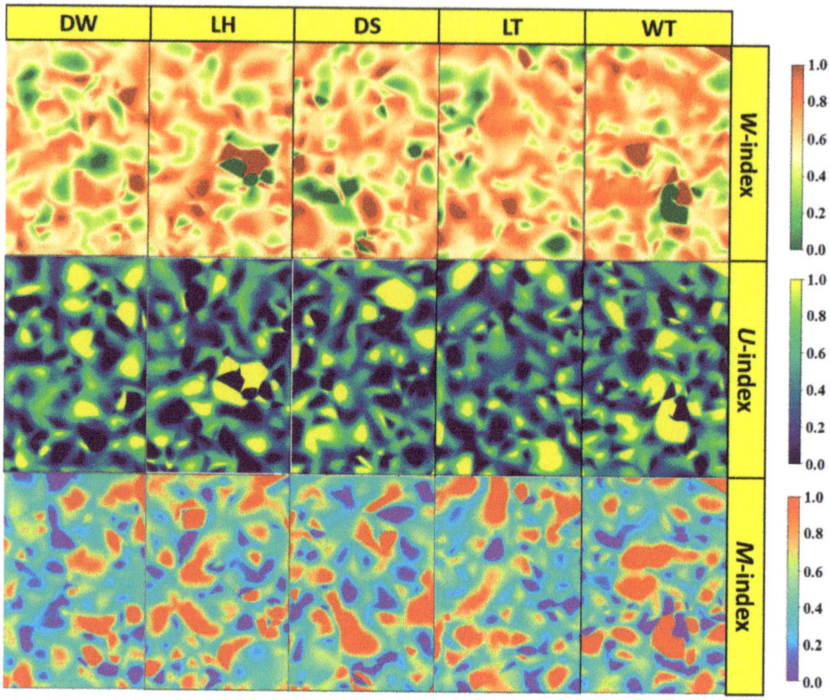

Figure 3. The distribution of W-index (ANGL), U-index (DOMI), and M-index (MING) for the five natural plots.

2.3. Construction of Dynamic Multi-Objective Optimization Model of Forest Spatial Structure

The multi-objective optimization of forest spatial structure is a dynamic optimization problem. Each goal is independent and affects the others. It is necessary to consider the tree species diversity, the inter- and intra-specific competition, and the tree distribution to make the overall forest structure healthy and stable. Therefore, no matter which forest management mode is chosen, stand spatial structure should be evaluated comprehensively instead of depending on a single structure index. Considering that tree species mixing, competition, and distribution pattern of trees are the main influencing factors of forest spatial structure, three objective functions of MING, DOMI, and ANGL are selected as the constraints of the evaluation method. In addition, the stand spatial structure optimization strategy first adjusts the horizontal distribution pattern of stands, and judges the horizontal distribution pattern of stands according to the average value of uniform angle index. It is generally believed that if the stand is not seriously disturbed, the horizontal distribution pattern should be random distribution after long-term forest development and succession [26–28]. Therefore, when making forest harvested strategy, we should first analyze the uniform angle index of the stand, and give priority to adjusting the horizontal distribution pattern of the stand from non-random distribution to random distribution.

The purpose of constructing the multi-objective optimization model is to optimize the stand structure, so the output is the target trees that affect the stand structure. The trees with small MING, large DOMI, and large ANGL significantly affect the overall stand structure [29,30], which needs to be removed to optimize the stand structure. Our study was based on utility theory [31,32], with the constraints of maximizing MING, minimizing DOMI and ANGL, and then constructing a multi-dimensional stand spatial structure index

(L) to evaluate the overall stand spatial structure after optimization. The L is constructed as follows:

$$\text{maximize } M_i = \frac{1}{n}\sum\nolimits_{j=1}^{n} v_{ij}, \tag{1}$$

$$\text{minimize } U_i = \frac{1}{n}\sum\nolimits_{j=1}^{n} k_{ij}, \tag{2}$$

$$\text{minimize } W_i = \frac{1}{n}\sum\nolimits_{j=1}^{n} Z_{ij}, \tag{3}$$

$$L_i = f(M_i, W_i, U_i), \tag{4}$$

$$L = \frac{1}{N}\cdot\frac{\omega_m}{\omega_w\cdot\omega_u}\sum\nolimits_{i=1}^{N}\frac{(1+M_i)\cdot(1+\sigma_{M_i})}{(1+W_i)\cdot(1+\sigma_{W_i})\cdot(1+U_i)\cdot(1+\sigma_{U_i})}, \tag{5}$$

where M_i, W_i and U_i are the species mingling, uniform angle, DBH dominance of the target tree i, respectively; N is the total number of trees in the stand; n is the number of the neighbors of the target tree i; σ_{M_i}, σ_{W_i} and σ_{U_i} are the standard deviation of species mingling, uniform angle, DBH dominance of the target tree i, respectively; ω_m, ω_w, and ω_u are the weights corresponding to the three optimization constraint objectives, respectively; L represents the overall level of stand spatial structure. The smaller L, the simpler the composition of tree species, the greater the competition among trees in certain areas, and the spatial distribution of stands is closer to non-random distribution.

The more ideal the stand spatial structure, the higher the degree of tree species mixing, the lower the degree of competition, and the more random the spatial distribution of trees. In addition, the thinning intensity is also an essential part of the forest harvest scheduling model. Therefore, quantitative evaluation of the impact of different harvesting intensities on the stand spatial structure characteristics plays an important role in the early stages of forest development and succession. In this study, according to the Technical Regulations on the Investigation and Design of Forest Cutting Areas in Hunan Province, the cutting intensity should be considered according to the comprehensive factors such as the forest management objectives, site conditions, and stand conditions (http://www.hunan.gov.cn, accessed on 2 December 2022). The tending cutting intensity was less than 20%~25%; the release cutting intensity was less than 40%, and the accretion cutting intensity was less than 30%. The cutting intensity varied in different regions. The plots in our study were in Hunan Province. In order to meet the actual thinning intensity and situation, we set four different levels of cutting intensity, namely 0% (T1), 15% (T2), 30% (T3), and 45% (T4), to explore the impact of harvesting intensity on the stand spatial structure.

2.4. Construction of Tree-Level Multi-Objective Forest Harvest Model (MO-PSO)

2.4.1. Particle Swarm Optimization

Particle swarm optimization (PSO) is an efficient and powerful population-based stochastic search technique for solving global optimization problems. PSO algorithm abstracts each solution of the problem to be solved as a particle in the solution space. The solution process is actually an iterative optimization process in the solution space by a particle swarm composed of m particles. Particles search for new positions by constantly updating their positions and speeds and determining whether the searched positions are optimal by using the fitness function.

$$v_{id} = \omega v_{id} + c_1 r_1 (p_{id} - x_{id}) + c_2 r_2 \left(p_{gd} - x_{id}\right), \tag{6}$$

$$x_{id} = x_{id} + v_{id} + min\sqrt{(x_i - x')^2 + (y_i - y')^2}, \tag{7}$$

where $X_i = (x_{i1}, x_{i2}, \cdots, x_{iD})$ represents the position of the ith particle in a D-dimensional search space, $V_i = (v_{i1}, v_{i2}, \cdots, v_{iD})$ is the velocity of ith particle, $P_{besti} = (p_{i1}, \ldots, p_{iD})$ is i^{th} particle's personal best position, $P_{gbest} = (p_{g1}, \ldots, p_{gD})$ is global best position. Acceleration constants c_1, c_2 and inertia weight ω are predefined by the user, and r_1 and r_2 are the uniformly generated random numbers in the range (0, 1).

When the particle position is updated according to Equation (7), it is possible that the particle position is not within the definition domain. Such particles are called transboundary-particles. When the transboundary-particles appear, we multiply the speed in Equation (7) by the adjustment coefficient β to pull the transboundary-particles back into the definition domain. The new location formula is shown in Equation (8). If the particles remain outside the boundary after adjustment, reduce the β value according to Equation (9) until the particles return to the decision space.

$$x_k^i = x_k^i + \beta v_k^i + min\sqrt{(x_i - x')^2 + (y_i - y')^2}, \tag{8}$$

$$\beta = 1/\left(\delta^2 + 1\right), \tag{9}$$

where δ is the number of adjustments.

2.4.2. The Method of MO-PSO Model Construction

In this study, each tree in the forest was regarded as a solution in the solution space, the forest space corresponded to the target solution space of the PSO algorithm, and the spatial coordinates of trees in the forest corresponded to the position of particles, and the Function was taken as the fitness function, thus transforming the multi-objective planning problem of stand spatial structure into an iterative optimization process of the particle swarm in the solution space.

For an ordinary particle swarm, N particles are randomly initialized, and then these particles are divided into M sub-swarm by K-means algorithm. In each sub-swarm, the particle with the shortest distance from other particles in the particle swarm is selected as the central particle of the particle swarm. The central particle is calculated by the following formula:

$$x_k^{Center} = arg\left(min\sum_{j=1}^{\frac{N}{M}} \| x_k^i - x_k^j \|\right),\ i \in \left[1, \frac{N}{M}\right], \tag{10}$$

$$dis\left(x_k^i, x_k^j\right) = \| x_k^i - x_k^j \| = \sqrt{\left(\sum_{l=1}^{n}\left(x_k^{il} - x_k^{jl}\right)^2\right)/n}, \tag{11}$$

in which x_k^{Center} represents the central position of each particle swarm, dis is the distance between i^{th} particle and j^{th} particle, x_k^i and x_k^j, respectively, represented as i^{th} particle and j^{th} particle in the particle swarm, and n is the dimension of the decision variable x. If the distance between the x_k^i and other particles in the same swarm is the smallest, then the central particle of the kth swarm $x_k^{Center} = x_k^i$.

After determining the center particle of the M sub-swarm, the distance between the sub-swarm k and other sub-swarms is calculated, respectively, and the dynamic particle group is constructed according to the results obtained. The upper limit and lower limit of the distance between particle swarms are D_{max} and D_{min}, respectively, and the particle swarm with the smallest distance from the particle swarm k is a particle swarm a. If $dis(x_k, x_a) > D_{max}$, a new particle swarm x_{M+1} is generated. If the minimum distance between a particle swarm and another particle swarm is less than D_{min}, a particle swarm is deleted. The calculation formula of the newly generated particle swarm is:

$$x_{M+1}^{il} = \left(x_k^{il} + x_a^{il}\right)/2 + c_1(-1)^{round(0.5+c_2)}\left|x_k^{il} - x_a^{il}\right|/2, \tag{12}$$

where x_{M+1}^{il} represents the l dimension component of the ith particle in the particle swarm $M+1$, c_1, c_2 is the random number in $[0,1]$, and round $(\cdot)$ represents the rounding function.

The particle swarm searches for trees that meet the optimization objectives in the stand space using a parallel mechanism, and the final output of the algorithm is the target tree that needs to be regulated or logged in the stand. The construction process of the multi-objective optimization harvesting model (MO-PSO) integrating neighborhood indices and PSO algorithm is shown in Figures 4 and 5. The simulation of the tree-level harvesting process was performed by c++ and Python.

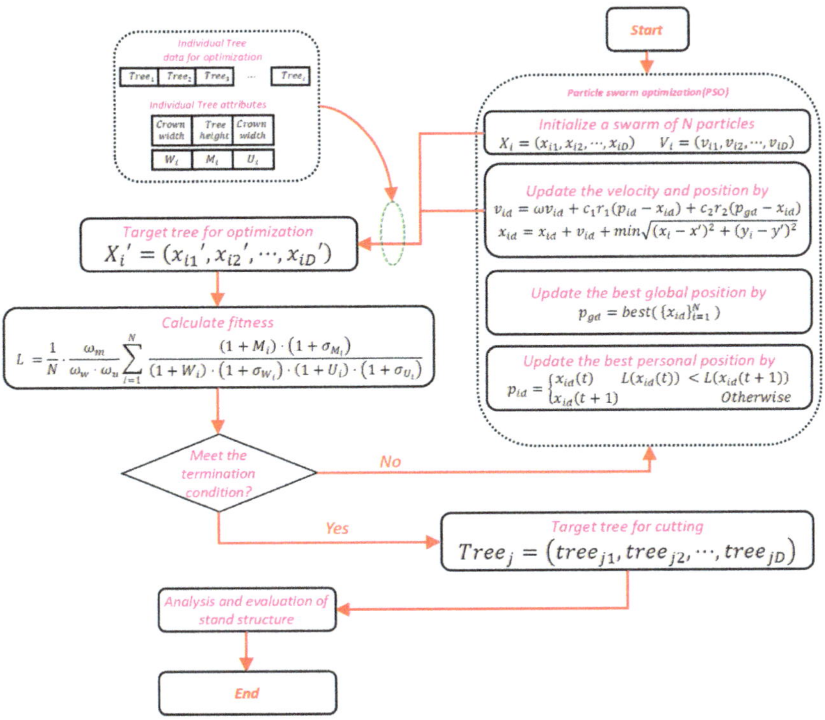

Figure 4. The architecture of the tree-level multi-objective forest harvest model (MO-PSO).

Step 1: Import spatial position information and individual tree attribute data of all trees Q, such as DBH, tree height, and crown width—import data.

Step 2: Randomly generate an initial population P with the size of N. Use the K-means algorithm to divide the randomly generated particles into M sub-swarm, and randomly generate the initial position and velocity of each particle—initial population.

Step 3: Calculate the Euclidean distance $\{D_i\}_{i=1}^{Q}$ from each particle to the tree, adjust the position of the particle to make it fall on the nearest tree, and calculate the individual fitness value according to the initial particle—Calculate distance.

Step 4: Initialize the external solution set p_k^i of ith particle in particle swarm k to be empty, the external set L_k of particle swarm k to be empty, and the external set G of the global optimal solution of all particles to be empty—initialize the external solution set.

Step 5: Store the non-dominated solution in the external set and update the optimal position—update particle position.

(1) The current position of each particle and the current position of all particles in each particle swarm are stored in the external set of each particle and the external set of the swarm, respectively, according to the non-dominated solution rules, and the optimal position $pbest_k^i(t)$ of ith particle at the time t and the optimal position $lbest_k(t)$ of the particle swarm k are calculated.

(2) The current positions of all particles are stored in the external set G of the global particle swarm according to non-dominated rules, and then the optimal particle of the population is found from the external set G of the global optimal solutions of all particles, and its position is the global optimal position $gbest(t)$.

Step 6: Update the speed and position of particles according to Equation (6) and Equation (7), adjust the position of particles, and calculate the current fitness value. If the new position exceeds the boundary of the definition domain, the out-of-boundary particles shall be treated according to Equation (8) and Equation (9)—calculate the current fitness value.

Step 7: Calculate the distance between particle swarm, insert or delete particle swarm according to the method described in Section 2.4.2—insert or delete particle swarm.

Step 8: For the inserted new particle swarm, initialize its particles and generate particle external set and group external set—update new particle swarm.

Step 9: Judge the change of L-index. If the L-index does not change, stop searching and output the target tree. Otherwise, return to Step 6 and continue to search for the optimal solution—judge the value of L-index.

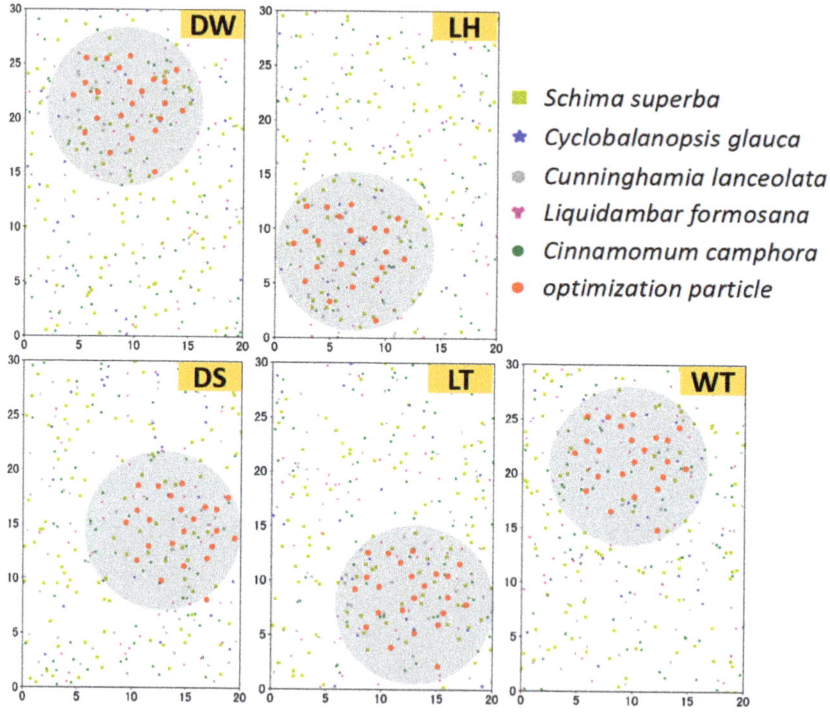

Figure 5. The schematic diagram of the MO-PSO model was used to search for target trees (optimal solution set) in five natural plots (DW, LH, DS, LT, and WT). Red point represents the optimization particle in the PSO algorithm. Five representative dominant tree species: Schima superba, Cyclobalanopsis glauca, Cunninghamia lanceolata, Liquidambar formosana, and Cinnamomum camphora.

3. Results

3.1. Model Performance

The simulated harvesting results in Table 3 show that the MO-PSO model was superior to the basic PSO model (PSO) and random thinning model Monte Carlo-based (RD-TH). For the uniform distribution, random distribution, and aggregated distribution stand, under the MO-PSO model, the average of the relative increased proportion ($\overline{RIP}$) were as large as 31.94%, 29.51%, and 24.71%, respectively, which were slightly larger than PSO model. However, they were far larger than RD-TH model. At the same time, the average iterations of MO-PSO (AITE) were about 26~43 times, which was far less than RD-TH (10,000 times simulating). As far as MO-PSO and PSO models, when the MO-PSO model integrated the multi-swarm objectives strategy, the ALIV and AITE were also evidently better than PSO. Obviously, our proposed approach has advantages in multi-dimensional stand spatial structure optimization.

Table 3. The model performance for the MO-PSO model, PSO model, and the RD-TH model.

Distribution Pattern	MO-PSO			PSO			RD-TH		
	AITE	ALIV	$\overline{\text{RIP}}$/%	AITE	ALIV	$\overline{\text{RIP}}$/%	RUNN	ALIV	$\overline{\text{RIP}}$/%
Uniform	26.9	0.697	31.94	34.2	0.664	30.34	100,000	0.631	27.01
Random	37.4	0.702	29.51	49.6	0.693	27.72	100,000	0.676	24.36
Aggregated	43.9	0.604	24.71	54.3	0.597	20.33	100,000	0.554	19.77

RIP: the relative increased proportion; $\overline{\text{RIP}}$: average of relatively increased proportion; AITE: average iterations; ALIV: average of L-index; RD-TH: random thinning model Monte Carlo-based; RUNN: random cutting times.

In scheduling individual tree harvests in our study, compared with the conventional single-objective optimization model, the MO-PSO model performed better in improving the stand multi-dimensional spatial structure before- and after-thinning. As shown in Figure 6 and Table 4, with the number of algorithm iterations increasing (Figure 6), the L index of five plots increased significantly, ranging from 1.0% to 11.3%. Under different thinning intensities, the average of MING increased by about 2.77% and showed an upward trend in all plots after thinning, and the variation in DS was the highest (about 4.86%). ANGL and DOMI decreased compared with those before thinning, and all five plots showed the same trend.

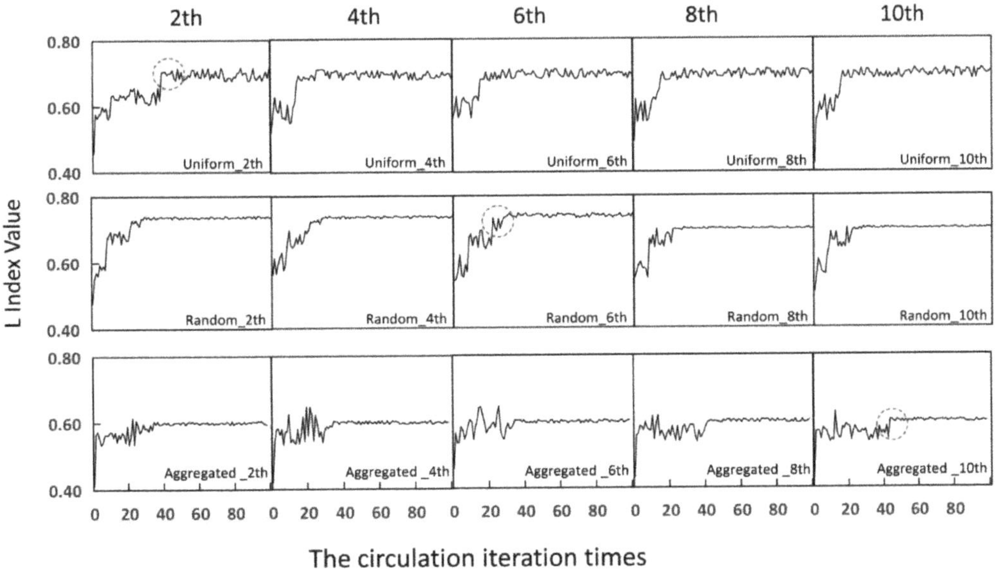

Figure 6. The L index change in the process of PSO algorithm solution in three types of simulated plots: Uniform: uniform distribution; Random: random distribution plot; Aggregated: aggregated distribution; 2nd, 4th, 6th, 8th, 10th represent 2th, 4th, 6th, 8th, 10th simulated cutting process.

Table 4. The statistical characteristics (mean (std)) for the five plots with alternative thinning intensity.

Plot Code	Intensity	DBH/cm	MING	RIP/%	DOMI	RIP/%	ANGL	RIP/%	L-Index	RIP/%
DW	0%	12.710 (3.143)	0.473 (0.223)		0.590 (0.351)		0.560 (0.259)		0.612 (0.225)	
	15%	13.107 (2.967)	0.491 (0.201)	3.81	0.425 (0.332)	−27.97	0.484 (0.246)	−13.57	0.671 (0.203)	9.60
	30%	13.412 (2.762)	0.502 (0.198)	2.24	0.336 (0.303)	−20.94	0.442 (0.246)	−8.68	0.719 (0.213)	7.19
	45%	13.430 (2.784)	0.515 (0.190)	2.59	0.303 (0.278)	−9.82	0.407 (0.224)	−7.92	0.749 (0.197)	4.15
LH	0%	10.360 (3.853)	0.572 (0.210)		0.485 (0.350)		0.550 (0.275)		0.594 (0.198)	
	15%	11.797 (3.521)	0.589 (0.199)	2.97	0.431 (0.338)	−11.13	0.514 (0.265)	−6.55	0.661 (0.176)	11.32
	30%	11.903 (3.326)	0.603 (0.192)	2.38	0.342 (0.314)	−20.65	0.474 (0.257)	−7.78	0.708 (0.188)	7.07
	45%	12.001 (3.117)	0.612 (0.189)	1.49	0.290 (0.283)	−15.20	0.447 (0.245)	−5.70	0.732 (0.172)	3.33
DS	0%	6.710 (2.976)	0.514 (0.223)		0.499 (0.357)		0.529 (0.250)		0.607 (0.215)	
	15%	7.286 (2.853)	0.539 (0.192)	4.86	0.447 (0.346)	−10.42	0.498 (0.240)	−5.86	0.672 (0.197)	10.63
	30%	7.074 (2.793)	0.548 (0.189)	1.67	0.350 (0.315)	−21.70	0.470 (0.246)	−5.62	0.708 (0.201)	5.44
	45%	7.295 (2.707)	0.564 (0.167)	2.92	0.311 (0.292)	−11.14	0.452 (0.242)	−3.83	0.738 (0.193)	4.23
LT	0%	11.320 (3.194)	0.604 (0.231)		0.493 (0.349)		0.531 (0.265)		0.605 (0.208)	
	15%	12.036 (3.013)	0.630 (0.203)	4.30	0.445 (0.339)	−9.74	0.493 (0.252)	−7.16	0.672 (0.184)	11.02
	30%	12.213 (2.936)	0.641 (0.204)	1.75	0.357 (0.316)	−19.78	0.449 (0.249)	−8.92	0.715 (0.194)	6.45
	45%	12.501 (2.903)	0.657 (0.189)	2.50	0.312 (0.291)	−12.61	0.431 (0.246)	−4.01	0.722 (0.177)	1.01
WT	0%	8.430 (3.723)	0.511 (0.231)		0.506 (0.356)		0.504 (0.274)		0.615 (0.225)	
	15%	8.989 (3.237)	0.533 (0.206)	4.35	0.456 (0.349)	−9.88	0.467 (0.262)	−7.34	0.681 (0.208)	10.75
	30%	9.035 (2.898)	0.540 (0.197)	1.16	0.370 (0.319)	−18.86	0.423 (0.262)	−9.42	0.721 (0.211)	5.85
	45%	9.127 (2.836)	0.549 (0.194)	1.83	0.312 (0.292)	−15.68	0.409 (0.261)	−3.31	0.746 (0.205)	3.49

RIP: the relative increased proportion; DBH: diameter at breast; W-index: uniform angle; U-index: DBH dominance; M-index: species mingling.

In addition, we also conducted simulated thinning on 30 simulated plots with uniform distribution (A), random distribution (B), and aggregated distribution (C). As shown in Figure 6, in the process of algorithm solution, the L index of A, B, and C all increased. The magnitude of fitness value (L index) change in A and B stands was slightly larger than that in C. The optimization time of A and B was shorter. The result showed that the optimized stand spatial structure was significantly better than before, and the PSO algorithm was feasible in solving the multi-objective optimization problem of stand spatial structure.

3.2. Thinning Intensities

As shown in Figure 7 and Table 4, the MING increased significantly with the thinning intensity increases for all the natural plots. However, the DOMI and ANGL values for all plots between two successive thinning intensities decreased significantly with the increase in thinning intensity. For the relatively increased proportion (RIP) of MING, DOMI, and ANGL, three similar inverse U-shaped tendencies were observed with the gradual increases in thinning intensity, L-index represents a multi-dimensional comprehensive stand spatial structure index, covering MING, DOMI, and ANG. L-index increased and the maximum values of the RIP of L-index usually appeared under the intensity of T2, indicating that the overall stand spatial structure was affected to varying degrees under different thinning intensities, and removing 15% of the trees (T2) was the optimal thinning intensity from the perspective of stand spatial structure overall optimization in our study.

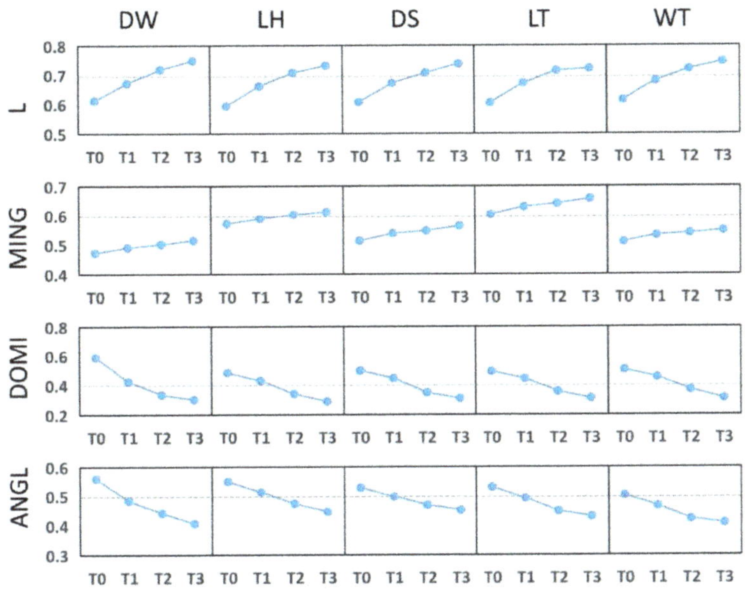

Figure 7. The Change trends of mean L index value under T0, T1, T2, and T3 thinning intensities: T0: 0% thinning intensity; T1: 15% thinning intensity; T2: 30% thinning intensity; T3: 45% thinning intensity.

3.3. Competition, Structure and Spatial Distribution Pattern

To explore the effects of MING, DOMI, and ANGL on the selection of harvested trees, we analyzed the distribution of MING, DOMI, and ANGL of harvested trees. As shown in Figure 8, the spatial structure indices distribution of harvested trees is different among the five plots. Of all five levels of the MING index distribution ($M_i = 0, 0.25, 0.5, 0.75, 1$), trees with lower values of MING index were more likely to be harvested ($M_i = 0$ or $M_i = 0.25$, that is, 1 or 0 of the nearest four neighboring trees are the same as the target tree), and the mean value of relative frequency was 33.7%. However, trees with higher or moderate MING index ($M_i = 0.75$ or $M_i = 1$) were relatively fewer. Similarly, trees with higher DBH dominance ($U_i = 0.75$ or $U_i = 1$) or the trees showed aggregated distribution ($W_i = 0.75$ or $W_i = 1$) were more likely to be harvested, and the mean values of the relative frequency of DOMI and ANGL were 47.3% and 35.6%, respectively. The results showed that the individual trees with higher DBH dominance, lower species mingling, and horizontal distribution aggregation were more likely to be harvested, indicating that these trees could have a negative influence on stand structure, tree species composition, or the intensification of tree competition.

In addition, changes in DOMI were the largest, indicating that thinning may significantly impact the competition among neighboring trees, followed by distribution pattern and tree species composition (Figure 9 and Table 4). Tree competition, tree diversity, and horizontal spatial distribution of trees should be fully considered in the forest spatial structure optimization.

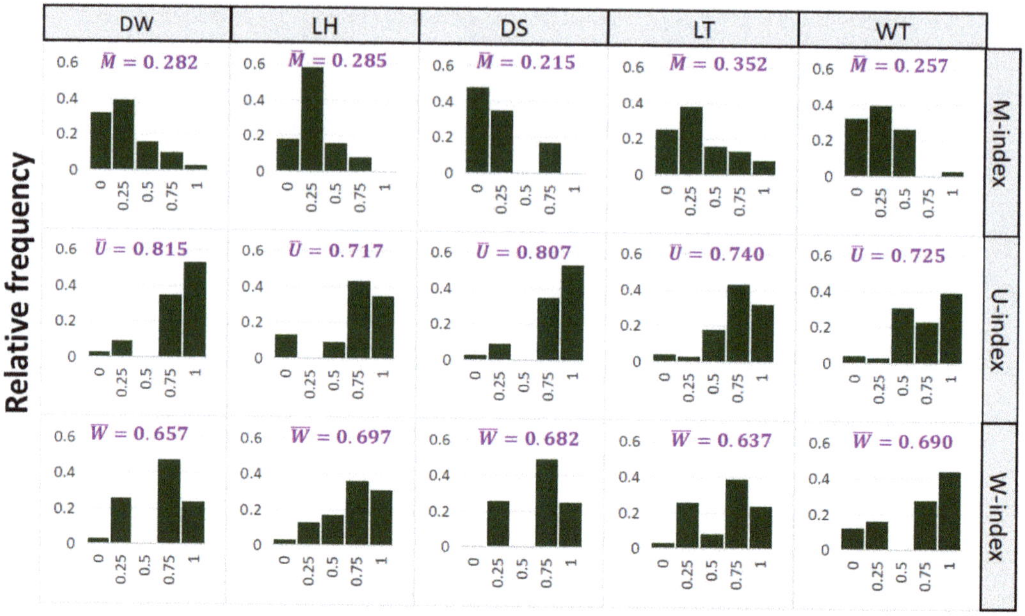

Figure 8. The frequency distribution of DOMI, ANGL, and MING of harvested trees.

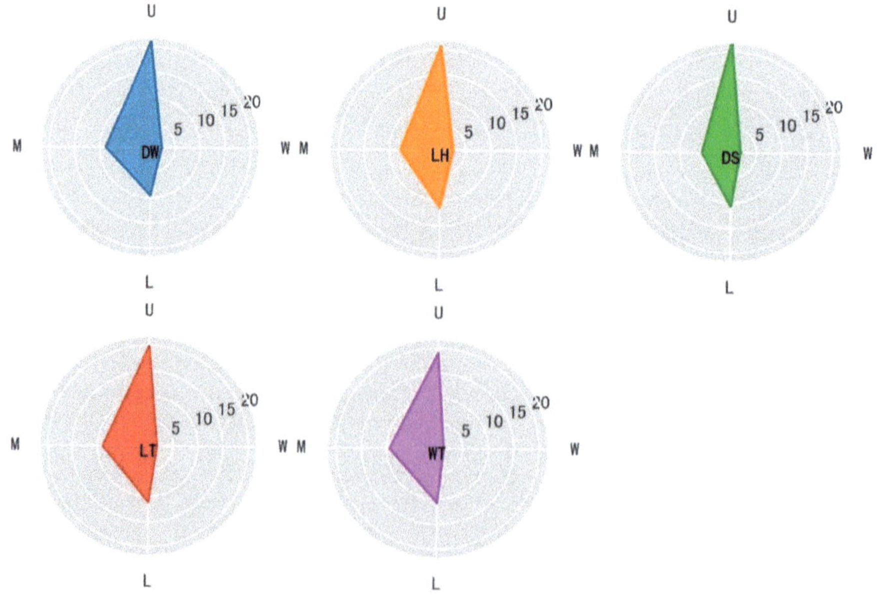

Figure 9. The RIP of DOMI, ANGL, MING, and L of five natural plots before and after thinning: RIP: average relative change rate; DOMI: DBH dominance; ANGL: uniform angle; MING: species mingling.

4. Discussion

In the mixed forest, the stand spatial distribution pattern, tree species mixing. and competition are important criteria in the harvested tree selection, and they are key endogenous constraints to tree growth. For instance, same- and different-species of trees through asymmetrical competition for light or symmetrical competition for water and nutrients affect the stand spatial structure and its succession process [33,34]. However, the traditional indices describing competition, stand distribution pattern. and tree species mixing, such as total basal area, tree density, Weibull distribution. and Gini coefficient, are all distance-independent, and cannot accurately quantify the impact of neighborhood interaction on the stand spatial structure [35,36]. Therefore, in this study, three distance-independent neighborhood indices, DBH dominance (DOMI), uniform angle (ANGL), and species mingling (MING), were selected to describe the degree of stand competition, spatial distribution of forest, and tree species diversity, respectively. DOMI measures the proportion of n ($n = 4$ in our study) nearest neighbors larger than the target tree, ANGL is used to describe the angular uniformity of the n nearest neighbors around the target tree, and MING is used to measure the proportion of n ($n = 4$) nearest neighbors not belonging to the same species as the target tree [10]. In order to explore the impact of thinning on forest spatial distribution pattern, mixing and competition, this study attempted to combine the PSO algorithm with multi-dimensional stand spatial structure indices (MING, DOMI, and ANGL) to construct a tree-level multi-objective forest planning model (MO-PSO), which is used to evaluate and simulate the impact of different thinning intensities on different stands. The results indicated that the MING, DOMI, and ANGL of different stands indices were better than those before thinning under alternative thinning intensities, and the multi-dimensional stand spatial structure (L-index) had been improved to varying degrees. Therefore, the model can provide guidance and help for multi-objects forest planning of natural secondary mixed forests.

In addition, the results indicated that DOMI, ANGL, and MING were better criteria to optimize tree selection. Obviously, the competition between target trees and neighboring trees and the degree of forest aggregation are important factors affecting forest growth. Trees in high aggregation and fierce competition environments have smaller tree size characteristics (DBH, height, crown) than those in free-growing (or less competitive) environments [30]. In our study, trees with larger size neighboring trees (DOMI > 0.5), highly aggregated distribution (ANGL > 0.5), and lower tree species mixing degree (MING < 0.5) were more likely to be harvested. The simulation thinning analysis results also showed that thinning had a greater impact on the degree of stand competition. Under different thinning intensities, with the improvement of the overall stand structure (L-index increases), the individual trees were more evenly distributed. Our results were consistent with some previous studies [11,12,37,38].

In previous studies, when the complexity of forest spatial management planning problems increased, showing nonlinear or discontinuous characteristics, the multiple optimization objectives conflicted. Researchers proposed to combine the heuristic algorithm with a single structure index for a solution [11,39], but the processing capacity and application scope of this technology is seriously limited. Therefore, in our study, the PSO algorithm was used to construct a multi-objective forest planning model, which quickly and fairly sought approximate optimal solutions and excellent local search ability. In addition, the initialization particle swarm generated randomly by the PSO algorithm avoids the subjectivity of harvesting tree selection and makes the solution more objective. The results indicated that the model had a significant effect on the optimization of multiple objectives, such as improving the degree of tree species mixing and reducing the aggregation degree of tree horizontal distribution and competition pressure, which are consistent with previous studies [12,40,41]. However, forest planning is affected by various factors such as topography, climate, and soil properties. Due to the limitations of research conditions and data accessibility, this study only incorporated the stand spatial structure characteristics into the model. At the same time, the logging gaps will produce naturally regenerated tree species over time [11], thus the model is unsuitable for monitoring and analyzing the dynamics in forest ecosystems. Therefore, further perfecting the forest planning model to improve the applicability and evaluation accuracy of the model under dynamic conditions will be the core and difficulty of the subsequent research.

Thinning reduces the number of trees, decreases the competition for resources in limited supply (e.g., light, nutrients, and water) among trees and expands the living space of individual trees [42]. However, the optimal thinning intensity varied greatly under different optimization objectives. In this study, the multi-dimensional stand spatial structure index (L-index) increased significantly under all four thinning intensities, and the value of some forest structural attributes showed certain fluctuations. Compared to the unharvested, L-index increased by 23.7% or more, but the relatively increased proportion (RIP) increased and then decreased significantly with increasing thinning intensity. When 15% of the trees were removed from the studied plots, the RIP values all reached the maximum, indicating that removing 30% of the trees might be the most effective thinning intensity to improve the multi-dimensional stand spatial structure. However, other studies have shown that the optimal thinning intensity to increase species diversity is about 60%, sustainable wood production is about 25~40%, promote stand growth and yield are 45% and 38%~52%, and improve soil physicochemical properties and microbial community are about 30%–45% [43–47]. Compared with other intensities, the optimal thinning intensity of stand structure optimization in this study was much less than other intensities. Therefore, whether removing 30% of the trees is the optimal thinning intensity, its impact on various aspects of the forest ecosystem and its differences, and whether it complies with relevant local forest management laws and regulations need further research.

In addition, some studies reported that advanced remote sensing technologies such as airborne laser scanning (ALS), ground laser scanning (TLS), and digital aerial photogrammetry (DAP) have been widely used in forest inventory and management. Integrating ALS and DAP data (data fusion) will help habitat modeling or drawing the forest attribute map [48], which might be of great help in calculating DOMI, ANGL, and MING. Combined with the MO-PSO model in this study, real-time tree-level forest planning decision-making will be possible. Although there are still many challenges in integrating ALS and DAP data to quantify three-dimensional spatial structure and attributes of stands [49], taking into account the synergy among MO-PSO, ALS, and DAP are crucial for making tree-level decisions, which will help to maintain or increase species diversity, bird habitat composition diversity, and habitat suitability.

5. Conclusions

Under the guidelines of the structure-considered forest management strategy, we presented an accurate, efficient, and comprehensive thinning model for the selection of harvesting trees by integrating the PSO algorithm and stand spatial structure indices with species diversity, competition, and spatial distribution pattern. The simulated cutting results showed that the MO-PSO model was superior to the basic PSO model (PSO) and random thinning model Monte Carlo-based (RD-TH). The analysis of the effects of MING, DOMI, and ANGL index on the selection of harvested trees showed that the individual trees with higher DBH dominance, lower species mingling, and horizontal distribution aggregation were more likely to be harvested. Thinning especially had a significant influence on tree competition from neighboring trees. A comprehensive analysis of four thinning intensities on the overall forest structure showed that removing 15% of the trees is the optimal thinning intensity in our study. The multi-dimensional stand spatial structure index increased by approximately 1.0% ~ 11.3% in the five natural plots. However, forest management and planning is a continuous, multi-factorial, and cyclic optimization process. The stand structure characteristics and research scope covered in our study were limited. Therefore, further research will consider more factors, such as the environment, temperature, and terrain.

Author Contributions: Conceptualization, H.Q. and H.Z.; formal analysis, K.L. and X.H.; funding acquisition, H.Z. and T.Y.; investigation, H.Q.; methodology, H.Q.; resources, T.Y.; software, X.H. and H.Q.; supervision, H.Z.; visualization, K.L. and H.Q.; writing—original draft, H.Q.; writing—review and editing, H.Q. and X.J. All authors have read and agreed to the published version of the manuscript.

Funding: This research was funded by the National Natural Science Foundation of China, grant number 32071681, 32271877 and the Foundation Research Funds of IFRIT, grant number CAFYBB2021ZE005-2.

Data Availability Statement: Not applicable.

Conflicts of Interest: The authors declare no conflict of interest.

References

1. Ryan, C.M.; Cerveny, L.K.; Robinson, T.L.; Blahna, D.J. Implementing the 2012 forest planning rule: Best available scientific information in forest planning assessments. *For. Sci.* **2018**, *64*, 159–169. [CrossRef]
2. Pukkala, T.; Lähde, E.; Laiho, O. Which trees should be removed in thinning treatments? *For. Ecosyst.* **2015**, *2*, 32. [CrossRef]
3. Borges, P.; Eid, T.; Bergseng, E. Applying simulated annealing using different methods for the neighborhood search in forest planning problems. *Eur. J. Oper. Res.* **2014**, *233*, 700–710. [CrossRef]
4. Borgesa, J.G.; Hoganson, H.M. Structuring a landscape by forestland classification and harvest scheduling spatial constraints. *For. Ecol. Manag.* **2000**, *130*, 269–275. [CrossRef]
5. Seiwa, K.; Eto, Y.; Hishita, M.; Masaka, K. Effects of thinning intensity on species diversity and timber production in a conifer (Cryptomeria japonica) plantation in Japan. *J. For. Res.* **2012**, *17*, 468–478. [CrossRef]
6. Bettinger, P.; Sessions, J. Spatial forest planning: To adopt, or not to adopt? *J. For.* **2003**, *101*, 24–29. [CrossRef]
7. Kangas, J.; Kuusipalo, J. Integrating biodiversity into forest management planning and decision-making. *For. Ecol. Manag.* **1993**, *61*, 1–15. [CrossRef]
8. Pohjanmies, T.; Triviño, M.; Le Tortorec, E.; Salminen, H.; Mönkkönen, M. Conflicting objectives in production forests pose a challenge for forest management. *Ecosyst. Serv.* **2017**, *28*, 298–310. [CrossRef]
9. Yousefpour, R.; Temperli, C.; Jacobsen, J.B.; Thorsen, B.J.; Meilby, H.; Lexer, M.J.; Lindner, M.; Bugmann, H.; Borges, J.G.; Palma, J.; et al. A framework for modeling adaptive forest management and decision making under climate change. *Ecol. Soc.* **2017**, *22*, 40. [CrossRef]
10. Pommerening, A. Evaluating structural indices by reversing forest structural analysis. *For. Ecol. Manag.* **2006**, *224*, 266–277. [CrossRef]
11. Bettinger, P.; Tang, M. Tree-level harvest optimization for structure-based forest management based on the species mingling index. *Forests* **2015**, *6*, 1121–1144. [CrossRef]
12. Dong, L.; Wei, H.; Liu, Z. Optimizing forest spatial structure with neighborhood-based indices: Four case studies from northeast China. *Forests* **2020**, *11*, 413. [CrossRef]
13. Knoke, T.; Stang, S.; Remler, N.; Seifert, T. Ranking the importance of quality variables for the price of high quality beech timber (*Fagus sylvatica* L.). *Ann. For. Sci.* **2006**, *63*, 399–413. [CrossRef]
14. Wing, B.M.; Boston, K.; Ritchie, M.W. A technique for implementing group selection treatments with multiple objectives using an airborne lidar-derived stem map in a heuristic environment. *For. Sci.* **2019**, *65*, 211–222. [CrossRef]
15. Vauhkonen, J.; Pukkala, T. Selecting the trees to be harvested based on the relative value growth of the remaining trees. *Eur. J. For. Res.* **2016**, *135*, 581–592. [CrossRef]
16. Packalen, P.; Pukkala, T.; Pascual, A. Combining spatial and economic criteria in tree-level harvest planning. *For. Ecosyst.* **2020**, *7*, 18. [CrossRef]
17. Kimmins, J.P.; Welham, C.; Seely, B.; Meitner, M.; Rempel, R.; Sullivan, T. Science in Forestry: Why does it sometimes disappoint or even fail us? *For. Chron.* **2005**, *81*, 723–734. [CrossRef]
18. Carnol, M.; Baeten, L.; Branquart, E.; Grégoire, J.-C.; Heughebaert, A.; Muys, B.; Ponette, Q.; Verheyen, K. Ecosystem services of mixed species forest stands and monocultures: Comparing practitioners' and scientists' perceptions with formal scientific knowledge. *For. An Int. J. For. Res.* **2014**, *87*, 639–653. [CrossRef]
19. Tóth, S.F.; McDill, M.E. Promoting large, compact mature forest patches in harvest scheduling models. *Environ. Model. Assess.* **2008**, *13*, 1–15. [CrossRef]
20. Yang, Q.; Wu, J.; Li, Y.; Li, W.; Wang, L.; Yang, Y. Using the particle swarm optimization algorithm to calibrate the parameters relating to the turbulent flux in the surface layer in the source region of the Yellow River. *Agric. For. Meteorol.* **2017**, *232*, 606–622. [CrossRef]
21. Fotakis, D.G.; Sidiropoulos, E.; Myronidis, D.; Ioannou, K. Spatial genetic algorithm for multi-objective forest planning. *For. Policy Econ.* **2012**, *21*, 12–19. [CrossRef]
22. Wang, L.; Zhou, Y.; Li, Q.; Zuo, Q.; Gao, H.; Liu, J.; Tian, Y. Forest Land Quality Evaluation and the Protection Zoning of Subtropical Humid Evergreen Broadleaf Forest Region Based on the PSO-TOPSIS Model and the Local Indicator of Spatial Association: A Case Study of Hefeng County, Hubei Province, China. *Forests* **2021**, *12*, 325. [CrossRef]

23. PrasannaVenkatesan, S.; Kumanan, S. Multi-objective supply chain sourcing strategy design under risk using PSO and simulation. *Int. J. Adv. Manuf. Technol.* **2012**, *61*, 325–337. [CrossRef]
24. Aguirre, O.; Hui, G.; von Gadow, K.; Jiménez, J. An analysis of spatial forest structure using neighbourhood-based variables. *For. Ecol. Manag.* **2003**, *183*, 137–145. [CrossRef]
25. Hui, G.; Zhang, G.; Zhao, Z.; Yang, A. Methods of forest structure research: A review. *Curr. For. Rep.* **2019**, *5*, 142–154. [CrossRef]
26. Zhang, J.; Chen, L.; Guo, Q.; Nie, D.; Bai, X.; Jiang, Y. Research on the change trend of dominant tree population distribution patterns during development process of climax forest communities. *Acta Bot. Sin.* **1999**, *23*, 256–268.
27. Chen, C.; Liu, J.; Yu, K.; Chen, J.; Ge, Y. Simulated Cutting for the Mixed Forest of Pinus massoniana and Broad-leaved Tree Species Based on Optimized Spatial Structure. *J. Northeast For. Univ.* **2010**, *30*, 29–32.
28. Hui, G.; HU, Y.; Zhao, Z. Research Progress of Structure-based Forest Management. *For. Res.* **2018**, *31*, 85–93.
29. Qiu, H.; Liu, S.; Zhang, Y.; Li, J. Variation in height-diameter allometry of ponderosa pine along competition, climate, and species diversity gradients in the western United States. *For. Ecol. Manag.* **2021**, *497*, 119477. [CrossRef]
30. Li, J.; Zhu, K.; Liu, S.; Li, D.; Zhang, G.; Liu, X.; Yang, W. Introducing tree neighbouring relationship factors in forest pattern spatial analysis: Weighted Delaunay triangulation method. *J. For. Res.* **2021**, *32*, 1941–1951. [CrossRef]
31. Weintraub, A.; Bare, B.B. New issues in forest land management from an operations research perspective. *Interfaces* **1996**, *26*, 9–25. [CrossRef]
32. Pukkala, T.; Ketonen, T.; Pykäläinen, J. Predicting timber harvests from private forests—A utility maximisation approach. *For. Policy Econ.* **2003**, *5*, 285–296. [CrossRef]
33. Stoll, P.; Weiner, J.; Muller-Landau, H.; Müller, E.; Hara, T. Size symmetry of competition alters biomass–density relationships. *Proc. R. Soc. B Biol. Sci.* **2002**, *269*, 2191–2195. [CrossRef] [PubMed]
34. Rewald, B.; Leuschner, C. Belowground competition in a broad-leaved temperate mixed forest: Pattern analysis and experiments in a four-species stand. *Eur. J. For. Res.* **2009**, *128*, 387–398. [CrossRef]
35. Fox, J.C.; Ades, P.K.; Bi, H. Stochastic structure and individual-tree growth models. *For. Ecol. Manag.* **2001**, *154*, 261–276. [CrossRef]
36. Li, Y.; Hui, G.; Wang, H.; Zhang, G.; Ye, S. Selection priority for harvested trees according to stand structural indices. *iForest-Biogeosci. For.* **2017**, *10*, 561–566. [CrossRef]
37. Pukkala, T.; Miina, J.; Kellomäki, S. Response to different thinning intensities in young Pinus sylvestris. *Scand. J. For. Res.* **1998**, *13*, 141–150. [CrossRef]
38. Lagergren, F.; Lindroth, A. Variation in sapflow and stem growth in relation to tree size, competition and thinning in a mixed forest of pine and spruce in Sweden. *For. Ecol. Manag.* **2004**, *188*, 51–63. [CrossRef]
39. Borges, J.G.; Hoganson, H.M.; Rose, D.W. Combining a decomposition strategy with dynamic programming to solve spatially constrained forest management scheduling problems. *For. Sci.* **1999**, *45*, 201–212.
40. Thomas, S.C.; Halpern, C.B.; Falk, D.A.; Liguori, D.A.; Austin, K.A. Plant diversity in managed forests: Understory responses to thinning and fertilization. *Ecol. Appl.* **1999**, *9*, 864–879. [CrossRef]
41. Li, Y.; Ye, S.; Hui, G.; Hu, Y.; Zhao, Z. Spatial structure of timber harvested according to structure-based forest management. *For. Ecol. Manag.* **2014**, *322*, 106–116. [CrossRef]
42. Forrester, D.I. Growth responses to thinning, pruning and fertiliser application in Eucalyptus plantations: A review of their production ecology and interactions. *For. Ecol. Manag.* **2013**, *310*, 336–347. [CrossRef]
43. Mäkinen, H.; Isomäki, A. Thinning intensity and growth of Norway spruce stands in Finland. *For. Int. J. For. Res.* **2004**, *77*, 349–364. [CrossRef]
44. Pérez-de-Lis, G.; García-González, I.; Rozas, V.; Arévalo, J.R. Effects of thinning intensity on radial growth patterns and temperature sensitivity in Pinus canariensis afforestations on Tenerife Island, Spain. *Ann. For. Sci.* **2011**, *68*, 1093. [CrossRef]
45. Dang, P.; Gao, Y.; Liu, J.; Yu, S.; Zhao, Z. Effects of thinning intensity on understory vegetation and soil microbial communities of a mature Chinese pine plantation in the Loess Plateau. *Sci. Total Environ.* **2018**, *630*, 171–180. [CrossRef]
46. Fang, S.; Lin, D.; Tian, Y.; Hong, S. Thinning intensity affects soil-atmosphere fluxes of greenhouse gases and soil nitrogen mineralization in a lowland poplar plantation. *Forests* **2016**, *7*, 141. [CrossRef]
47. Settineri, G.; Mallamaci, C.; Mitrović, M.; Sidari, M.; Muscolo, A. Effects of different thinning intensities on soil carbon storage in Pinus laricio forest of Apennine South Italy. *Eur. J. For. Res.* **2018**, *137*, 131–141. [CrossRef]
48. Vogeler, J.C.; Yang, Z.; Cohen, W.B. Mapping post-fire habitat characteristics through the fusion of remote sensing tools. *Remote Sens. Environ.* **2016**, *173*, 294–303. [CrossRef]
49. Vastaranta, M.; Wulder, M.A.; White, J.C.; Pekkarinen, A.; Tuominen, S.; Ginzler, C.; Kankare, V.; Holopainen, M.; Hyyppä, J.; Hyyppä, H. Airborne laser scanning and digital stereo imagery measures of forest structure: Comparative results and implications to forest mapping and inventory update. *Can. J. Remote Sens.* **2013**, *39*, 382–395. [CrossRef]

forests MDPI

Article

A Progressive Plane Detection Filtering Method for Airborne LiDAR Data in Forested Landscapes

Shangshu Cai [1], Xinlian Liang [1,*] and Sisi Yu [2,3,4]

[1] State Key Laboratory of Information Engineering in Surveying, Mapping and Remote Sensing, Wuhan University, Wuhan 430070, China
[2] Wuhan Botanical Garden, Chinese Academy of Sciences, Wuhan 430074, China
[3] Sino-Africa Joint Research Center, Chinese Academy of Sciences, Wuhan 430074, China
[4] Department of Public Administration, Law School, Shantou University/Institute of Local Government Development, Shantou University, Shantou 515063, China
* Correspondence: xinlian_liang@hotmail.com

Abstract: Ground filtering is necessary in processing airborne light detection and ranging (LiDAR) point clouds for forestry applications. This study proposes a progressive plane detection filtering (PPDF) method. First, the method uses multi-scale planes to characterize terrain, i.e., the local terrain with large slope variations is represented by small-scale planes, and vice versa. The planes are detected in local point clouds by the random sample consensus method with decreasing plane sizes. The reliability of the planes to represent local terrain is evaluated and the planes with optimal sizes are selected according to evaluation results. Then, ground seeds are identified by selecting the interior points of the planes. Finally, ground points are iteratively extracted based on the reference terrain, which is constructed using evenly distributed neighbor ground points. These neighbor points are identified by selecting the nearest neighbor points of multiple subspaces, which are divided from the local space with an unclassified point as center point. PPDF was tested in six sites with various terrain and vegetation characteristics. Results showed that PPDF was more accurate and robust compared to the classic filtering methods including maximum slope, progressive morphology, cloth simulation, and progressive triangulated irregular network densification filtering methods, with the smallest average total error and standard deviation of 3.42% and 2.45% across all sites. Moreover, the sensitivity of PPDF to parameters was low and these parameters can be set as fixed values. Therefore, PPDF is effective and easy-to-use for filtering airborne LiDAR data.

Keywords: ground filtering; LiDAR; forestry applications; terrain

1. Introduction

Airborne light detection and ranging (LiDAR) can penetrate vegetation to capture accurate and detailed three-dimensional structure information of forests and is less affected by weather conditions than optical sensors [1–10]. It has become a mainstream tool in forest investigation and management [11–13].

Ground filtering, i.e., classifying ground and non-ground points, is a pre-requisite step of LiDAR data processing in many applications [14]. Therefore, various types of ground filtering methods have been developed to automatically separate LiDAR point clouds into ground and non-ground points. However, the performance of these methods is usually unreliable in forested landscapes especially containing dense vegetation and complex terrain (e.g., highly steep slopes, break lines, and ridges) [15–18]. In addition, these methods are not easy to use, because they often require tuning many parameters to achieve satisfactory results [19].

Existing ground filtering methods can be mainly categorized as slope-based [20–25], mathematical morphology-based [26–34], and surface-based methods [19,35–41]. The implementation of these three types of methods can be summarized in two steps: ground seed (i.e., the initial ground point) identification and ground point extraction. In general, ground seeds are identified based on the feature that the ground point has the smallest height in a local region [41,42]. Ground points are iteratively extracted based on the feature that the height differences of the ground points to reference ground points or reference terrain are relatively small [43]. The slope-based methods provide an innovation to the ground point extraction step. The ground points are extracted by comparing the height differences between unclassified points and their nearest neighbor ground points (i.e., reference ground points). The threshold of height differences is set based on the maximum terrain slope in a landscape [20]. The greater the terrain slope, the greater the threshold. This innovation can reduce the risk of ground points being misclassified as non-ground points on steep slopes [44]. The mathematical morphology-based methods present an innovation in the ground seed identification step. Ground seeds are identified by selecting the lowest points within sliding windows. The window sizes are determined adaptively according to the sizes of the non-ground objects in a landscape through morphological opening operation [26,45]. The method can identify more ground seeds compared to the largest size of windows [14,16,46]. The surface-based methods propose an innovation to the ground point extraction step. Ground points are extracted by judging the height differences between unclassified points and the reference terrain constructed by neighboring ground points [35,41]. Compared to the slope-based methods, this innovation can improve filtering accuracy on undulating terrain due to the use of contextual information [14,47].

However, these methods still have some limitations when applied in forested landscapes. First, it is difficult to identify ground seeds on raised terrain, since the height of raised terrain is the local maximum [48]. The lack of ground seeds leads to poor filtering results on the terrain. Second, the window sizes in urban landscapes are usually determined by manually measuring the largest buildings. Nevertheless, the window sizes in forested landscapes are extremely difficult to be determined, because vegetation would not always be penetrated by laser pulses and the largest non-penetrated vegetation is not easily identified by users [15]. Third, the number of laser pulses penetrating vegetation to reach the ground is inconsistent due to the varying density of the canopy level, resulting in uneven ground points. The reference terrain for ground point extraction may be unreliable since the neighbor ground points searched are unevenly distributed. This also reduces the filtering accuracy.

To overcome these limitations, we propose a progressive plane detection filtering (PPDF) method, which is expected to effectively filter out non-ground points while preserving the terrain's details. Moreover, the proposed method is universally applicable, i.e., the users without much experience do not need to carefully adjust the parameters to obtain satisfactory filtering results in different forested landscapes. Compared with the existing methods, the main contributions of the proposed method are presented as follows:

(1) A progressive plane detection method is proposed to characterize terrain. This method quantitatively evaluates the reliability of the planes with different sizes to represent local terrain and adopts the planes with optimal sizes according to evaluation results. Finally, terrain is characterized by multi-scale planes, that is, the local terrain with large slope variations is represented by small-scale planes and vice versa. This method provides a high-quality reference for ground point extraction on various terrains. More importantly, this method does not require setting the window size, improving its utility.

(2) An improved surface-based filtering method is developed to extract ground points. The method uses the interior points of the multi-scale planes as ground seeds. Compared with the local minimum method, it can identify more ground seeds on various terrains (e.g., raised terrain) and is resistant to negative outliers. In addition, the neighbor ground points are extracted from multiple subspaces, ensuring the spatial uniform distribution of the selected neighbor ground points. These improvements increase the accuracy of ground point extraction.

2. Data and Method

The research framework consists of three parts, including data preparation, method design, and performance evaluation (Figure 1). First, the data with various terrain, vegetation, and point cloud features are prepared. Then, our method is designed, which can be divided into point extraction under forest canopy, progressive random sample consensus (RANSAC) plane detection, and ground point extraction. Point extraction under forest canopy aims to minimize the interference of non-ground points on the plane detection on the ground by eliminating the points with large height in a neighborhood. Based on the points under forest canopy, terrain is characterized based on multi-scale planes using progressive RANSAC plane detection. Ground points are extracted by selecting the points with small distance between them and the reference terrain characterized by planes. Finally, based on the prepared data, the performance of the proposed method is evaluated by adopting the accuracy metrics and comparing with classical filtering methods.

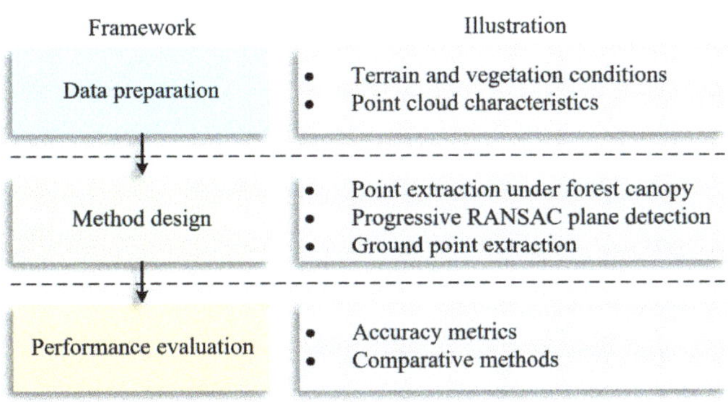

Figure 1. Research framework.

2.1. Data Description

We selected six 300 × 300 m forested sites to evaluate the performance of the proposed method (Figure 2). LiDAR data for these sites is supplied by OpenTopography (California, America). The details of LiDAR data collection are summarized in Table 1. The data in Site 1 and Sites 2–6 were collected by Leica ALS50 (Leica, Gallen, Switzerland) and Optech GEMINI (Optech, Ontario, Canada) airborne laser terrain mapper, respectively. The collection was performed using flight altitudes of 600–900 m above the ground, scan frequencies of 40–83 Hz, scan angles of 14–25° from nadir, and adjacent swath overlaps of 50%–100%, generating average point densities of 2.16–21.74 points/m^2.

Table 1. LiDAR data collection information for the six forested sites.

Site	Location	Collection Date	System	Flying Height (m)	Scan Frequency (Hz)	Scan Angle (°)	Overlap (%)	Mean Density (Points/m^2)
1	Lake Tahoe, Sierra Nevada	August, 2010	Leica ALS50	900	83	14	100	21.74
2	North, Wasatch	July, 2008	Optech GEMINI ALTM	700	70	20	50	7.9
3	East, Modesto	August, 2010	Optech GEMINI ALTM	600	40	21	50	7.82
4	East, Ephrain	July, 2010	Optech GEMINI ALTM	600	40	21	50	6.73
5	Southeast, Butte	August, 2010	Optech GEMINI ALTM	600	40	21	50	8.1
6	West Reno	July, 2007	Optech GEMINI ALTM	700	40	25	50	2.16

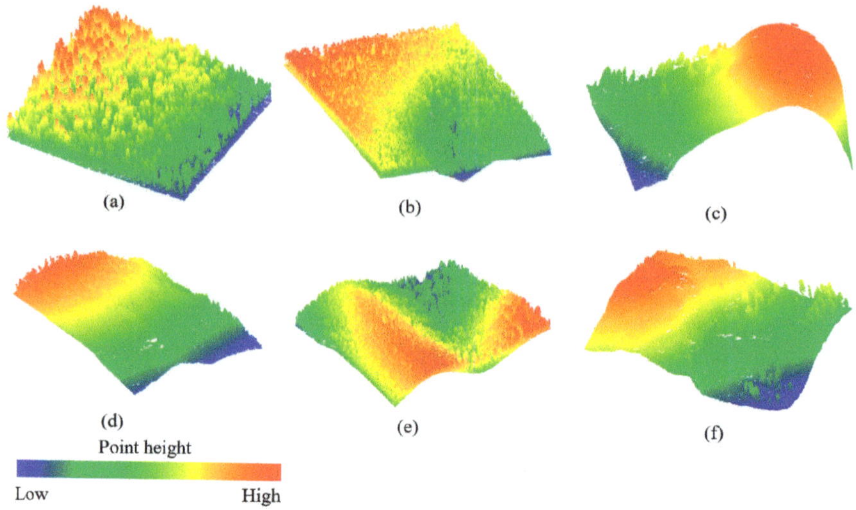

Figure 2. LiDAR point clouds of six forested sites: (**a**–**f**) Sites 1–6.

The statistics of terrain and vegetation conditions for six forested sites are shown in Figure 3. They were calculated based on reference ground and non-ground points, which were generated through human–computer interaction. The slope of local terrain in each plot was obtained by calculating the angles between horizontal planes and the local terrain surfaces that were constructed by fitting planes based on the neighbor ground points of each ground point. The 10 m pixel-level canopy cover was calculated based on canopy height models (CHMs). The 1 m resolution CHMs were generated from LiDAR point clouds and, then, the canopy cover of each 10 m grid cell was estimated by calculating the percentage of the pixels that are higher than 2 m [16,49]. Canopy height was counted from the pixels with height higher than 2 m in the 1 m resolution CHMs. The terrain slope, canopy cover, and canopy height maps of a sample site are given in Figure 4. These sites contain various terrain conditions (i.e., gentle slopes, steep slopes, hilltops, valleys, ridges, and break lines) with an average terrain slope of 18.3–37.11°. Furthermore, vegetation conditions vary widely with an average canopy cover of 7.27%–84.67% and canopy height of 7.83–23.6 m.

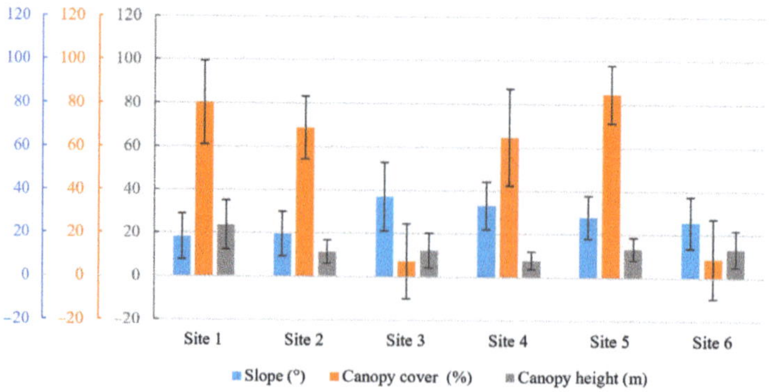

Figure 3. Statistics for the six forested sites.

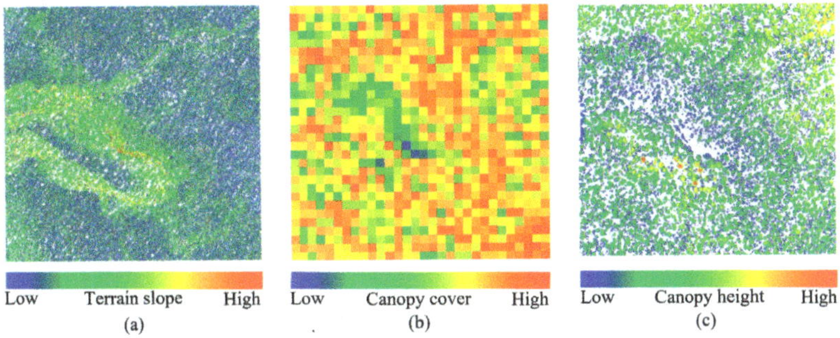

Figure 4. (**a**) Terrain slope, (**b**) canopy cover, and (**c**) canopy height maps for a sample site.

2.2. Overview of the Proposed Filtering Method

Irregular terrain can be characterized by multi-scale planes. The larger the slope variations of local terrain, the smaller the plane size should be (Figure 5). How to automatically detect the planes on the ground and determine optimal plane size are two core problems. Specifically, detecting planes on the ground of unfiltered LiDAR point clouds is difficult due to the negative influence of non-ground points. As a result, it is necessary to reduce non-ground points as much as possible before detecting planes and to adopt a plane detection method that is robust to noise (i.e., non-ground points). In terms of setting plane size, a small-scale plane can represent local terrain well, but the plane is often incorrectly detected in vegetation, since the number of ground points may not be dominant compared to vegetation points in a small region. In contrast, a large-scale plane tends to smooth local terrain with large slope variations, such as ridges and break lines. Thus, it is necessary to evaluate the reliability of the plane to represent local terrain and to determine optimal plane size based on evaluation results.

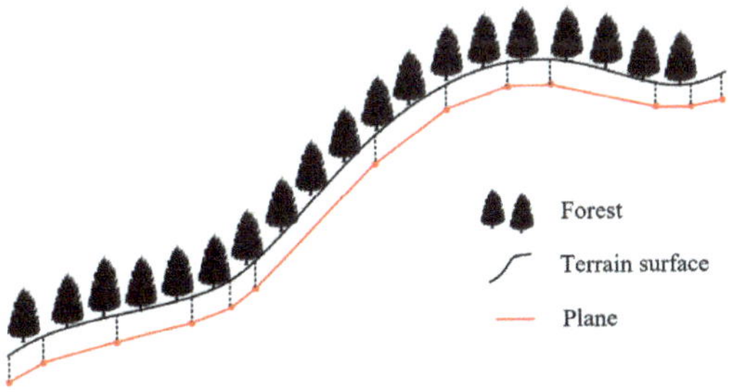

Figure 5. An illustration of irregular terrain characterized by multi-scale planes. The local terrain with larger slope variations is represented by the plane with smaller size.

To solve the two problems, a progressive plane detection method is proposed to characterize terrain. First, the points under forest canopy, which consist of ground and understory vegetation points, are extracted in unfiltered point clouds based on the feature that canopy points are higher than the points under forest canopy. Second, the points under forest canopy are divided into multiple regions and RANSAC plane detection method is adopted to detect planes in each region. It is because RANSAC has the advantage of being resistant to noise and, thus, has a greater potential to detect planes on the ground [50]. Next, the reliability of the plane representing local terrain is evaluated based on the maximum distance from the points under the detected plane to the plane. The smaller the distance is, the more reliable the plane is, and vice versa. If the reliability is low, the points in the region are further divided into multiple subregions with smaller size. The plane detection and reliability evaluation are iteratively performed, until no local terrain can be represented by planes. Finally, based on the terrain characterized by multi-scale planes, an improved surface-based filtering method is developed to extract ground points. The method identifies ground seeds by selecting the interior points of the detected planes and, then, extracts ground points according to the distances of the points to the reference terrain, which is constructed with the neighbor ground points with spatially uniform distribution.

The proposed method is decomposed into three parts: point extraction under forest canopy, progressive RANSAC plane detection, and ground point extraction. In the following sections, the principle, implementation steps, and parameter settings in each part are described in detail, and the results of each part are verified by taking Site 2 as an example.

2.3. Point Extraction under Forest Canopy

The negative effects of non-ground points (mainly from forest canopy) require to be reduced as much as possible to detect planes on the ground. Otherwise, planes may be detected in non-ground objects even using anti-noise plane detection methods. The points under forest canopy are extracted according to the feature that these points are lower than canopy points in a local region. The points (P_{ufc}) under the canopy are expressed as:

$$P_{ufc} = \left\{ p |\, h_p - h_0 < T_h \right\} \tag{1}$$

where p is an unclassified point; h_p is the height of p; h_0 is the height of the local minimum point; and T_h is the height threshold.

The implementation steps include regional division, seed identification under forest canopy, and point extraction under forest canopy. First, an unfiltered point cloud is gridded and a grid cell represents a local region. Then, the lowest points in each cell are identified as the seeds under forest canopy. Finally, the points under forest canopy are extracted based on the seeds under forest canopy cell by cell. In a cell, a point is identified as the point under forest canopy if the height difference between the point and the seed is less than a threshold.

This part includes two parameters, i.e., cell size and height difference. The cell size was set to 2 m, considering that the lowest point within 2 m is usually located under the forest canopy. The height difference was set to 5 m, since canopy points usually are 2–5 m above the ground according to the studies of canopy cover estimation [49,51]. To ensure that the ground points are completely preserved, the height difference was conservatively set to 5 m.

The points under forest canopy using the above procedure in a sample site (Site 2) are shown in Figure 6. Ground points are not only completely preserved but also most of forest canopy points (gray points) are accurately removed in gentle slopes and steep slopes (see enlarged figures A and B). Thus, the extracted points under forest canopy can provide an important basis for subsequent progressive RANSAC plane detection.

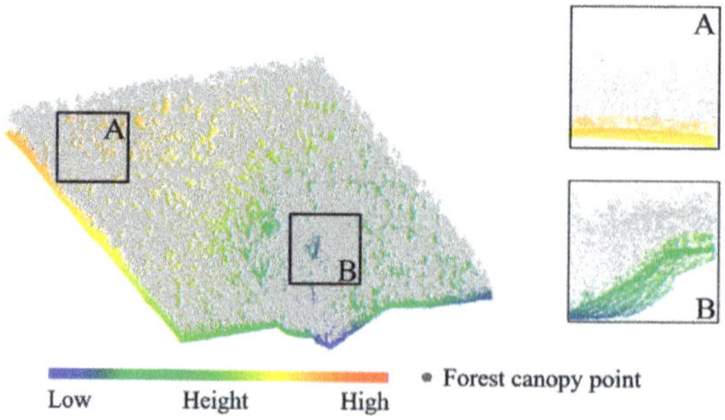

Figure 6. The results of point extraction under forest canopy in a sample site (Site 2). The proposed method can effectively extract the points under forest canopy in gentle slopes and steep slopes, as shown in the enlarged figures A and B.

2.4. Progressive RANSAC Plane Detection

Based on the points under forest canopy, the RANSAC plane detection method is adopted to detect planes on the ground, since this method is robust to noise (including understory vegetation points and negative outliers). However, determining the optimal plane size is critical to describe the terrain in detail and accurately. To solve this problem, we propose a progressive RANSAC plane detection method with decreasing plane sizes. The method detects a plane in a local region and then evaluates the reliability of the plane to represent the terrain in the local region by calculating the maximum distance from points below the detected plane to the plane using Equation (2).

$$\frac{|ax_m+by_m+cz_m+d|}{\sqrt{a^2+b^2+c^2}} < T_d \quad (2)$$

where a, b, c, and d are the coefficients of the detected plane; (x_m, y_m, z_m) is the coordinate of the point p_m below the plane farthest from the plane; and T_d is the distance threshold. In the region with low reliability, the plane with smaller size is detected until the terrain cannot be characterized by a plane.

The implementation steps include regional division, plane detection, reliability evaluation of plane to represent local terrain, and regional subdivision. First, the points under forest canopy are gridded. The cell size of grid is the initial plane size for plane detection. Second, a plane is detected based on the points in a cell using RANSAC. The detected plane is preserved if the number of interior points for the plane is less than a threshold. Third, the maximum distance from the points under the detected plane to the plane is calculated. If the distance is lower than a division distance threshold, the ground in the cell can be represented with the current size plane. Fourth, the points in the cell with low reliability are further gridded with 0.5× the cell size. Steps 2–4 are repeated until no planes can be detected.

The progressive RANSAC plane detection method has three parameters, including initial plane size, number of interior points, and division distance. These parameters were set to 10 m, 20 points, and 1 m empirically in this study.

The plane detection results in the sample site are shown in Figure 7. The local terrain is represented using planes of decreasing sizes as the slope variation increases. Specifically, gentle slopes, steep slopes, and discontinuous terrain are represented by large-, medium-, and small-scale planes (Figures 7a, 7b and 7c), respectively. This confirms that the proposed reliability evaluation method is effective. Thus, the terrain characterized by multi-scale planes can provide a complete and accurate reference for subsequent ground point extraction.

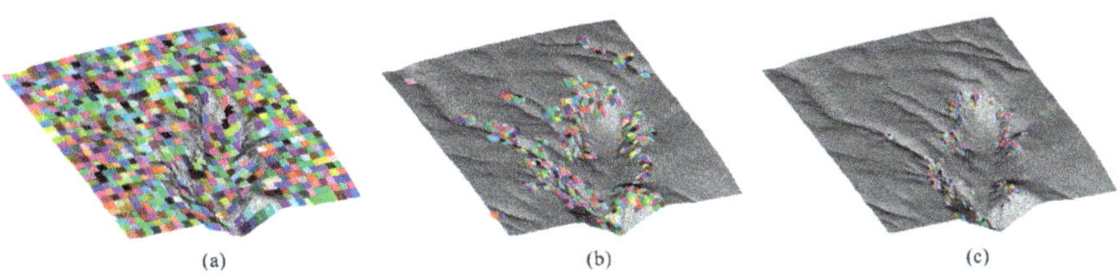

Figure 7. The results of progressive RANSAC plane detection in the sample site. (**a**) Gentle slopes, (**b**) steep slopes, and (**c**) discontinuous terrain are represented by large-, medium-, and small-scale planes, respectively. Colored square and gray surface indicate detected planes and true terrain, respectively.

2.5. Ground Point Extraction

After the terrain is characterized by multi-scale planes, ground points are extracted using an improved surface-based filtering method. The main improvements of the method include two aspects. On the one hand, ground seeds are identified by selecting the points within buffers of detected planes, rather than the lowest points within sliding windows.

On the other hand, the reference terrain for ground point extraction is constructed using the neighbor ground points with spatially uniform distribution. Specifically, the method divides the two-dimensional local space into four subspaces with an unclassified point as center point (Figure 8) and, then, the ground points (P_n) nearest to the center point in each subspace are extracted to construct the local reference terrain using Equation (3).

$$P_n = \begin{cases} p_1, \text{ if } x_p > x_o, y_p > y_o \text{ and } \arg\min\|\vec{op}\|_2 \\ p_2, \text{ if } x_p < x_o, y_p > y_o \text{ and } \arg\min\|\vec{op}\|_2 \\ p_3, \text{ if } x_p < x_o, y_p < y_o \text{ and } \arg\min\|\vec{op}\|_2 \\ p_4, \text{ if } x_p > x_o, y_p < y_o \text{ and } \arg\min\|\vec{op}\|_2 \end{cases} \quad (3)$$

where p_1, p_2, p_3, and p_4 are the points selected in the four subspaces, respectively; (x_o, y_o) and (x_p, y_p) are the coordinates of the center point o and the neighbor point p, respectively; and $\vec{op}$ is the vector between points o and p.

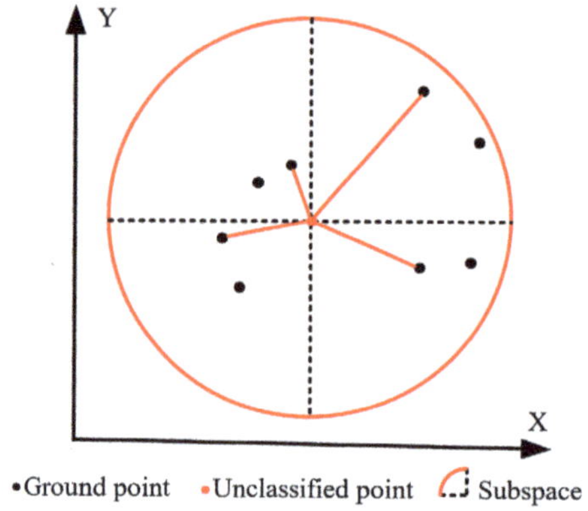

Figure 8. Schematic diagram of the identification for evenly distributed neighbor ground points.

The implementation steps include point cloud attribute initialization, ground seed identification, and ground point extraction. First, the attributes of all points in the point cloud are marked as unclassified points. Second, unclassified points are regarded as ground seeds if the distances between the points and the detected planes are less than a buffer difference threshold. The attributes of identified ground seeds are modified to ground points. Third, the neighbor ground points of an unclassified point are searched from multiple subspaces. The local reference terrain is constructed based on the searched ground points using the least-squares plane fitting method. The unclassified point is marked as ground point if the distance between the point and the local reference terrain is less than a classification distance threshold. Step 3 is repeated until no points are classified as ground points.

The part includes two parameters, i.e., buffer difference and classification distance. They were set to 0.5 m and 1.5 m empirically in this study.

The results of ground seed identification and ground point extraction in the sample are shown in Figures 9 and 10, respectively. Compared with the local minimum method (Figure 9a), the proposed method can identify more ground points on various terrain details (Figure 9b), which provides a more detailed reference for ground point extraction. In addition, the proposed method obtains more accurate ground filtering results than the traditional k-nearest neighbor method in regions with uneven ground points (see enlarged figures A and B in Figure 10).

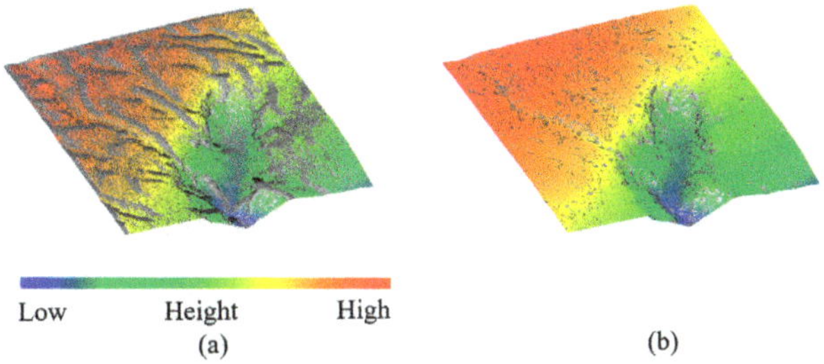

Figure 9. Ground seeds identified by (**a**) the local minimum method and (**b**) the proposed method. Gray surface indicates true terrain.

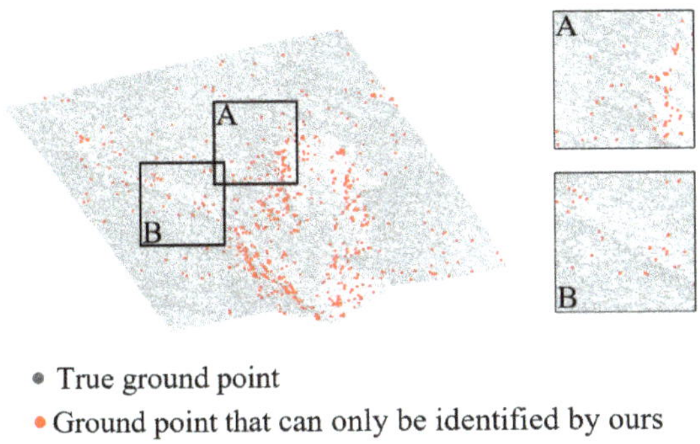

Figure 10. The ground points that can be identified only by the proposed method compared to the traditional *k*-nearest neighbor method. The proposed method obtains more accurate ground filtering results in the regions with uneven ground points, as shown in the enlarged figures A and B.

2.6. Experimental Setup

The experiments were carried out on a laptop computer with 40 GB RAM and an AMD Ryzen 7 5800 H with Radeon Graphics @ 3.20 GHz. The details of the experiments, including accuracy metrics and comparative methods, are given in this section.

2.6.1. Accuracy Metrics

We quantitatively evaluated the performance of ground filtering methods using type I error, type II error, and total error for all sites [52]. Type I error indicates the proportion of ground points being misclassified into non-ground points (Equation (4)). Type II error represents the proportion of opposite condition (Equation (5)). Total error is the proportion of all misclassified points (Equation (6)).

$$\text{Type } I = \frac{a}{c} \tag{4}$$

$$\text{Type II} = \frac{b}{d} \tag{5}$$

$$\text{Total} = \frac{a+b}{c+d} \tag{6}$$

where a is the number of ground points misclassified as non-ground points, b is the number of non-ground points misclassified as ground points, and c and d are the number of reference ground points and non-ground points.

2.6.2. Comparative Methods

The proposed method was compared with four well-known filtering methods, including maximum slope filtering (MSF) [20], progressive morphology filtering (PMF) [26], cloth simulation filtering (CSF) [19], and progressive triangular irregular network densification filtering (PTDF) methods [35]. As described in Section 1, MSF and PMF belong to the slope-based and mathematical morphology-based methods, respectively, while CSF and PTDF belong to the surface-based methods. MSF was implemented using the C++ programming language. PMF was implemented using the C++ programming language and the PCL library [53]. CSF was implemented using CloudCompare software v2.11.3 (Daniel Girardeau, Grenoble, France) [54]. PTDF was implemented using Terrisolid software V8i (Terrasolid, Espoo, Finland) [55]. Table 2 lists the parameter settings of the comparative methods. These parameters were set by visually determining whether most ground points and non-ground points were accurately distinguished. Specifically, the filtered point cloud was segmented into multiple strips with the width of 50 m, because ground filtering results can be better checked in cross-section view. We observed the results from different views in CloudCompare software v2.11.3 (Daniel Girardeau, Grenoble, France) and selected the parameter settings corresponding to the optimal results.

Table 2. Parameter settings of the comparative methods. MSF, PMF, CSF, and PTDF represent the four comparative methods, i.e., maximum slope filtering, progressive morphology filtering, cloth simulation filtering, and progressive triangular irregular network densification filtering methods.

Method	Parameter			
MSF	Cell size: 1 m	Window size: 10 m	Slope: 15°	Maximum height: 3 m
PMF	Window size: 10 m	Slope: 15°	Initial height: 0.5 m	Maximum height: 3 m
CSF	Rigidness: 3	Cloth resolution: 0.3 m	Max iterations: 500	Classification threshold: 0.5 m
PTDF	Window size: 10 m	Terrain angle: 88°	Iteration angle: 15 °	Iteration distance: 1.5 m

3. Results

The accuracy comparison of the proposed method (i.e., PPDF) and the comparative methods (i.e., MSF, PMF, CSF, and PTDF) are given in Figure 11. PPDF had the lowest average total error of 3.42% and standard deviation of 2.45% across all sites, which include various terrain (e.g., terrain slope and height) and vegetation conditions (e.g., canopy cover and height). PPDF also achieved high accuracy with less than 5% total error even in sites (e.g., Sites 2 and 6) containing dense vegetation and various discontinuous terrain. The smallest type I errors were achieved by the proposed method in five out of six sites, and the total error in the remaining sample (i.e., Site 1) was rather close to the optimal result, with a difference of 0.34%. In terms of type II errors, the proposed method obtained optimal or near-optimal results in most sites.

The digital terrain models (DTMs) constructed by the proposed method were highly similar to the reference DTMs in all forested sites. The proposed method completely preserved various terrain features, including the gentle slopes in Site 1 (Figure 12a), valleys in Site 2 (Figure 12b), steep slopes in Site 3 (Figure 12c), hilltops in Site 4 (Figure 12d), ridges in Site 5 (Figure 12e), and break lines in Site 6 (Figure 12f). Meanwhile, vegetation points with different cover and height values were effectively filtered out.

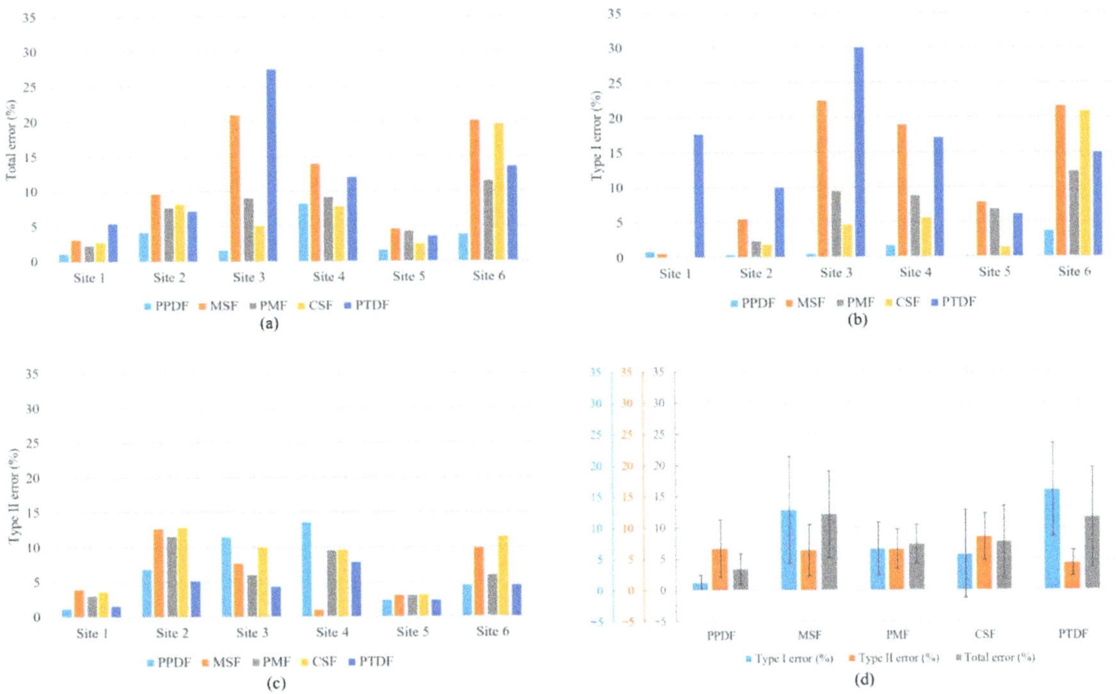

Figure 11. Accuracy comparison of the proposed method and the comparative methods: (**a**) total error, (**b**) type I error, (**c**) type II error, and (**d**) average and standard deviation values of the three errors. MSF, PMF, CSF, and PTDF represent the four comparative methods, i.e., maximum slope filtering, progressive morphology filtering, cloth simulation filtering, and progressive triangular irregular network densification filtering methods.

Figures 13 and 14 show the DTMs of the proposed and well-known ground filtering methods for two representative sites (Sites 2 and 6), which are characterized by high canopy cover and various discontinuous terrain, respectively. In Site 2, the proposed method generated DTM that are closer to the reference result. In comparison, MSF failed to preserve highly steep slopes and ridges (see ellipses A and B in Figure 13) and accurately filtered out vegetation points (see ellipse C in Figure 13). PMF often smoothed ridges (see ellipse C in Figure 13). CSF tended to misclassify vegetation points as ground points (see ellipse B in Figure 13). PTDF often smoothed highly steep slopes (see ellipse A in Figure 13). In Site 6, various morphological break lines were preserved more realistically by the proposed method compared to other methods, as shown in ellipses A, B, and C in Figure 14.

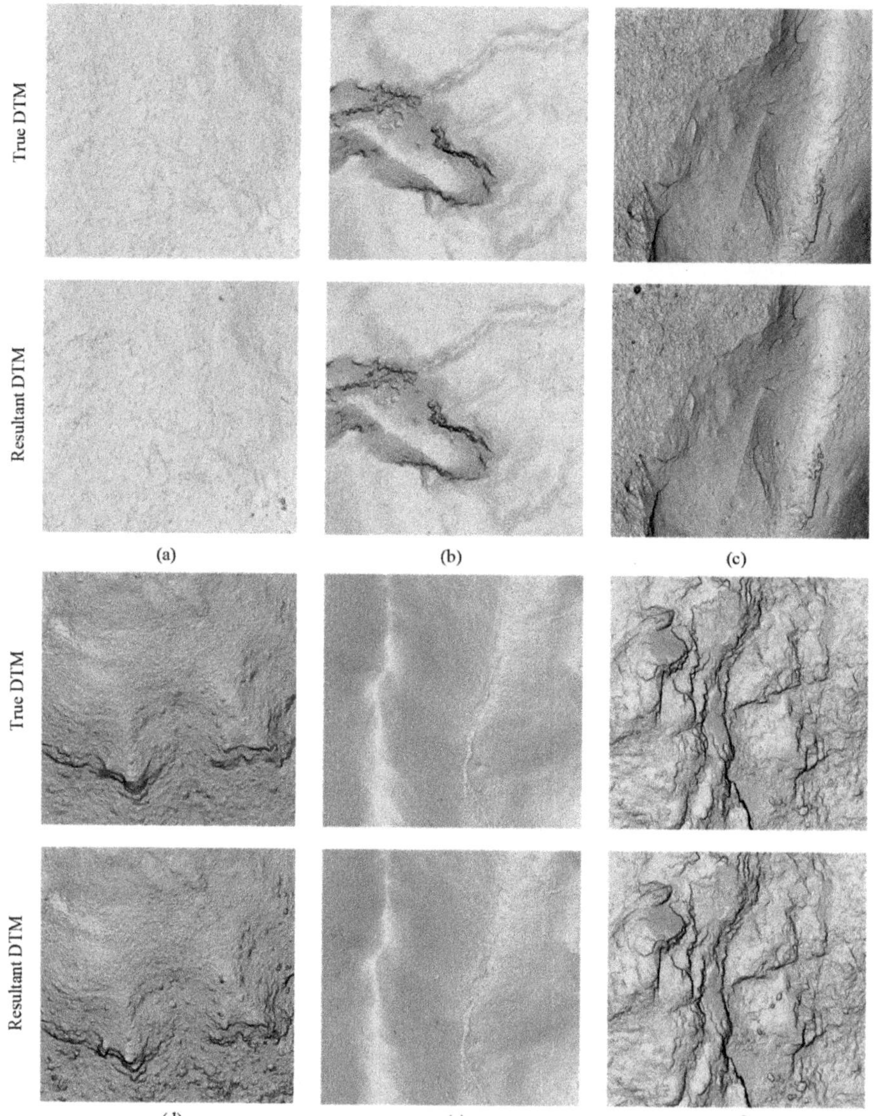

Figure 12. DTMs of the proposed method in all sites: (**a–f**) Sites 1–6.

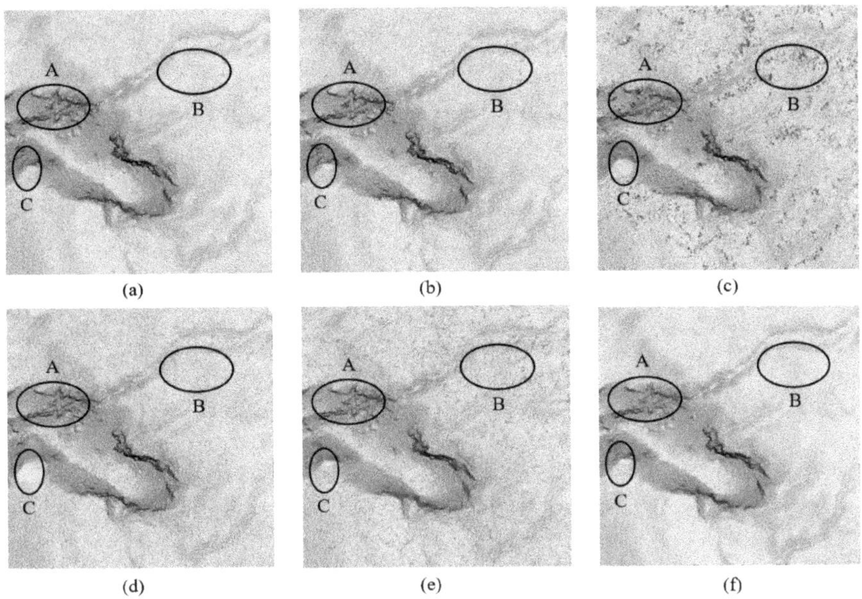

Figure 13. DTMs of (**a**) reference, (**b**) PPDF, (**c**) MSF, (**d**) PMF, (**e**) CSF, and (**f**) PTDF in Site 2. MSF, PMF, CSF, and PTDF represent the four comparative methods, i.e., maximum slope filtering, progressive morphology filtering, cloth simulation filtering, and progressive triangular irregular network densification filtering methods.

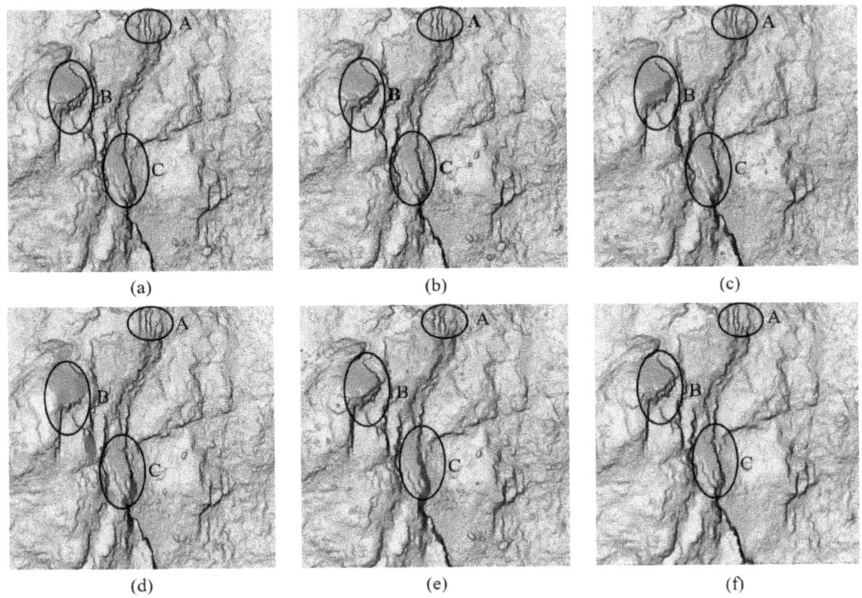

Figure 14. DTMs of (**a**) reference, (**b**) PPDF, (**c**) MSF, (**d**) PMF, (**e**) CSF and (**f**) PTDF in Site 6. MSF, PMF, CSF and PTDF represent the four comparative methods, i.e., maximum slope filtering, progressive morphology filtering, cloth simulation filtering, and progressive triangular irregular network densification filtering methods.

The computational efficiency comparison between the proposed method and well-known methods is given in Figure 15. The proposed method was comparable to PMF and CSF, and significantly higher than PMF. In addition, the computational efficiency of the proposed method was negatively related to the point density. Site 1, Sites 2–5, and Site 6 had high, medium, and low point densities and exhibited the same trend in runtime. Note that the computational efficiency of PTDF was not evaluated, since Terrsolid software V8i (Terrsolid, Espoo, Finland) cannot provide runtime information.

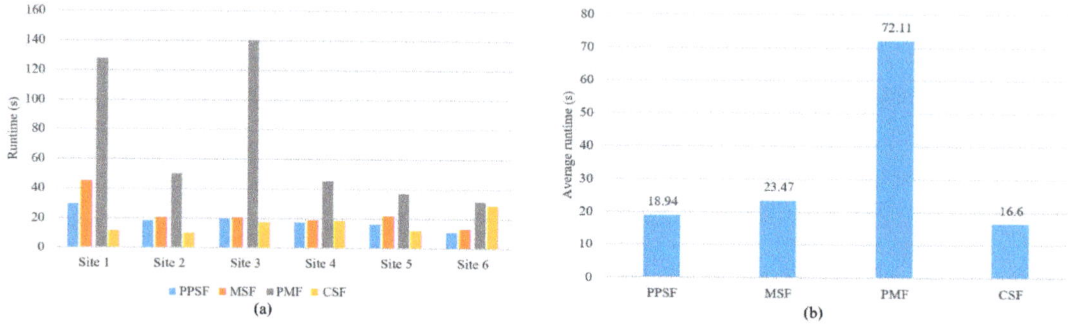

Figure 15. Runtime comparison between the proposed and well-known ground filtering methods in terms of (**a**) each forested site and (**b**) their average values. MSF, PMF, CSF, and PTDF represent the four comparative methods, i.e., maximum slope filtering, progressive morphology filtering, cloth simulation filtering, and progressive triangular irregular network densification filtering methods.

4. Discussion

4.1. Performance Analysis

Compared with other methods, PPDF was more accurate and more robust even in the landscapes containing dense vegetation and complex terrain, with the smallest average total error and standard deviation (Figure 11). This is attributed to the contributions described in Section 1. To test the performance of the proposed method with and without each contribution, we performed two tests in all sites. The accuracy comparison between the proposed method with and without the two contributions is given in Figure 16. Results showed that both contributions helped to improve ground filtering accuracy in all sites. For the average total errors of the proposed method with both contributions, only the first contribution and the second contribution were 3.42%, 3.75%, and 9.12%, respectively. In comparison, the contribution of the progressive plane detection method was greater especially in the sites containing the terrain with large slope variations, e.g., Site 6.

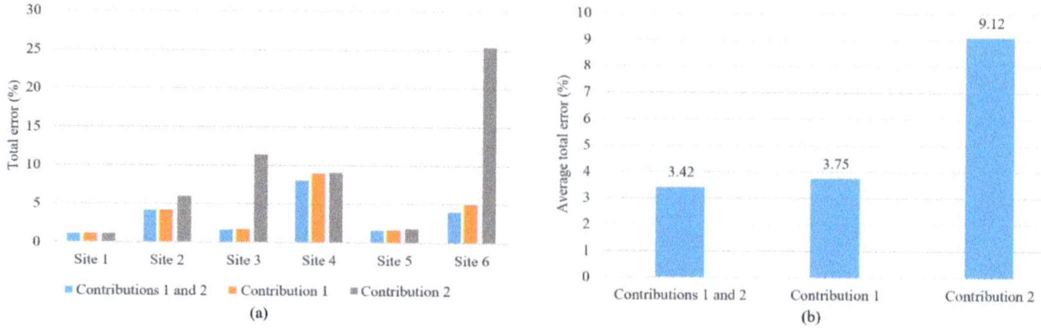

Figure 16. Accuracy comparison of the proposed method with and without each contribution in terms of (**a**) each forested site and (**b**) their average values.

PPDF had obviously higher a ground filtering accuracy than MSF (Figure 11). The possible reasons include two aspects. On the one hand, MSF fails to obtain ground seeds on raised terrain and negative outliers may also be misjudged as ground seeds. On the other hand, MSF judges the attributes of unclassified points only based on the nearest neighbor ground points. In contrast, PPDF can obtain more complete ground seeds on various terrains based on the terrain characterized by multi-scale planes. The misjudgment of negative outliers as ground seeds can also be avoided due to the introduction of the feature that the local terrain is flat. Additionally, ground points are extracted based on the reference terrain which uses more contexts. As a result, the proposed method outperforms MSF. Although PMF can obtain more ground seeds than MSF, it still cannot select enough ground seeds on raised terrain, causing the terrain to be smoothed (Figures 13d and 14d). Nevertheless, PPDF overcomes this limitation by identifying ground seeds based on the feature that the local terrain is flat, rather than the local minimum (Figures 13b and 14b). CSF often produces filtering errors probably because it only extracts ground points based on the nearest neighbor ground points (Figure 13e). However, PPDF reduces these errors by using a reference terrain that contains more context. PTDF tends to ignore terrain details, especially in the sites with steep slopes and break lines (Figures 13f and 14f). Although PTDF cannot obtain sufficient ground seeds on the terrain, PTDF extracts ground points through iterative upward densification, that is, ground points are extracted from points above the reference terrain in each iteration [16,39]. However, PPDF identifies ground points by iteratively upward and downward densification, thus, better preserving the terrain's details. Overall, PPDF has advantages over other methods in terms of accuracy and reliability.

4.2. Parameter Sensitivity Analysis

The sensitivity of PPDF to key parameters was evaluated, considering that the parameter settings in different landscapes are a core factor affecting the ease of use for the method. We iteratively executed the proposed method by replacing one parameter with a certain interval while keeping other parameters at default settings. The influence of each parameter on filtering the results is given in Figure 17. The effect of initial plane size on the total error was limited even using a particularly large plane size (Figure 17a), because the optimal plane size can be automatically and accurately determined by evaluating the reliability of the plane to represent the local terrain. The number of interior points had insignificant influences on filtering the results (Figure 17b). There was no obvious accuracy difference with the change in the division distance threshold (Figure 17c). The effect of the classification distance on filtering the results was small in a large interval (Figure 17d), probably because a more detailed reference terrain weakens the dependence of surface-based filtering methods on the classification of the distance threshold [14,16,47]. In general, these results suggested that the proposed method was insensitive to the settings of the initial plane size, the number of interior points, the division distance, and the classification distance, and these parameters can be set to fixed values.

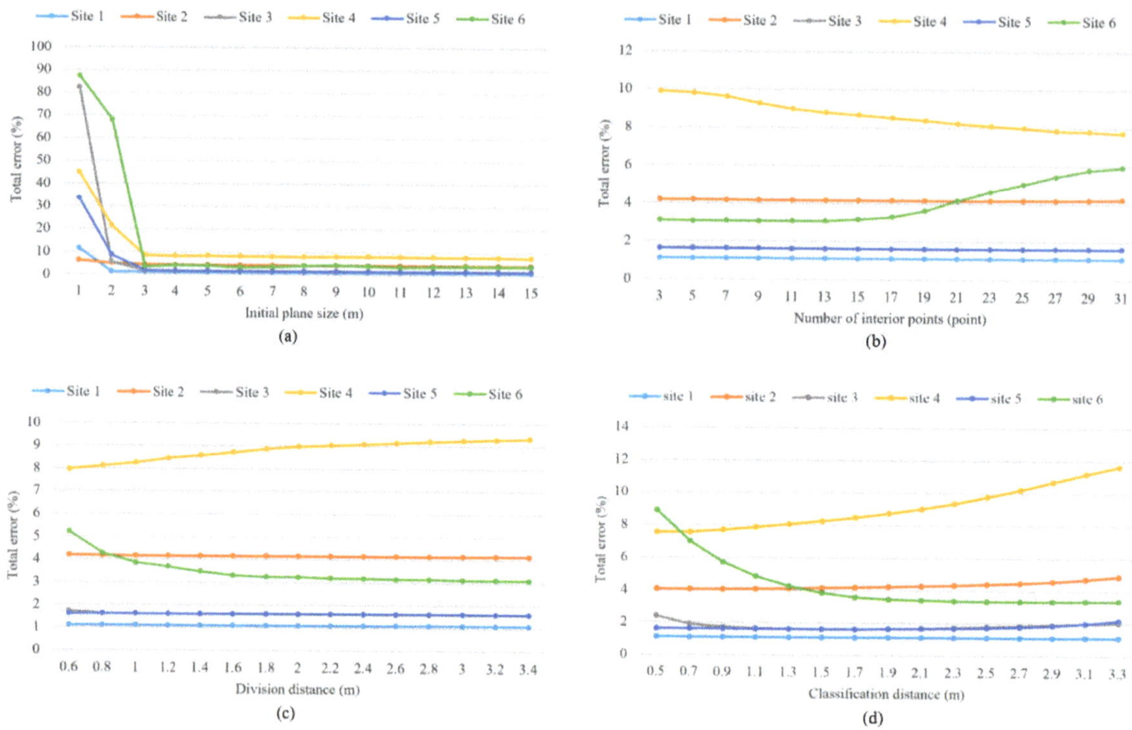

Figure 17. The sensitivity of the proposed method to the settings of (**a**) initial plane size, (**b**) number of interior points, (**c**) division distance, and (**d**) classification distance. Experiments were performed in each site. Only one parameter was adjusted in each execution, while other parameters were kept at default settings.

5. Conclusions

To improve the effectiveness and ease of use of ground filtering in forested landscapes containing dense vegetation and complex terrain, a progressive plane detection filtering method for airborne LiDAR point clouds is proposed in this study. The novelty of the proposed method lies within its use of progressive plane detection to identify ground seeds as completely as possible on various terrains and the identification of evenly distributed neighbor ground points to construct a reference terrain to increase the accuracy of ground point extraction in the regions with uneven ground points. We demonstrated the effectiveness of the proposed method by quantitatively and qualitatively comparing four classic filtering methods, which belong to different types of filtering methods that have been published. Meanwhile, the ease of use of the proposed method was verified by testing the sensitivity of the proposed method to the parameters. There are two major conclusions.

(1) The proposed method can improve the effectiveness of ground filtering in the forested landscapes. It achieved the smallest average total error and standard deviation compared to other methods and the preservation of terrain details was greatly improved especially in regions with large terrain slope variations (e.g., steep slopes, break lines, and ridges).

(2) The proposed method has the advantage of ease of use. It was insensitive to parameters. Therefore, these parameters can be set as fixed values, which makes it easier for the users with insufficient experience to execute ground filtering in their own applications.

Author Contributions: Conceptualization, S.C.; data curation, S.C.; formal analysis, S.C. and S.Y.; funding acquisition, X.L.; investigation, S.C.; methodology, S.C. and X.L.; project administration, S.C. and X.L.; resources, S.C. and X.L.; software, S.C.; supervision, S.C. and X.L.; validation, S.C. and S.Y.; visualization, S.C.; writing—original draft, S.C.; writing—review and editing, S.C., X.L. and S.Y. All authors have read and agreed to the published version of the manuscript.

Funding: This research was funded by the Natural Science Fund of China, grant number 32171789, 32211530031; Wuhan University, grant number WHUZZJJ202220; Open Fund of Key Laboratory of Mine Environmental Monitoring and Improving around Poyang Lake, Ministry of Natural Resources, grant number MEMI-2021-2022-03; International S&T Cooperation Program of Hubei Province, grant number 2022EHB048; National Key R&D Program of China, grant number 2022YFB3903800; Open Fund of Key Research Base of Philosophy and Social Science of Higher Education in Guangdong Province - Local Government Development Research Institute of Shantou University, grant number 07422002.

Data Availability Statement: The authors do not have permission to share data.

Conflicts of Interest: The authors declare no conflict of interest.

References

1. Drake, J.B.; Dubayah, R.O.; Clark, D.B.; Knox, R.G.; Blair, J.B.; Hofton, M.A.; Chazdon, R.L.; Weishampel, J.F.; Prince, S. Estimation of tropical forest structural characteristics using large-footprint lidar. *Remote Sens. Environ.* **2002**, *79*, 305–319. [CrossRef]
2. Brandtberg, T. Classifying individual tree species under leaf-off and leaf-on conditions using airborne lidar. *ISPRS J. Photogramm. Remote Sens.* **2007**, *61*, 325–340. [CrossRef]
3. Popescu, S.C. Estimating biomass of individual pine trees using airborne lidar. *Biomass Bioenergy* **2007**, *31*, 646–655. [CrossRef]
4. Hyyppä, J.; Yu, X.; Hyyppä, H.; Vastaranta, M.; Holopainen, M.; Kukko, A.; Kaartinen, H.; Jaakkola, A.; Vaaja, M.; Koskinen, J.; et al. Advances in Forest Inventory Using Airborne Laser Scanning. *Remote Sens.* **2012**, *4*, 1190–1207. [CrossRef]
5. Hyyppä, J.; Hyyppä, H.; Leckie, D.; Gougeon, F.; Yu, X.; Maltamo, M. Review of methods of small-footprint airborne laser scanning for extracting forest inventory data in boreal forests. *Int. J. Remote Sens.* **2008**, *29*, 1339–1366. [CrossRef]
6. Wulder, M.A.; White, J.C.; Nelson, R.F.; Næsset, E.; Ørka, H.O.; Coops, N.C.; Hilker, T.; Bater, C.W.; Gobakken, T. Lidar Sampling for Large-Area Forest Characterization: A Review. *Remote Sens. Environ.* **2012**, *121*, 196–209. [CrossRef]
7. Ferraz, A.; Bretar, F.; Jacquemoud, S.; Gonçalves, G.; Pereira, L.; Tomé, M.; Soares, P. 3-D mapping of a multi-layered Mediterranean forest using ALS data. *Remote Sens. Environ.* **2012**, *121*, 210–223. [CrossRef]
8. Shi, Y.; Wang, T.; Skidmore, A.K.; Heurich, M. Important LiDAR metrics for discriminating forest tree species in Central Europe. *ISPRS J. Photogramm. Remote Sens.* **2018**, *137*, 163–174. [CrossRef]
9. Mielcarek, M.; Stereńczak, K.; Khosravipour, A. Testing and evaluating different LiDAR-derived canopy height model generation methods for tree height estimation. *Int. J. Appl. Earth Obs. Geoinf.* **2018**, *71*, 132–143. [CrossRef]
10. Dai, W.; Guan, Q.; Cai, S.; Liu, R.; Chen, R.; Liu, Q.; Chen, C.; Dong, Z. A Comparison of the Performances of Unmanned-Aerial-Vehicle (UAV) and Terrestrial Laser Scanning for Forest Plot Canopy Cover Estimation in *Pinus massoniana* Forests. *Remote Sens.* **2022**, *14*, 1188. [CrossRef]
11. Hartling, S.; Sagan, V.; Sidike, P.; Maimaitijiang, M.; Carron, J. Urban Tree Species Classification Using a WorldView-2/3 and LiDAR Data Fusion Approach and Deep Learning. *Sensors* **2019**, *19*, 1284. [CrossRef]
12. Liang, X.; Wang, Y.; Pyörälä, J.; Lehtomäki, M.; Yu, X.; Kaartinen, H.; Kukko, A.; Honkavaara, E.; Issaoui, A.E.I.; Nevalainen, O.; et al. Forest in situ observations using unmanned aerial vehicle as an alternative of terrestrial measurements. *For. Ecosyst.* **2019**, *6*, 20. [CrossRef]
13. Liang, X.; Kukko, A.; Balenovic, I.; Saarinen, N.; Junttila, S.; Kankare, V.; Holopainen, M.; Mokros, M.; Surovy, P.; Kaartinen, H.; et al. Close-Range Remote Sensing of Forests: The state of the art, challenges, and opportunities for systems and data acquisitions. *IEEE Geosci. Remote Sens. Mag.* **2022**, *10*, 32–71. [CrossRef]
14. Chen, C.; Wang, M.; Chang, B.; Li, Y. Multi-Level Interpolation-Based Filter for Airborne LiDAR Point Clouds in Forested Areas. *IEEE Access* **2020**, *8*, 41000–41012. [CrossRef]
15. Hui, Z.; Jin, S.; Xia, Y.; Nie, Y.; Xie, X.; Li, N. A mean shift segmentation morphological filter for airborne LiDAR DTM extraction under forest canopy. *Opt. Laser Technol.* **2021**, *136*, 106728. [CrossRef]
16. Zhao, X.; Guo, Q.; Su, Y.; Xue, B. Improved progressive TIN densification filtering algorithm for airborne LiDAR data in forested areas. *ISPRS J. Photogramm. Remote Sens.* **2016**, *117*, 79–91. [CrossRef]
17. Véga, C.; Durrieu, S.; Morel, J.; Allouis, T. A sequential iterative dual-filter for Lidar terrain modeling optimized for complex forested environments. *Comput. Geosci.* **2012**, *44*, 31–41. [CrossRef]
18. Bigdeli, B.; Amirkolaee, H.A.; Pahlavani, P. DTM extraction under forest canopy using LiDAR data and a modified invasive weed optimization algorithm. *Remote Sens. Environ.* **2018**, *216*, 289–300. [CrossRef]
19. Zhang, W.; Qi, J.; Wan, P.; Wang, H.; Xie, D.; Wang, X.; Yan, G. An Easy-to-Use Airborne LiDAR Data Filtering Method Based on Cloth Simulation. *Remote Sens.* **2016**, *8*, 501. [CrossRef]

20. Vosselman, G. Slope Based Filtering of Laser Altimetry Data. *Int. Arch. Photogramm. Remote Sens.* **2000**, *33*, 935–942.
21. Meng, X.; Wang, L.; Silván-Cárdenas, J.L.; Currit, N. A multi-directional ground filtering algorithm for airborne LIDAR. *ISPRS J. Photogramm. Remote Sens.* **2009**, *64*, 117–124. [CrossRef]
22. Susaki, J. Adaptive Slope Filtering of Airborne LiDAR Data in Urban Areas for Digital Terrain Model (DTM) Generation. *Remote Sens.* **2012**, *4*, 1804–1819. [CrossRef]
23. Shan, J.; Aparajithan, S. Urban DEM Generation from Raw Lidar Data. *Photogramm. Eng. Remote Sens.* **2005**, *71*, 217–226. [CrossRef]
24. Chen, Z.; Gao, B.; Devereux, B. State-of-the-Art: DTM Generation Using Airborne LIDAR Data. *Sensors* **2017**, *17*, 150. [CrossRef]
25. Wan, P.; Zhang, W.; Skidmore, A.K.; Qi, J.; Jin, X.; Yan, G.; Wang, T. A simple terrain relief index for tuning slope-related parameters of LiDAR ground filtering algorithms. *ISPRS J. Photogramm. Remote Sens.* **2018**, *143*, 181–190. [CrossRef]
26. Zhang, K.; Chen, S.-C.; Whitman, D.; Shyu, M.-L.; Yan, J.; Zhang, C. A progressive morphological filter for removing nonground measurements from airborne LIDAR data. *IEEE Trans. Geosci. Remote Sens.* **2003**, *41*, 872–882. [CrossRef]
27. Pingel, T.J.; Clarke, K.C.; McBride, W.A. An improved simple morphological filter for the terrain classification of airborne LIDAR data. *ISPRS J. Photogramm. Remote Sens.* **2013**, *77*, 21–30. [CrossRef]
28. Li, Y.; Yong, B.; Wu, H.; An, R.; Xu, H. An Improved Top-Hat Filter with Sloped Brim for Extracting Ground Points from Airborne Lidar Point Clouds. *Remote Sens.* **2014**, *6*, 12885–12908. [CrossRef]
29. Mongus, D.; Lukač, N.; Žalik, B. Ground and building extraction from LiDAR data based on differential morphological profiles and locally fitted surfaces. *ISPRS J. Photogramm. Remote Sens.* **2014**, *93*, 145–156. [CrossRef]
30. Yang, B.; Huang, R.; Dong, Z.; Zang, Y.; Li, J. Two-step adaptive extraction method for ground points and breaklines from lidar point clouds. *ISPRS J. Photogramm. Remote Sens.* **2016**, *119*, 373–389. [CrossRef]
31. Chen, Q.; Gong, P.; Baldocchi, D.; Xie, G. Filtering Airborne Laser Scanning Data with Morphological Methods. *Photogramm. Eng. Remote Sens.* **2007**, *73*, 175–185. [CrossRef]
32. Chen, D.; Zhang, L.; Wang, Z.; Deng, H. A mathematical morphology-based multi-level filter of LiDAR data for generating DTMs. *Sci. China Inf. Sci.* **2012**, *56*, 1–14. [CrossRef]
33. Tan, Y.; Wang, S.; Xu, B.; Zhang, J. An improved progressive morphological filter for UAV-based photogrammetric point clouds in river bank monitoring. *ISPRS J. Photogramm. Remote Sens.* **2018**, *146*, 421–429. [CrossRef]
34. Liang, X.; Zhang, J.; Li, H. An Adaptive Morphological Filter for LIDAR Data Filtering in Urban Area. *J. Remote Sens.* **2007**, *2*, 276–281. [CrossRef]
35. Axelsson, P. DEM Generation from Laser Scanner Data Using Adaptive TIN Models. *Int. Arch. Photogramm. Remote Sens.* **2000**, *33*, 111–118.
36. Zhang, J.; Lin, X. Filtering airborne LiDAR data by embedding smoothness-constrained segmentation in progressive TIN densification. *ISPRS J. Photogramm. Remote Sens.* **2013**, *81*, 44–59. [CrossRef]
37. Hu, H.; Ding, Y.; Zhu, Q.; Wu, B.; Lin, H.; Du, Z.; Zhang, Y.; Zhang, Y. An adaptive surface filter for airborne laser scanning point clouds by means of regularization and bending energy. *ISPRS J. Photogramm. Remote Sens.* **2014**, *92*, 98–111. [CrossRef]
38. Evans, J.S.; Hudak, A.T. A Multiscale Curvature Algorithm for Classifying Discrete Return LiDAR in Forested Environments. *IEEE Trans. Geosci. Remote Sens.* **2007**, *45*, 1029–1038. [CrossRef]
39. Nie, S.; Wang, C.; Dong, P.; Xi, X.; Luo, S.; Qin, H. A revised progressive TIN densification for filtering airborne LiDAR data. *Measurement* **2017**, *104*, 70–77. [CrossRef]
40. Chen, C.; Chang, B.; Li, Y.; Shi, B. Filtering airborne LiDAR point clouds based on a scale-irrelevant and terrain-adaptive approach. *Measurement* **2021**, *171*, 108756. [CrossRef]
41. Chen, C.; Li, Y.; Li, W.; Dai, H. A multiresolution hierarchical classification algorithm for filtering airborne LiDAR data. *ISPRS J. Photogramm. Remote Sens.* **2013**, *82*, 1–9. [CrossRef]
42. Meng, X.; Lin, Y.; Yan, L.; Gao, X.; Yao, Y.; Wang, C.; Luo, S. Airborne LiDAR Point Cloud Filtering by a Multilevel Adaptive Filter Based on Morphological Reconstruction and Thin Plate Spline Interpolation. *Electronics* **2019**, *8*, 1153. [CrossRef]
43. Meng, X.; Currit, N.; Zhao, K. Ground Filtering Algorithms for Airborne LiDAR Data: A Review of Critical Issues. *Remote Sens.* **2010**, *2*, 833–860. [CrossRef]
44. Chen, C.; Guo, J.; Wu, H.; Li, Y.; Shi, B. Performance Comparison of Filtering Algorithms for High-Density Airborne LiDAR Point Clouds over Complex LandScapes. *Remote Sens.* **2021**, *13*, 2663. [CrossRef]
45. Hui, Z.; Wang, L.; Ziggah, Y.Y.; Cai, S.; Xia, Y. Automatic morphological filtering algorithm for airborne lidar data in urban areas. *Appl. Opt.* **2019**, *58*, 1164–1173. [CrossRef]
46. Wei, L.; Yang, B.; Jiang, J.; Cao, G.; Wu, M. Vegetation filtering algorithm for UAV-borne lidar point clouds: A case study in the middle-lower Yangtze River riparian zone. *Int. J. Remote Sens.* **2016**, *38*, 2991–3002. [CrossRef]
47. Cai, S.; Zhang, W.; Liang, X.; Wan, P.; Qi, J.; Yu, S.; Yan, G.; Shao, J. Filtering Airborne LiDAR Data Through Complementary Cloth Simulation and Progressive TIN Densification Filters. *Remote Sens.* **2019**, *11*, 1037. [CrossRef]
48. Hui, Z.; Hu, Y.; Yevenyo, Y.Z.; Yu, X. An Improved Morphological Algorithm for Filtering Airborne LiDAR Point Cloud Based on Multi-Level Kriging Interpolation. *Remote Sens.* **2016**, *8*, 35. [CrossRef]
49. Cai, S.; Zhang, W.; Jin, S.; Shao, J.; Li, L.; Yu, S.; Yan, G. Improving the estimation of canopy cover from UAV-LiDAR data using a pit-free CHM-based method. *Int. J. Digit. Earth* **2021**, *14*, 1477–1492. [CrossRef]

50. Daghigh, H.; Tannant, D.D.; Daghigh, V.; Lichti, D.D.; Lindenbergh, R. A critical review of discontinuity plane extraction from 3D point cloud data of rock mass surfaces. *Comput. Geosci.* **2022**, *169*, 105241. [CrossRef]
51. Tang, H.; Song, X.-P.; Zhao, F.A.; Strahler, A.H.; Schaaf, C.L.; Goetz, S.; Huang, C.; Hansen, M.C.; Dubayah, R. Definition and measurement of tree cover: A comparative analysis of field-, lidar- and landsat-based tree cover estimations in the Sierra national forests, USA. *Agric. For. Meteorol.* **2019**, *268*, 258–268. [CrossRef]
52. Sithole, G.; Vosselman, G. Experimental comparison of filter algorithms for bare-Earth extraction from airborne laser scanning point clouds. *ISPRS J. Photogramm. Remote Sens.* **2004**, *59*, 85–101. [CrossRef]
53. Point Cloud Library. Available online: https://pointclouds.org/ (accessed on 15 November 2022).
54. CloudCompare Software and Introduction. Available online: https://www.danielgm.net/cc/ (accessed on 15 November 2022).
55. Terrisolid Software and Introduction. Available online: http://www.terrasolid.com/home.php (accessed on 10 November 2022).

forests

MDPI

Article

New Coupled Canopy–Light Model (CCLM) to Improve Visual Polymorphism Simulation of Fir Morphology

Yuanqing Zuo [1,2,3], Huaiqing Zhang [1,2,3,*], Zeyu Cui [1,2,3], Yang Liu [1,2,3], Kexin Lei [1,2,3], Xingtao Hu [4], Hanqing Qiu [1,2,3], Jiansen Wang [1,2,3], Jing Zhang [1,2,3] and Tingdong Yang [1,2,3]

1 Institute of Forest Resource Information Techniques, Chinese Academy of Forestry, Beijing 100091, China
2 Key Laboratory of Forest Management and Growth Modelling, NFGA, Beijing 100091, China
3 National Long Term Scientific Research Base of Huangfengqiao Forest Monitoring and Simulation in Hunan Province, Beijing 100091, China
4 School of Geography and Environmental Sciences, Guizhou Normal University, Guiyang 550025, China
* Correspondence: zhang@ifrit.ac.cn; Tel.: +86-1-365-116-6672

Abstract: Environmental factors substantially influence the growth of trees. The current studies on tree growth simulation have mainly focused on the effect of environmental factors on diameter at breast height and tree height. However, the influence of environmental factors, especially light, on canopy morphology has not been considered, hindering the accurate understanding of the range of characteristics of tree morphology that occur due to environmental changes. To solve this problem, this study investigated the influence of light on the changes in canopy morphology and constructed a coupled canopy–light model (CCLM) to visually simulate the polymorphism of fir morphology. Using the Huangfengqiao Forestry Farm in You County, Hunan Province, China, as the study area, we selected a typical sample plot. Field surveys of the fir trees in the sample plot were conducted for three consecutive years to obtain longitudinal data of fir tree canopy shape. We constructed the canopy curves using a cubic uniform B-spline to construct 3D models of the fir trees in different years. The topographic and spatial location distribution data of the fir trees were used to construct a 3D scene of the sample plot in the UE4 3D engine, and the light distribution for each part of the canopy was calculated in a 3D scene by using the annual average photosynthetically active radiation (PAR) as the light parameter, which we combined with the ray-tracing algorithm. This study constructed the CCLM from the fir diameter using the breast-height growth model (BDGM) and the height–diameter curve model (HDCM), the fir trees' canopy shape description from two years, and the light distribution data. We compared the canopy data obtained from canopy simulations using the CCLM with those obtained using a growth model based on spatial structure (GMBOSS) and those obtained from field surveys to identify any difference in the effectiveness of the canopy simulations using the CCLM and GMBOSS. Based on the BDGM and HDCM, we constructed the CCLM of firs with a determination coefficient (R^2) of 0.829, combining data on canopy shape descriptions obtained from two years of field surveys and the light distribution data of each part of the canopy obtained through the ray-tracing algorithm. The Euclidean distance between the canopy description data obtained using the CCLM and the canopy description data obtained from the field survey was 15.561; that between the GMBOSS and the field survey was 23.944. A virtual forest stand environment was constructed from the survey data, combining ray-tracing algorithms to construct the CCLM model of fir in a virtual forest stand environment for growth visualization and simulation. Compared with the canopy description data obtained using the GMBOSS, the canopy description data obtained using the CCLM better fit the canopy description data obtained from the field survey, and the Euclidean distance decreased from 23.944 to 15.561.

Keywords: canopy-light model; polymorphism modeling; canopy morphology; visual simulation

1. Introduction

Forest tree canopy shapes are diverse, and many factors contribute to their diversity. In addition to the genetic factors of the forest tree itself, the growing environment of the forest tree is also crucial for canopy shape diversity [1,2]. Many environmental factors affect the canopy morphology of forest trees, including fire, water, light, temperature, soil, elevation, spatial structure, species interaction, and so on [3,4]. At the scale of the sample plot, differences in soil, temperature, and elevation within the same sample plot are not significant [5–7]. However, competition for spatial resources in forest stands can lead to significant differences in spatial structure between forest trees and their neighboring trees, resulting in significant differences in light conditions for forest trees within forest stands [8–10]. The growth of the canopy varies under different light conditions.

Therefore, it is important to consider the effect of light on the canopy shape, and there are some scholars who have already started such research [11,12]. Shanin et al. proposed a new model, which operates with the 3D-representation of tree canopies and light transmission through the canopy using discrete spatial and temporal resolution, paying special attention to the simulation of asymmetry in canopy shape as a result of competition for light [13]. Duchemin et al. considered the growth of the tree canopy as a continuous front propagation process and propose a model which only requires two parameters to describe the shape of the crown: the intensities of phototropic and gravitropic growing responses [14]. Hu et al. found that with the increase of light, the canopy of seedlings developed from wide and loose to relatively tight and narrow [15]. He et al. proposed that shading can increase the leaf area of ginkgo seedlings, prolong their growth period, and promote the growth of long branches and plant height [16]. Xu et al. believed that hard light promoted the differentiation of branches by inhibiting the growth of the trunk, and that shading promoted the rebranching of branches by inhibiting the growth of first-order branches [17]. The canopy shape of a forest tree is closely related to its ability to obtain light resources, which affects its light distribution. The light distribution of the canopy also affects the canopy shape of the forest tree [18], so a mutual influence exists between the two. When the light intensity is below or above a certain value, branch growth stagnation or even death may occur. Between these limits, the growth rate of branches first increases and then decreases with the increase in light intensity [19–21].

To research the influence of light on the canopy shape, it is necessary to calculate the canopy light distribution [22–24]. The methods used for calculating plant canopy light distribution are constantly advancing with the development of technology, from the direct observation of plants to measuring plants with the aid of instruments and, more recently, to the digital simulation of plants [25–27]. In the process, the accuracy and efficiency of model calculations are constantly increased [28–30]. Initially, canopy analyzers were used to capture photos of plants in different orientations, and physical indicators such as light spots and light transmission were analyzed based on the acquired images [31]. Alternatively, a light meter can be used to determine the light distribution of a plant canopy [32], a method that relies on hardware devices to measure the light distribution information of plants, which is a cumbersome and inefficient process. Lao [33] and Wang [34] used a 3D digitizer to measure the 3D digital morphology of plant canopies and analyzed the canopy light distribution. However, 3D digitizers cannot rapidly acquire data and do not accurately capture the actual light distribution.

With the continuous development of information technology, hardware and software are more often being combined to construct mathematical physical models to study the light distribution in plant canopies and to construct 3D models of plants [35,36]. Simulations are mainly performed using various illumination models. Illumination models, also known as light–dark effect models, are mathematical models that replace real physical models of light irradiation on the surface of an object. The two main types of illumination models are local [37] and global [38]. Local illumination models can be subdivided into the Lambert, Gourand, and Phong illumination models [39,40]. Global illumination models can be subdivided into the ray-tracing, radiance, and photon mapping algorithms [41–43].

Among them, the most widely studied is the ray-tracing algorithm. Zhang studied the key technology of the apple-tree canopy illumination model and implemented a GPU-based method to calculate plant canopy light [44]. Hua studied tree canopy light distribution based on the 3D spatial structure and ray-tracing algorithm, scanned and reconstructed the 3D structure of poplars, and implemented the canopy light distribution algorithm based on the GPU [45].

In terms of studies on the effect of light on the canopy morphology of forest trees, Shanin et al. and Duchemin et al. quantitatively analyzed the effect of light on canopy morphology at the single-tree scale. Hu et al. qualitatively described light intensity using the words "forest gap", "full light", and "shade", and then qualitatively studied the effect of light on canopy morphology at the forest-stand scale. However, quantitative study of the mathematical relationship between light intensity and canopy width at the forest-stand scale are lacking. Regarding plant canopy light distribution, the above-mentioned researchers have tended to focus on the canopy light distribution of a single tree, ignoring the light distribution in the whole forest stand. No further studies have been conducted on the relationship between light distribution and canopy morphology.

As the main management tree species of Huangfengqiao Forest Farm, fir trees have a wide range of ecological, economic and medicinal values [46]. In this study, fir was selected as the research object. It is worth noting that the method in this paper also can be applied to other non-fir species. The purpose of our study was to put forward a new coupled canopy–light model (CCLM) based on light distribution data and canopy morphological data. The CCLM expresses the mathematical relationship between light intensity and canopy width and was used to improve the visual polymorphism simulation of fir morphology. The development of the CCLM involved four steps: (1) constructing a 3D scene in UE4 by forestry survey data, (2) obtaining the light distribution data of the fir tree canopy in a 3D scene with a ray-tracing algorithm, (3) building a coupled canopy-light model (CCLM) of fir and performing a visual simulation, and (4) comparing the CCLM with the method proposed by Ma et al. [47] and calculating the accuracy of the CCLM in visual polymorphism simulation of fir morphology.

2. Materials and Methods

2.1. Study Area

We selected the Huangfengqiao Forestry Farm in China as the study area, which is in a zonal distribution located at the east and west border of You County in Hunan Province, with a latitude and longitude range of 113°04′ to 113°43′ E and 26°43′ to 27°06′ N. The terrain is dominated by low and medium mountainous landscapes, with slopes ranging from 20 to 35 degrees and elevation from 115 and 1270 m. The climate is subtropical monsoonal, with an average annual temperature of 17.8 °C, an annual average precipitation of 1410.8 mm, a frost-free period of 292 days, and an average annual sunshine time of 1612 h. The forest management area is 152,438 acres, with 891,262 cubic meters of forest wood storage. The forest is rich in trees and species, with more than 430 species of woody plants, including more than 10 species of rare and endangered trees under national protection, such as *Taxus chinensis* var. *mairei*, ginkgo, flame tree, and tulip tree [48] (Figure 1).

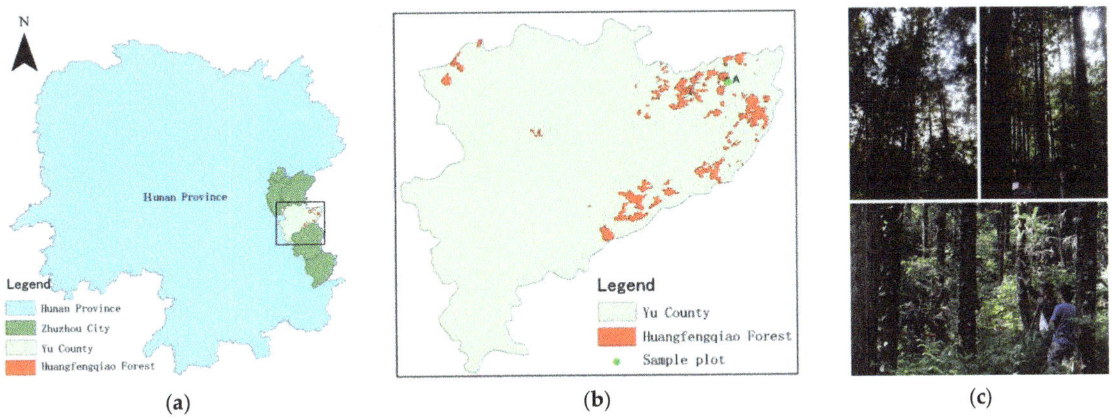

Figure 1. Description of study area: (**a**) Location of study area in Hunan Province. (**b**) Enlargement of inset in (**a**) showing details of the Huangfengqiao Forest and the sample area localization in Yu Country. (**c**) Example of the vegetation structure with forest canopy aspects.

2.2. Data Sampling

Considering factors such as site conditions, altitude, slope, aspect, difficulty of data collection, etc., we selected a typical sample plot with an area of 40 m by 40 m and a total of 230 fir trees in the Huangfengqiao Forestry Farm. We used a UAV orthophoto to obtain the spatial topographic elevation of the sample plot. An electronic total station was used to obtain the relative coordinates (X, Y) of the roots of the fir trees to represent the distribution of the fir trees in the sample plot (Figure 2). The fir trees in the sample plot were surveyed for three consecutive years from 2015 to 2017 with conventional field surveys; leveling poles and laser altimeters were used to measure tree crown morphology data, and we recorded the tree height, canopy width, height at maximum canopy width, and under-living branch height in four quadrants (as shown in Table 1). The canopy width was calculated in one-meter intervals for the corresponding height in the four quadrants. We collected the global long-series high-resolution photosynthetically active radiation (PAR) data from the National Tibetan Plateau Science Data Center [49].

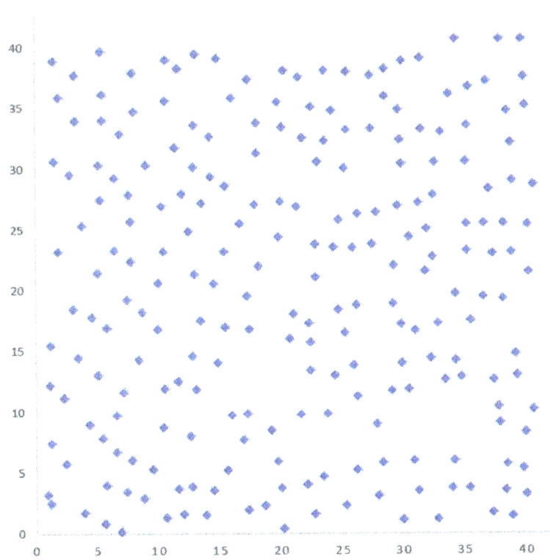

Figure 2. Spatial distribution of fir trees in sample plot. The blue points are the coordinate points representing the relative position of the fir.

Table 1. Basic information of fir trees in sample site.

Year	Item	Tree Height (m)	Diameter at Breast Height (cm)	Canopy Width (m)	Under-Living Branch Height (m)
2015	Maximum	18	28.6	7.75	9.3
	Minimum	9.1	8	1.05	1.2
	Average	11.09	16.81	2.83	5.24
2016	Maximum	19.1	31.3	8	10.2
	Minimum	9.5	8.1	1.4	1.8
	Average	12.15	17.30	3.27	6.06
2017	Maximum	19.88	31.96	9.15	12
	Minimum	9.9	8.4	1.86	1.92
	Average	12.64	17.60	3.51	6.64

2.3. Canopy Shape Description and Light Parameters

The commonly used canopy shape descriptors include under-living branch height, height at maximum canopy width, canopy width, and so on. When using these indicators to identify similarities in canopy shapes, the accuracy is insufficient. We therefore addressed this issue by measuring the width of the canopy at one-meter intervals to obtain a more detailed description of the morphological canopy characteristics. When we compared the two groups of canopy morphology models, the canopy of each tree in the first group of models was recorded at one-meter intervals in the four quadrants to obtain a set of canopy widths x_n. The canopy widths of the corresponding trees, as well as their directions and heights, were recorded for the second group of models to acquire a set of canopy widths y_n (Figure 3). We compared the two sets of data x_n and y_n using the Euclidean distance to identify the similarity between the two groups in terms of the canopy shapes. The smaller the Euclidean distance, the more similar the canopy shapes between the two groups were.

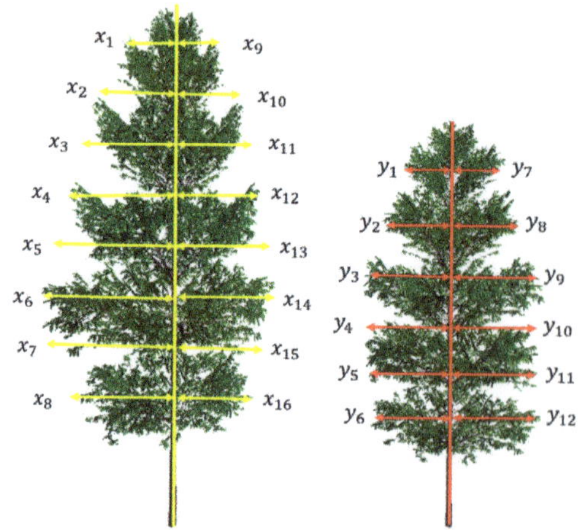

Figure 3. Description of canopy shape and evaluation of fit. The x and y represent the distance from the canopy edge to the trunk; the arrows with two directions are used to identify this distance.

There are two common methods of measuring light intensity: the radiant illuminance, which represents the radiant flux per unit area in watts per square meter (W/m^2), indicating the amount of energy received by the surface of the object; and the light intensity, which indicates the light flux per unit area in lux, indicating the amount of illumination of the surface of the object. Light intensity is the human visual perception of light, indicating light and darkness, whereas the perception of light by plants can be measured by radiation illumination. Solar radiation is not always effective for plant photosynthesis; the component of solar radiation that is effective for plant photosynthesis is called photosynthetically active radiation (PAR), which has a wavelength range of 380 to 710 nm. The annual average PAR is used as the light parameter in the ray-tracing algorithm. Because the light condition of a canopy is affected by the angle of solar irradiation, the angle of solar irradiation and PAR vary at different times. In a 3D scene, the changes in solar orientation and PAR are simplified by dividing them into seven moments (6:00, 8:00, 10:00, 12:00, 14:00, 16:00, and 18:00), and the sum of PAR at these seven moments at the canopy measurement point is calculated to represent the PAR at that point.

2.4. Global Illumination Algorithm Based on Ray Tracing

The ray-tracing algorithm is a global illumination algorithm that can accurately simulate global illumination. Unlike local illumination models, which only consider a single reflection or refraction from a light source to the human eye or camera, global illumination models account for the effect of other nonlight surfaces on the result. The ray-tracing algorithm simulates light in reality and tracks the path of light propagation. In a scene, after the light and object collide, a secondary light is generated due to reflection and refraction, and the secondary light enters the scene for collision testing until the number of iterations reaches the expected value, or the light is finally shot into the camera. Considering the probability of the light eventually entering the camera, combined with the principle of reversibility of the optical path, light is emitted from the camera to the light source and then backtracked to calculate the loss of energy, which is finally colored on a light screen. This operation is repeated for each pixel to obtain a 2D representation of a 3D scene. In this study, we used real survey data to construct a virtual forest stand environment, and we then used a ray-tracing algorithm to calculate the light distribution of each part of the canopy to provide light distribution data for the construction of the CCLM (Figure 4).

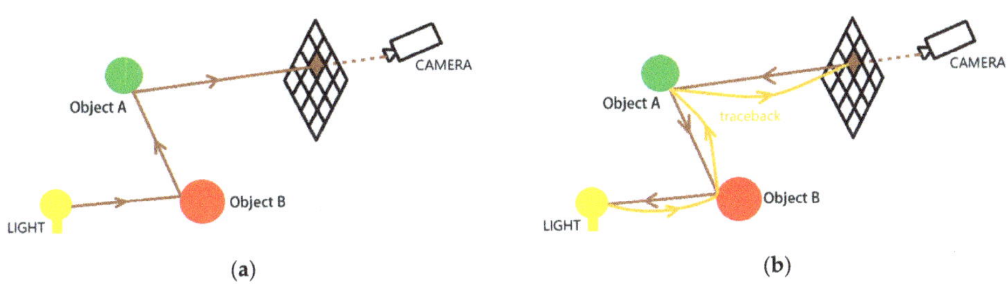

Figure 4. Schematic diagram of ray-tracing algorithm: (**a**) light starts from light source and eventually hits a camera; (**b**) light shines out of the camera and backtracks.

2.5. Construction of Coupled Canopy–Light Model (CCLM)

The Coupled canopy–light model is shown in Figure 5. From the field surveys, we obtained base canopy morphological data (Base_CMDD), expressed as a set (i, d, h, w), which indicates the width of the canopy for the corresponding tree (number i), direction (direction d), and height (height h) as w. We used points (h, w) as the type value points (data points describing the geometry of the curve) Q_i of the cubic uniform B-spline curve [50]. The control points P_i were solved by the type value points. The system of equations to solve for the control points is shown in Equation (1).

$$\begin{bmatrix} 0.5 & 1 & & & & 0 \\ 1 & 4 & 1 & & & \\ & 1 & 4 & 1 & & \\ & & \cdots & & & \\ & & & 1 & 4 & 1 \\ 0 & & & & 1 & -1 \end{bmatrix} \begin{bmatrix} P_1 \\ P_2 \\ P_3 \\ \cdots \\ P_n \\ P_{n+1} \end{bmatrix} = 6 \begin{bmatrix} (3Q_1 - Q_1')/12 \\ Q_1 \\ Q_2 \\ \cdots \\ Q_{n-1} \\ (Q_{n-1} - Q_n)/6 \end{bmatrix} \quad (1)$$

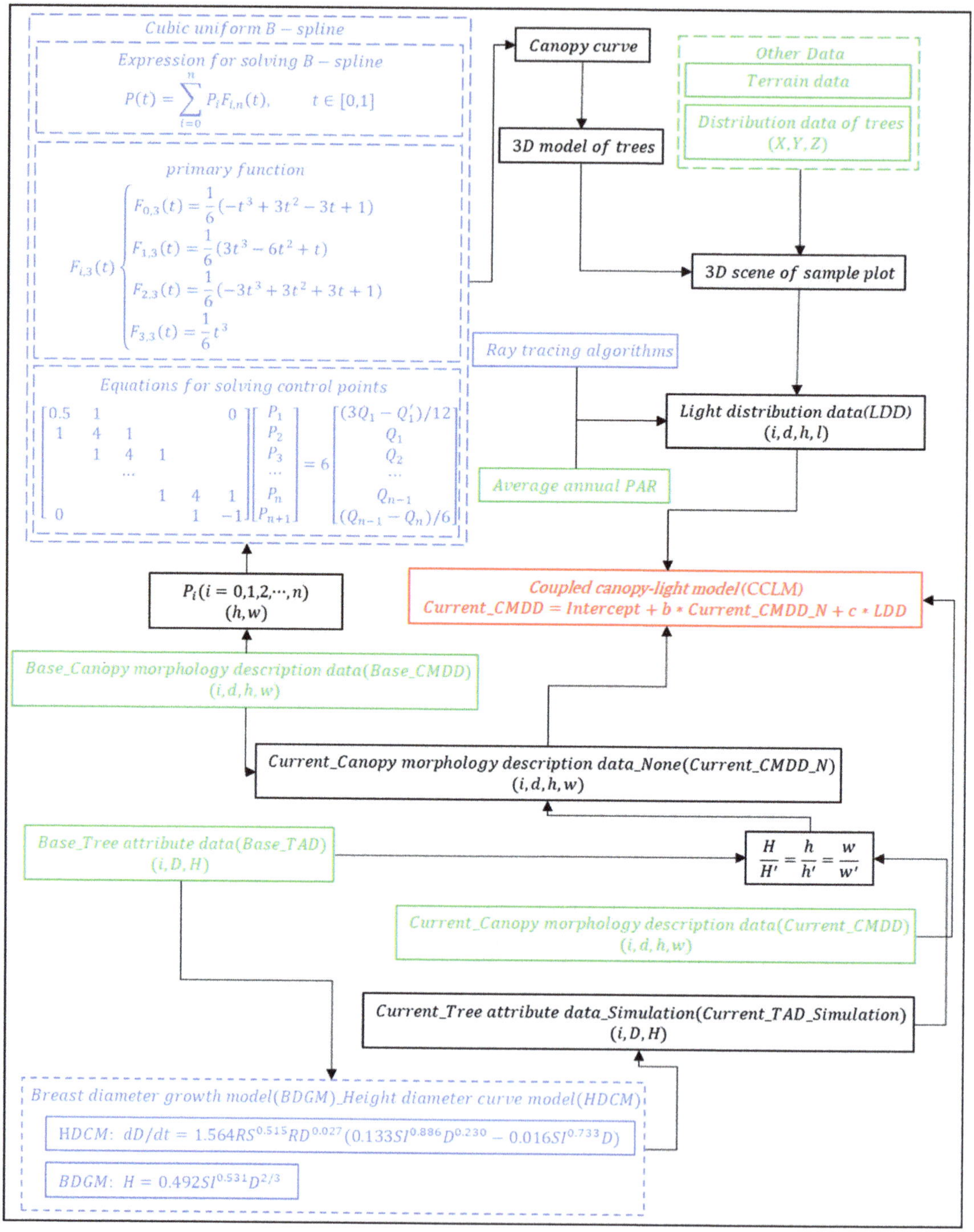

Figure 5. Coupled canopy–light model (CCLM).

The control points P_i were counted based on the system of equations in (1), and the B-sample curve of the canopy curve was calculated using P_i with the expression:

$$P(t) = \sum_{i=0}^{n} P_i F_{i,n}(t), t \in [0,1] \tag{2}$$

For the cubic uniform B-spline curve, the basis function is $F_{i,3}(t)$:

$$\begin{cases} F_{0,3}(t) = \frac{1}{6}(-t^3 + 3t^2 - 3t + 1) \\ F_{1,3}(t) = \frac{1}{6}(3t^3 - 6t^2 + t) \\ F_{2,3}(t) = \frac{1}{6}(-3t^3 + 3t^2 + 3t + 1) \\ F_{3,3}(t) = \frac{1}{6}t^3 \end{cases} \tag{3}$$

The canopy curves of each tree in the four quadrants could be obtained, and the 3D model of the tree was constructed from the canopy curves in each quadrant, which we combined with the topographic data and the distribution information of the fir trees to construct the 3D scene of the sample plot. The light distribution data (LDD), expressed as (i, d, h, l), was calculated by combining the ray-tracing algorithm with the annual average PAR as the light parameter in the 3D scene.

The base period tree attribute data (Base_TAD) data are expressed as (i, D, H), where i represents the tree number, D represents the diameter at breast height, and H represents the tree height. We obtained the current tree attribute data, also expressed as (i, D, H), from a simulation (Current_TAD_Simulation) using the diameter in the breast- diameter growth model (BDGM) and the height-diameter curve model (HDCM). The BDGM and HDCM were based on a previous study [51], where *RS* is the relative plant distance, *RD* is the relative dominance, *D* is the diameter at breast height, and *SI* is the status index, reflecting the combination of topography, soil, and climate.

$$HDCM : dD/dt = 1.564RS^{0.515}RD^{0.027}\left(0.133SI^{0.886}D^{0.230} - 0.016SI^{0.733}D\right) \tag{4}$$

$$BDGM : H = 0.492SI^{0.531}D^{2/3} \tag{5}$$

A quantitative relationship exists:

$$\frac{H}{H'} = \frac{h}{h'} = \frac{w}{w'} \tag{6}$$

from which the current canopy morphological data without the effect of light (Current_CMDD_N) is obtained, expressed as (i, d, h, w), where H denotes H in Base_TAD; H′ denotes H in Current_TAD_Simulation; and h and w denote h and w in Base_CMDD, respectively; h′ and w′ denote h and w in Current_CMDD_N, respectively.

The current canopy morphological data (Current_CMDD) were obtained through field surveys, and the CCLM was finally constructed using statistical methods, expressed as the equation of Current_CMDD (i, d, h, w), Current_CMDD_N (i, d, h, w), and LDD (i, d, h, l).

$$Current\_CMDD = Intercept + b * Current\_CMDD\_N + c * LDD \tag{7}$$

2.6. Contrast Validation Model

After we constructed the CCLM, we incorporated the LDD and Current_CMDD_N into the CCLM to obtain the effects of light on the current canopy morphology, which is the current canopy morphological data obtained from the simulation using the CCLM (Current_CMDD_CCLM).

We then combined the data on under-living branch height (H_b), height at maximum canopy width (H_c), canopy width (C), age (Age), and height of trees (H) from Base_TAD with the growth model based on spatial structure [33] (GMBOSS) to obtain the current tree attribute data based on spatial structure (Base_TAD_SS).

We constructed a canopy curve using the under-living branch height, height at maximum canopy width, canopy width, and height of trees in Base_TAD_SS as the type value points of the B-spline curve, and the current canopy morphological data based on the spatial structure (Current_CMDD_SS) was obtained from the canopy shape curve.

The GMBOSS is expressed as Equations (8)–(10), where P_v denotes the vertical spatial structure parameter, and P_h denotes the horizontal spatial structure parameters.

$$H_b = -3.020 + 0.409 * Age + 2.601 * P_v \quad (8)$$

$$H_c = -3.036 + 0.469 * Age + 2.690 * P_v \quad (9)$$

$$C = 0.660 + 0.037 * Age + 0.160 * P_h \quad (10)$$

Current_CMDD, Current_CMDD_CCLM, and Current_CMDD_SS are all canopy morphological data, which can be represented by a set (i, d, h, w). The distance is calculated with Equation (10), where x_i and y_i are the canopy widths at the corresponding locations, and n denotes the number of canopy description data.

$$D = \sqrt{\sum_{i=1}^{n}(x_i - y_i)^2} \quad (11)$$

We calculated the Euclidean distance D_CCLM between Current_CMDD and Current_CMDD_CCLM and the Euclidean distance D_SS between Current_CMDD and Current_CMDD_SS, and we compared the magnitudes of D_CCLM and D_SS to determine the degree of superiority of the two fitting effects. The contrast validation model is shown in Figure 6.

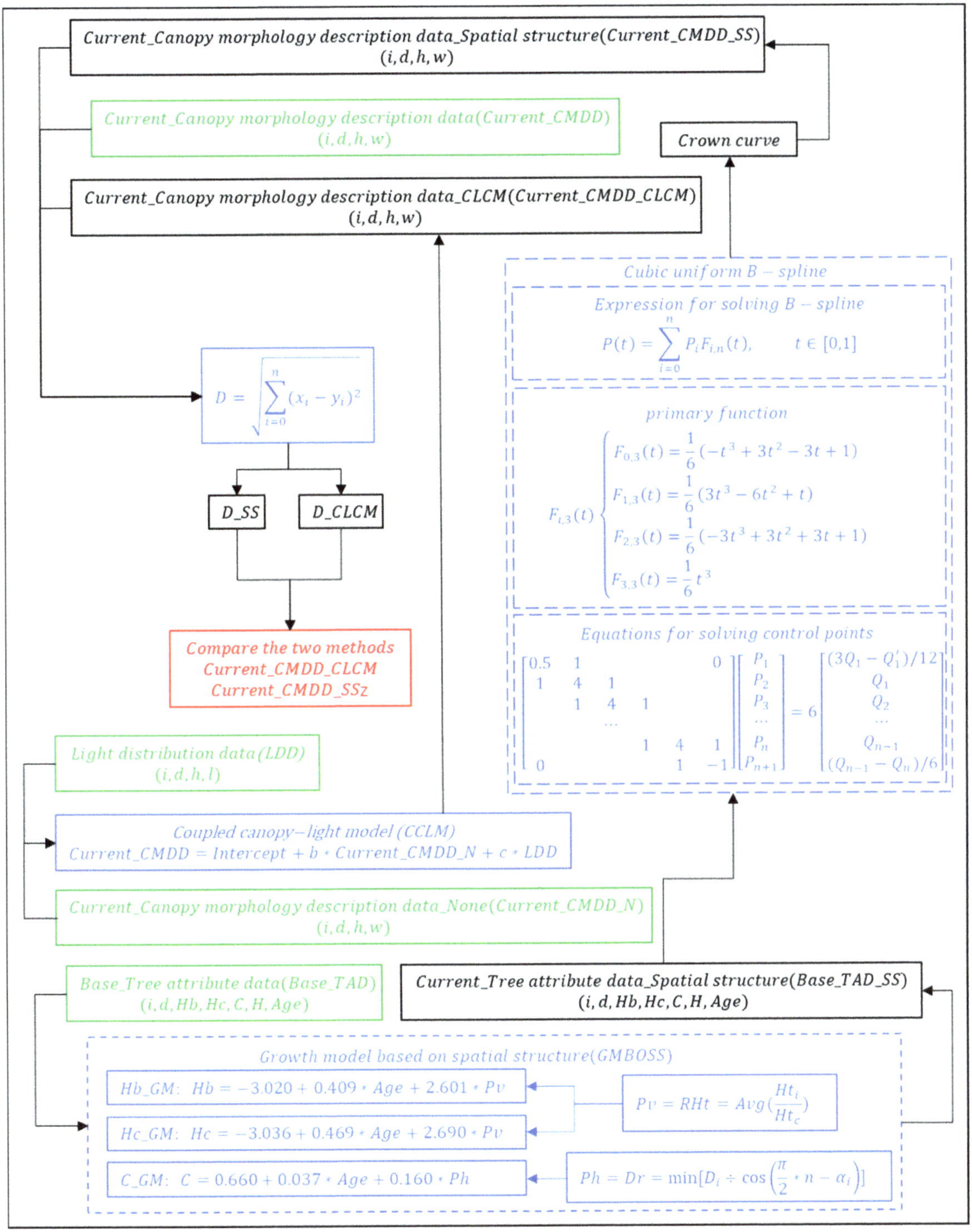

Figure 6. Contrast validation model.

3. Results

3.1. Calculation of PAR for Fir Tree Canopies Based on Ray-Tracing Algorithm

The annual average PAR in You County, Hunan Province, was 119.13 and 121.62 W/m^2 in 2015 and 2016, respectively. The daily variation in PAR showed a single-peak curve, which gradually increased from morning and reached the daily maximum around noon, and then began to gradually decrease. We combined the annual average PAR and the daily variation curve of the PAR, and we estimated the annual average values of the seven moments. The PAR intensity at each measurement point was obtained with the ray-tracing algorithm to calculate the summation to represent the PAR intensity at that point. A measurement point was taken at one-meter intervals for each fir tree in the four directions of the crown shape, and the PAR intensity (LDD) of the measurement point was calculated for a total of two years. The specific results are shown in Table 2. The maximum value of PAR at each point roughly positively correlated with the annual average PAR of the year. Due to the presence of forest stand closure, some points that receive little sunlight are always present. The average value of PAR at each point showed a decreasing trend, with little difference in the annual average PAR between the two years. Perhaps with the growth of the forest stand, the forest stand gradually becomes closed and light is blocked at more points in the forest stand, causing a decrease in the mean value.

Table 2. Photosynthetically active radiation (PAR) at measurement points in 2015 and 2016.

Project	Maximum	Minimum	Average	Standard Deviation
LDD15	674.48	6.23	338.26	113.73
LDD16	683.23	5.75	330.75	125.34

3.2. Construction and Validation of CCLM for Fir

3.2.1. Model Fitting

We used the 2015 survey data as Base_CMDD and the 2016 survey data as Current_CMDD; at present, the LDD is LDD15. All these data were used as the basis to obtain Current_CMDD_N with the BDGM and HDCM combined. The specific results are shown in Table 3: the maximum, minimum, mean, and standard deviation of Current_CMDD were 3.60, 0.20, 1.97, and 0.84, respectively. The maximum, minimum, mean, and standard deviation of Current_CMDD_N were 3.72, 0.15, 1.89, and 0.79, respectively.

Table 3. Canopy morphological data from field surveys and from BDGM and HDCM simulations in 2016.

Project	Maximum	Minimum	Average	Standard Deviation
Current_CMDD	3.60	0.20	1.97	0.84
Current_CMDD_N	3.72	0.15	1.89	0.79

Current_CMDD describes the fir canopy morphological data from field surveys in 2016. We used Current_CMDD_N for the fir canopy morphological data from BDGM and HDCM simulations in 2016, which could indirectly represent the age factor. Each measurement point had a corresponding height, but the height was not used as a model parameter because the canopy width of the same height could be different for different fir trees. We used the multiple stepwise regression method, and Current_CMDD, Current_CMDD_N, and LDD were analyzed in SPSS. The results are shown in Table 4.

Table 4. Results of CCLM fitting.

Independent Variable	Model	Variable by Stepwise Regression	R^2	Rc^2	SEE
Current_CMDD	1	Current_CMDD_N	0.813	0.806	0.3823
	2	Current_CMDD_N, LDD	0.829	0.821	0.3491

With the introduction of Current_CMDD_N and LDD as the independent variables to the model, the fit of the model increased to some extent, with the coefficient of determination (R^2) increasing from 0.813 to 0.829, the corrected coefficient of determination (Rc^2) increasing from 0.806 to 0.821, and the standard estimation error decreasing from 0.3823 to 0.3491. The results indicated that Current_ CMDD_N and LDD had a stronger influence on the actual canopy width at the measurement point. The model with a better fit was selected as the final model, with a larger Rc^2 and a smaller standard estimation error (SEE).

The specific model parameters are shown in Table 5. For the CCLM, the *p*-values for the intercept, Current_CMDD_N, and LDD were 0.34, 0.018, and 0.025, respectively. The *p*-values of the three variables entering the model were all less than 0.05, showing significant differences. The final analysis determined that the fir CCLM was

$$Current\_CMDD = -40.418 + 0.927 * Current\_CMDD\_N + 0.012 * LDD \quad (12)$$

Table 5. Statistics of the CCLM.

Independent Variable	Variable	Parameter	Standard Deviation	*t*-Test	*p*-Value
Current_CMDD	Intercept	−40.418	18.391	−29.194	0.034
	Current_CMDD_N	0.927	0.327	4.294	0.018
	LDD	0.012	0.001	13.623	0.025

3.2.2. Model Testing

As described in the above sections, based on the data from field surveys in 2015 and 2016, Current_CMDD, Current_CMDD_N, and LDD were fitted to obtain the fir CCLM. We then used the data from the field surveys in 2016 and 2017 to validate the model. We used the 2016 survey data as Base_CMDD and the 2017 survey data as Current_CMDD, at which time the LDD was LDD16. The Current_CMDD_N could be obtained by using the BDGM and HDCM. The LDD and Current_CMDD_N were incorporated into the fir CCLM to obtain the Current_CMDD_CCLM. Using the GMBOSS, we obtained the Current_CMDD_SS. We performed the t-test on the results, and the results are shown in Table 6. The table shows that the results of both the Current_CMDD_CCLM and Current_CMDD_SS were larger than that of the Current_CMDD, but the result of Current_CMDD_SS was larger. The *p*-values were all greater than 0.05, indicating a lack of significant difference between Current_CMDD_SS and the measured value, with no significant difference between Current_CMDD_CCLM and the measured value.

Table 6. Results of t-test of CCLM and GMBOSS simulations.

Independent Variable	Variable	Mean Value	Standard Deviation	*t*-Test	*p*-Value
CMDD	Current_CMDD	0.823	0.318		
	Current_CMDD_SS	0.857	0.392	1.846	0.821
	Current_CMDD_CCLM	0.843	0.334	1.623	0.835

3.2.3. Model Comparison

Current_CMDD_CCLM, Current_CMDD_SS, and Current_CMDD are canopy morphological descriptions, all of which are a series of sets of canopy widths at measurement points from corresponding fir trees, directions, and heights. The 3D spatial coordinates of the measurement points were obtained by the coordinates of the fir roots (X, Y, Z), the directions of the measurement points, and the canopy widths of the measurement points, which we represented in the 3D space (Figure 7). The Euclidean distance D_SS between Current_CMDD_SS and Current_CMDD was 23.944, and the D_CCLM between Current_CMDD_CCLM and Current_CMDD was 15.561. The smaller the distance, the more accurate the simulation. This showed that the simulation using the CCLM is better than that using the GMBOSS at the forest-stand scale.

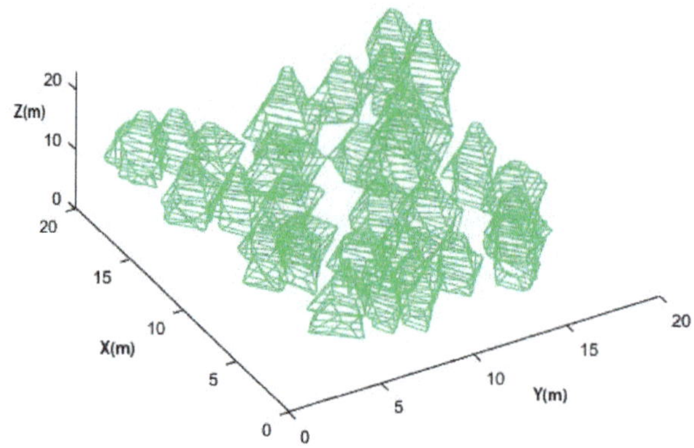

(a)

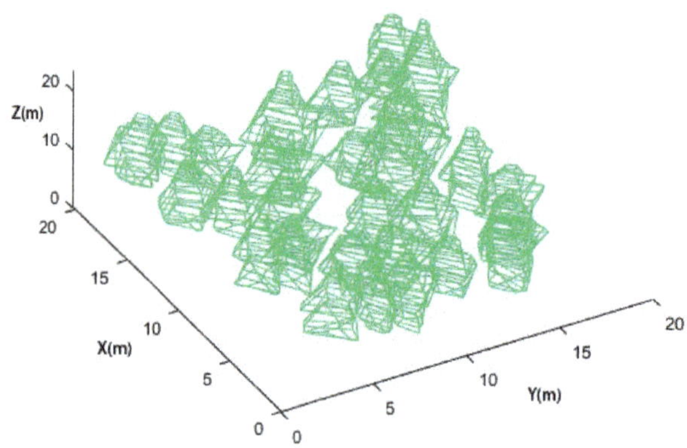

(b)

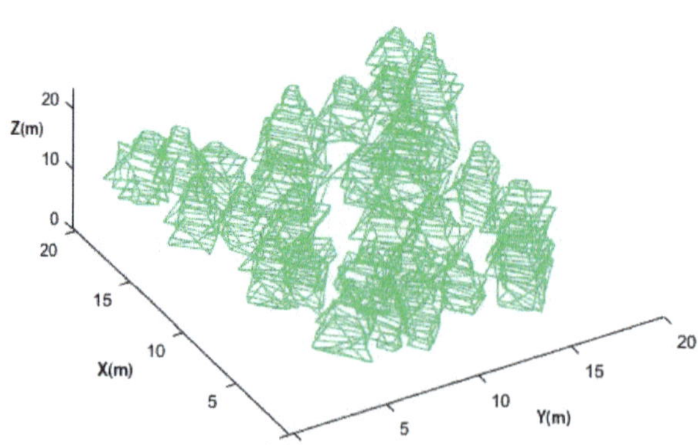

(c)

Figure 7. Schematic diagrams of canopy morphological data in 3D space: (**a**) Current_CMDD; (**b**) Current_CMDD_CCLM; and (**c**) Current_CMDD_SS.

We attempted to identify differences in the canopy width (w) between Current_CMDD_CCLM and Current_CMDD (Figure 8a) and between Current_CMDD_SS and Current_CMDD (Figure 8b). The simulation effect shown in Figure 8a is better than that in Figure 8b, demonstrating that the simulation using the CCLM is better than that using the GMBOSS.

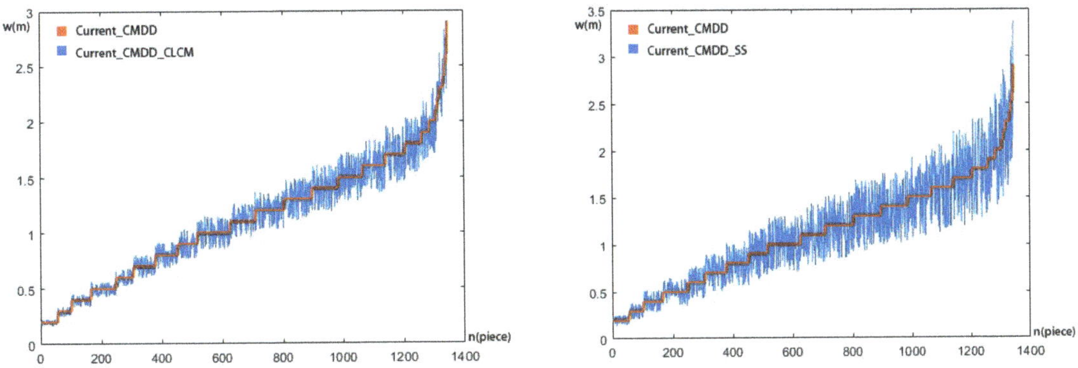

(a) (b)

Figure 8. Comparison of canopy width in canopy morphological data: (**a**) Current_CMDD and Current_CMDD_CCLM; (**b**) Current_CMDD and Current_CMDD_SS.

The height of the fir trees in the sample plot ranged from 9.1 to 18.5 m. We randomly selected one fir tree each with a height of 10 m, 12 m, 14 m, 16 m, and 18 m. The canopy curves of the fir trees were plotted according to Current_CMDD_CCLM, Current_CMDD_SS, and Current_CMDD in two quadrants (east–west and north–south), as shown in Figure 9. And according to the canopy curves, three-dimensional models of the fir trees is constructed (Figure 10). Because the tree heights in Current_CMDD_CCLM and Current_CMDD_SS were both obtained from BDGM and HDCM simulations, the tree heights obtained from the simulations using both methods were the same. The tree heights obtained from the simulations were slightly higher than the measured tree heights, except for the fir tree with a height of 10 m.

As for the under-living branch height, height at maximum canopy width, and canopy width, the values of under-living branch height of the 10 m fir tree in the west and north, the height at the maximum canopy width of the 12 m fir tree in the west, and the height at maximum canopy width of the 18 m fir tree in the east obtained using GMBOSS simulations were closer to the measured values. For the rest of the values, the results obtained using CCLM simulations were closer to the measured values.

For the fit of the overall canopy curve, the GMBOSS simulation used the under-living branch height, height at maximum canopy width, canopy width, and height as the type value points of the canopy morphology and lacked control over other parts of the canopy morphological, so the fit between the canopy morphology obtained from the CCLM simulation and the actual canopy morphology was better than that between the GMBOSS simulation and actual canopy morphology. Therefore, at the single-tree scale, the results obtained using CCLM simulations were better than those obtained using GMBOSS simulations in the overall canopy curve or canopy morphological descriptors such as under-living branch height, height at maximum canopy width, and canopy width.

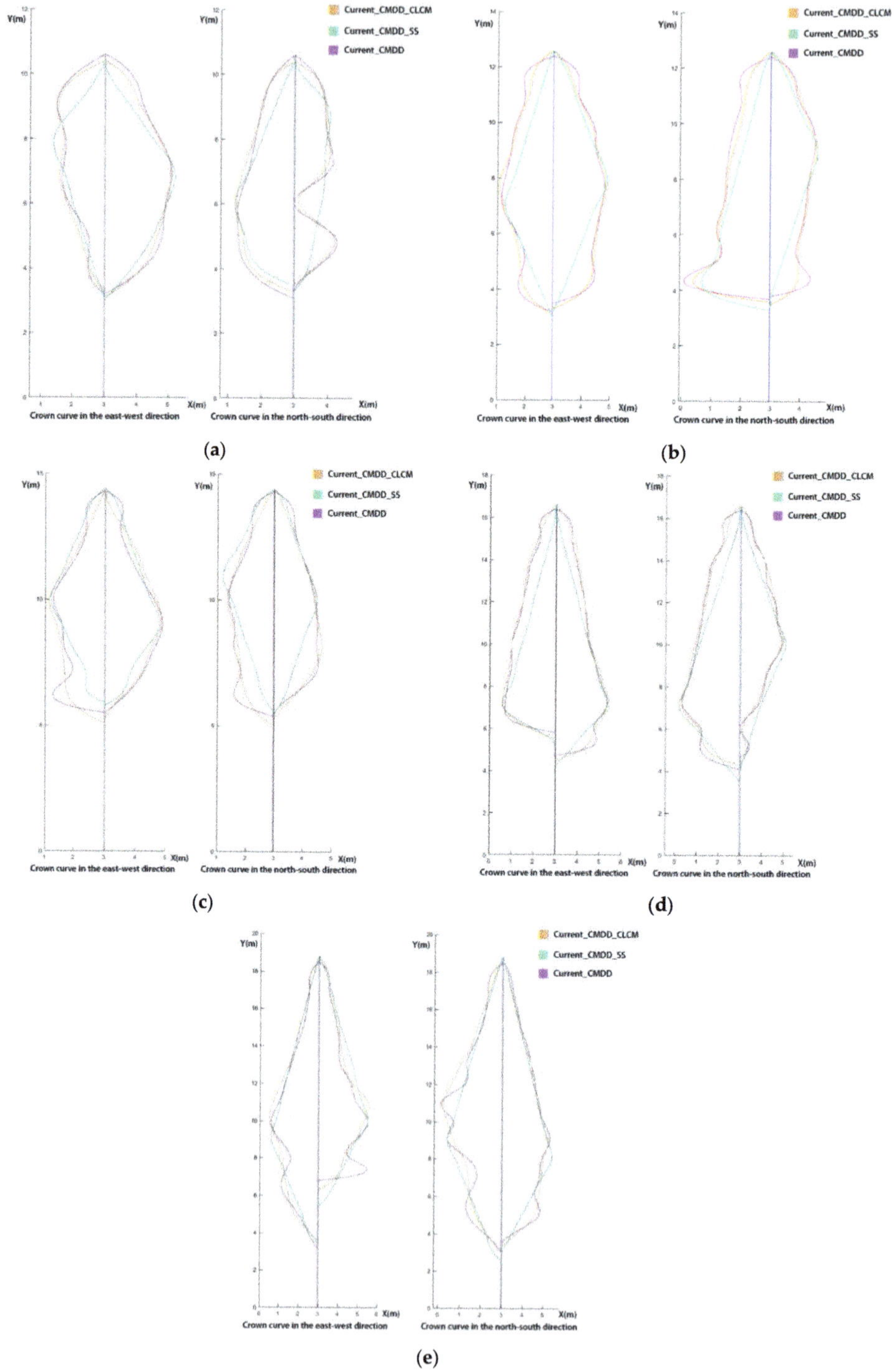

Figure 9. Identifying differences in fir canopy curves from different methods with a height of (**a**) 10 m; (**b**) 12 m; (**c**) 14 m; (**d**) 16 m; and (**e**) 18 m.

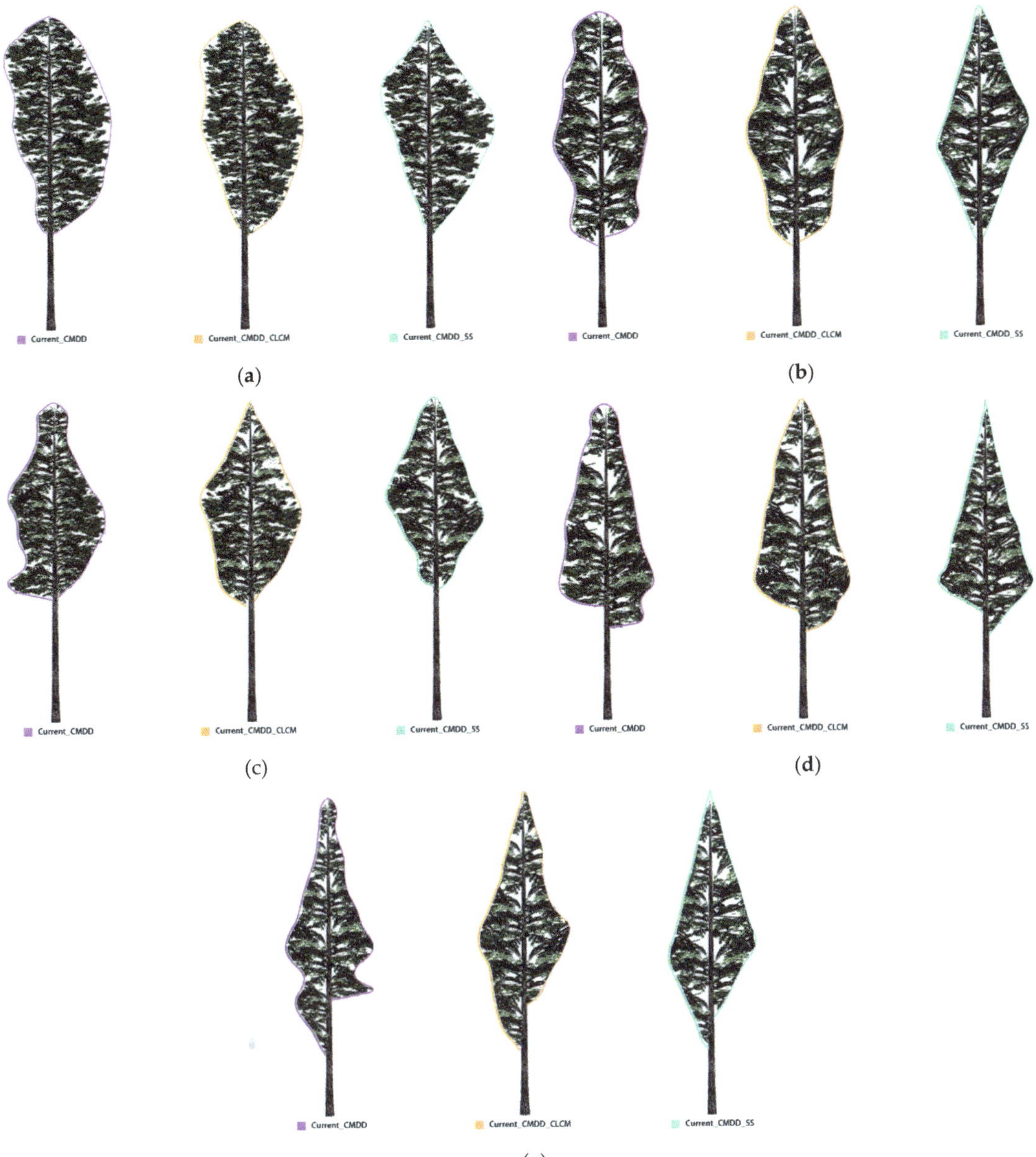

Figure 10. Identifying differences in 3D fir models from different methods with a height of (**a**) 10 m; (**b**) 12 m; (**c**) 14 m; (**d**) 16 m; and (**e**) 18 m.

4. Discussion

In the traditional forest growth simulation process based on the BDGM and HDCM, the changes in diameter at breast height and tree height are considered, whereas the changes in canopy morphology are not. However, the environment has an important influence on canopy morphology. Some researchers have considered the influence of spatial structure and visually simulated fir tree canopies based on spatial structure. One of the most important factors for plant growth is light, and the influence of spatial structure on plant canopy morphology largely occurs because spatial structure affects the light distribution in the plant canopy, which in turn affects the photosynthesis of the plant and thus the canopy morphology. In studies on the effect of light on plant canopy morphology, researchers have quantitatively examined the effect of light on canopy morphology at the scale of a single tree and qualitatively studied this at the scale of the forest stands by describing light intensity through words such as "stand gap", "full light", and "shade". However, quantitative research on the effect of light on the canopy morphology at the forest-stand scale is lacking. In this study, based on the BDGM and HDCM, we introduced a ray-tracing algorithm from the computer field to calculate the light distribution in each part of the canopy in a virtual stand and combined this with field survey data, which simplified the acquisition of canopy light distribution data in a forest stand. The light distribution data were then introduced into the BDGM and HDCM to construct the CCLM, which quantitatively evaluated the effect of light on the canopy morphology at the forest-stand scale.

The canopy width in Huangfengqiao Forestry Farm was measured at one-meter intervals in four quadrants to describe the canopy morphology; in addition, we measured the under-living branch height, height at maximum canopy width, and canopy width of the fir trees in the four quadrants in the sample plot. A 3D scene was constructed with the actual data, and a ray-tracing algorithm was used in the 3D scene to calculate the light distribution data in each part of the tree canopy. Using the stepwise regression method, the relationships between canopy morphological data and age (the canopy morphological data obtained from the BDGM and HDCM were used to represent age) and light distribution data were analyzed. A CCLM of the fir tree was constructed with a coefficient of determination (R^2) of 0.829. The model is expressed in Equation (12). Current_CMDD was the current canopy morphological data obtained from the field surveys, Current_CMDD_N was the canopy morphological data obtained from the BDGM and HDCM simulations without adding any environmental factors, and LDD was the light distribution data. The value of the intercept was −40.418 ($p = 0.034$). The value of the Current_CMDD_N was 0.927 ($p = 0.018$). The value of the LDD was 0.012 ($p = 0.025$).

The canopy data from the field surveys, CCLM simulation, and GMBOSS simulation were compared to evaluate the fit of the canopy simulation using the CCLM and GMBOSS. For the forest-stand scale, the Euclidean distance between the overall canopy data was calculated to evaluate the fit with actual canopy morphology. The results indicated a divergence of 23.944 in the Euclidean distance between the canopy data from the field surveys and that from the GMBOSS simulation. The Euclidean distance between the canopy data from the field surveys and that from the CCLM simulation was 15.561. This showed that the simulation results using the CCLM were 8.383 times more accurate than those obtained using the GMBOSS. The fitting effect of the CCLM was better than that of the GMBOSS at the forest-stand scale.

For the single-tree scale, we randomly selected one 10 m high fir tree, one 12 m high fir tree, one 14 m high fir tree, one 16 m high fir tree, and one 18 m high fir tree, according to the height range of fir trees. The canopy curves of the fir trees were separately plotted based on the canopy data from the field surveys, CCLM simulation, and GMBOSS simulation to compare and analyze the fitting effect of the canopy curves of single trees of different heights. Because the tree heights obtained from the CCLM and GMBOSS simulations were simulated by the BDGM and HDCM, the height obtained from the simulations using the two methods were the same. As for the under-living branch height, height at maximum canopy width, and canopy width, the values of under-living branch height of the 10 m high fir tree in the west and north, the height at the maximum canopy width of the 12 m high fir tree in the west, and the height at maximum canopy width of the 18 m high fir tree in the east obtained using GMBOSS simulations were closer to the measured values. For the rest of the values, the results obtained using the CCLM simulation were closer to the measured values. For the fit of the canopy curve of a single tree, the GMBOSS simulation used under-living branch height, height at maximum canopy width, canopy width, and height as the type value points of the canopy morphology, which lacked control over other parts of the canopy morphology, so the fit between the canopy morphological data from the CCLM simulation and the actual canopy morphological data was better than that between the canopy morphology from GMBOSS simulation and the actual canopy. Therefore, at the single-tree scale, the results obtained from the CCLM simulations were more accurate than those obtained from GMBOSS simulations in the overall canopy curve or canopy morphological descriptors such as under-living branch height, height at maximum canopy width, and canopy width.

This study is based on the BDGM and HDCM by introducing the factor of light to construct the CCLM, so the accuracy of the CCLM is increased based on the accuracy of the BDGM and HDCM. As such, the accuracy of the selected BDGM and HDCM strongly influenced on the accuracy of results. The light distribution data in this study were calculated by using the annual average PAR with the ray-tracing algorithm. Although the ray-tracing algorithm can accurately simulate realistic light situations, the study is not free from errors, which may also have had an impact on the experimental results. Other factors that may affect the canopy morphology include broken tips, pests, and diseases, thereby influencing the accuracy of the study results. Additionally, both tree growth and solar movement are continuously changing processes. In this study, the initial state of the stand was used as the basis, and the solar position at seven moments was selected to calculate the light distribution of the canopy. These factors deserve more attention in future research.

5. Conclusions

In this study, we considered the effect of light on the canopy shape and constructed a new coupled canopy–light model (CCLM) based on light distribution data and canopy morphological data, which was used to improve the visual polymorphism simulation of fir morphology. The model is expressed in Equation (12) with a coefficient of determination (R2) of 0.829. The value of the intercept was −40.418 (p = 0.034). The value of the Current_CMDD_N was 0.927 (p = 0.018). The value of the LDD was 0.012 (p = 0.025). For the forest-stand scale, the Euclidean distance between the canopy data from the field surveys and that from the CCLM simulation was 15.561. The CCLM was compared with the GMBOSS proposed by Ma et al., and the Euclidean distance between the canopy data from the field surveys and that from the GMBOSS simulation was 23.944. The results indicated that the simulation results using the CCLM were 8.383 times smaller than those obtained using the GMBOSS. This means that the CCLM was 35.0% more accurate than the GMBOSS in the visual polymorphism simulation of fir morphology, and the fitting effect of the CCLM was better than that of the GMBOSS at the forest-stand scale. The computer graphics technology was closely combined with constructing tree canopy shape and improved visual polymorphism simulation of fir morphology, which will enable us to provide more accurate tree canopy models for forestry research. However, to make visual polymorphism simulation more accurate, we will consider the impact of other environmental factors in the future.

Author Contributions: Conceptualization, H.Z. and Y.Z.; methodology, Y.Z.; software, Y.Z. and K.L.; formal analysis, Y.L., H.Q. and Z.C.; resources, X.H., J.Z. and J.W.; writing—original draft preparation, Y.Z.; writing—review and editing, Y.Z. and H.Z.; visualization, Y.Z.; supervision, H.Z. and T.Y. All authors have read and agreed to the published version of the manuscript.

Funding: This research was funded by the National Natural Science Foundation of China, grant number 32071681, 32271877.

Institutional Review Board Statement: Not applicable.

Informed Consent Statement: Not applicable.

Data Availability Statement: Not applicable.

Conflicts of Interest: The authors declare that there are no conflict of interest.

References

1. Soualiou, S.; Wang, Z.; Sun, W.; de Reffye, P.; Collins, B.; Louarn, G.; Song, Y. Functional-structural plant models mission in advancing crop science: Opportunities and prospects. *Front. Plant Sci.* **2021**, *12*, 747142. [CrossRef] [PubMed]
2. Sierra-de-Grado, R.; Moulia, B.; Fournier, M.; Ricardo, A.; Rafael, D.-B. Genetic control of stem form in Pinus pinaster Ait. seedlings exposed to lateral light. *Trees* **1997**, *11*, 455–461. [CrossRef]
3. Maciel, E.A.; Martins, V.F.; de Paula, M.D.; Huth, A.; Guilherme, F.A.G.; Fischer, R.; Giles, A.; Barbosa, R.I.; Cavassan, O.; Martins, F.R. Defaunation and changes in climate and fire frequency have synergistic effects on aboveground biomass loss in the brazilian savanna. *Ecol. Model.* **2021**, *454*, 109628. [CrossRef]
4. Goyal, A.; Szarzynska, B.; Fankhauser, C. Phototropism: At the crossroads of light-signaling pathways. *Trends Plant Sci.* **2013**, *18*, 393–401. [CrossRef]
5. Wang, Y.; Wei, X.L. Research progress on the effects of different light levels on plant growth, physiology, biochemistry and morphological structure. *Chin. J. Mt. Agric. Biol.* **2010**, *29*, 353–359.
6. Henke, M.; Buck-Sorlin, G. Using a full spectral raytracer for calculating light microclimate in functional-structural plant modelling. *Comput. Inform.* **2017**, *36*, 1492–1522. [CrossRef]
7. Mishra, P.; Lohumi, S.; Khan, H.A.; Nordon, A. Close-range hyperspectral imaging of whole plants for digital phenotyping: Recent applications and illumination correction approaches. *Comput. Electron. Agric.* **2020**, *178*, 105780. [CrossRef]
8. Harja, D.; Vincent, G.; Mulia, R.; van Noordwijk, M. Tree shape plasticity in relation to canopy exposure. *Trees* **2012**, *26*, 1275–1285. [CrossRef]
9. Guo, Y.; Fourcaud, T.; Jaeger, M.; Zhang, X.; Li, B. Plant growth and architectural modelling and its applications. *Ann. Bot.* **2011**, *107*, 723–727. [CrossRef]
10. Li, Z.W.; Huai, Y.J.; Fu, H. Virtual plant growth simulation based on illumination model. *J. Beijing For. Univ.* **2013**, *35*, 81–86.
11. Allen, W.A.; Richardson, A.J. Interaction of light with a plant canopy. *JOSA* **1968**, *58*, 1023–1028. [CrossRef]
12. Verhagen AM, W.; Wilson, J.H.; Britten, E.J. Plant production in relation to foliage illumination. *Ann. Bot.* **1963**, *27*, 627–640. [CrossRef]
13. Shanin, V.; Grabarnik, P.; Shashkov, M.; Ivanova, N.; Bykhovets, S.; Frolov, P.; Stamenov, M. Canopy asymmetry and niche segregation as an adaptation of trees to competition for light: Conclusions from simulation experiments in mixed boreal stands. *Math. Comput. For. Nat. Resour. Sci.* **2020**, *12*, 26.
14. Duchemin, L.; Eloy, C.; Badel, E.; Moulia, B. Tree canopies grow into self-similar shapes controlled by gravity and light sensing. *J. R. Soc. Interface* **2018**, *15*, 20170976. [CrossRef]
15. Hu, X.S.; Hong, W.; Wu, C.Z.; Hong, T.; Fan, H.L.; Song, P. Plasticity response of canopy structure in seedlings of Mucuna pruriens under different light conditions. *J. Plant Resour. Environ.* **2006**, *15*, 55–59.
16. He, B.H.; Zhong, Z.C. Effects of different light intensities and drought stress on the growth of Ginkgo biloba components. *J. Guangxi Norm. Univ. Nat. Sci. Ed.* **2005**, *23*, 66–69.
17. Xu, C.Y. Plastic response of canopy structure of young purple linden trees under different light conditions. *J. Appl. Ecol.* **2001**, *12*, 339–343.
18. Wang, Q.; Chen, J.L.; Sun, Z.Q. Application of LAI-2000 canopy analyzer in the study of light distribution characteristics of different plant populations. *Sci. Agric. Sin.* **2006**, *39*, 922–927.
19. Barthélémy, D.; Caraglio, Y. Plant architecture: A dynamic, multilevel and comprehensive approach to plant form, structure, and ontogeny. *Ann. Bot.* **2007**, *99*, 375–407. [CrossRef]
20. Luft, T.; Balzer, M. *Expressive Illumination of Foliage Based on Implicit Surfaces*; The Eurographics Association: Saarbrücken, Germany, 2006.
21. Svenning, J.C. Canopy illumination limits the population growth rate of a neotropical understorey palm (Geonoma macrostachys, Arecaceae). *Plant Ecol.* **2002**, *159*, 185–199. [CrossRef]
22. Gao, Y.; Gao, S.; Jia, L.; Dai, T.; Wei, X.; Duan, J.; Liu, S.; Weng, X. Canopy characteristics and light distribution in Sapindus mukorossi Gaertn. are influenced by crown architecture manipulation in the hilly terrain of Southeast China. *Sci. Hortic.* **2018**, *240*, 11–22. [CrossRef]
23. de Visser, P.H.B.; Buck-Sorlin, G.H.; van der Heijden, G.W.A.M. Optimizing illumination in the greenhouse using a 3D model of tomato and a ray tracer. *Front. Plant Sci.* **2014**, *5*, 48. [CrossRef] [PubMed]
24. Lewis, P. Three-dimensional plant modelling for remote sensing simulation studies using the Botanical Plant Modelling System. *Agronomie* **1999**, *19*, 185–210. [CrossRef]
25. Morita, K.; Yonemori, T.; Urashita, M.; Ogawa, Y.; Matoda, K.; Taharazako, S. Effects of intermittent illumination on the plant growth. *Mem. Fac. Agric. Kagoshima Univ.* **1996**, *32*, 79–91.
26. Burgess, A.J.; Retkute, R.; Herman, T.; Murchie, E.H. Exploring relationships between canopy architecture, light distribution, and photosynthesis in contrasting rice genotypes using 3D canopy reconstruction. *Front. Plant Sci.* **2017**, *8*, 734. [CrossRef]
27. Kull, O.; Broadmeadow, M.; Kruijt, B.; Meir, P. Light distribution and foliage structure in an oak canopy. *Trees* **1999**, *14*, 55–64. [CrossRef]
28. Cournède, P.H.; Mathieu, A.; Houllier, F.; Barthélémy, D.; de Reffye, P. Computing competition for light in the GREEN LAB model of plant growth: A contribution to the study of the effects of density on resource acquisition and architectural development. *Ann. Bot.* **2008**, *101*, 1207–1219. [CrossRef] [PubMed]

29. Vicent, M.J.; Rosell, V.; Vivó, R. A polar-plate-based method for natural illumination of plants and trees. *Comput. Graph.* **2005**, *29*, 203–208. [CrossRef]
30. LAZÁR, D. Chlorophyll a fluorescence rise induced by high light illumination of dark-adapted plant tissue studied by means of a model of photosystem II and considering photosystem II heterogeneity. *J. Theor. Biol.* **2003**, *220*, 469–503. [CrossRef]
31. Sassenrath-Cole, G.F. Dependence of canopy light distribution on leaf and canopy structure for two cotton (Gossypium) species. *Agric. For. Meteorol.* **1995**, *77*, 55–72. [CrossRef]
32. Yan, T.F.; Wang, H.T.; Geng, B.; Zhang, Y.T.; Shu, H.R. Canopy light energy distribution patterns and seasonal changes in 25-year-old 'Fuji' apple orchards. *Chin. Agric. Bull.* **2011**, *27*, 200–205.
33. Lao, C.L. A Three-Dimensional Plant Canopy Radiative Transfer Model Based on the Monte Carlo Ray-Tracing Method. Ph.D. Thesis, China Agricultural University, Beijing, China, 2005.
34. Wang, X.P. Construction and Validation of a Three-Dimensional Spatial Distribution Model of Photosynthetically Active Radiation in Maize Canopy. Ph.D. Thesis, China Agricultural University, Beijing, China, 2004.
35. Měch, R.; Prusinkiewicz, P. Visual models of plants interacting with their environment. In Proceedings of the 23rd Annual Conference on Computer Graphics and Interactive Techniques, New York, NY, USA, 1 August 1996; pp. 397–410.
36. Van Haevre, W.; Fiore, F.D.; Bekaert, P.; Van Reeth, F. A ray density estimation approach to take into account environment illumination in plant growth simulation. In Proceedings of the 20th Spring Conference on Computer Graphics, New York, NY, USA, 2–6 August 2004; pp. 121–131.
37. Akenine-Mo, T.; Haines, E.; Hoffman, N. *Real-Time Rendering*; A K Peters/CRC Press: Natick, MA, USA, 2018.
38. Whitted, T. An improved illumination model for shaded display. In Proceedings of the 6th Annual Conference on Computer Graphics and Interactive Techniques, Chicago, IL, USA, 8–10 August 1979; p. 14.
39. Whitted, T. Origins of global illumination. *IEEE Comput. Graph. Appl.* **2020**, *40*, 20–27. [CrossRef]
40. Phong, B.T. Illumination for computer generated pictures. *Commun. ACM* **1975**, *18*, 311–317. [CrossRef]
41. Cohen, M.F.; Wallace, J.R.; Hanrahan, P. *Radiosity and Realistic Image Synthesis*; Morgan Kaufmann: Burlington, MA, USA, 1993.
42. Heckbert, P.S. Adaptive radiosity textures for bidirectional ray tracing. In Proceedings of the 17th Annual Conference on Computer Graphics and Interactive Techniques, New York, NY, USA, 1 September 1990; pp. 145–154.
43. Wallace, J.R.; Elmquist, K.A.; Haines, E.A. A ray tracing algorithm for progressive radiosity. In Proceedings of the 16th Annual Conference on Computer Graphics and Interactive Techniques, New York, NY, USA, 1 July 1989; pp. 315–324.
44. Zhang, J. Research on Key Technologies for Canopy Light Modeling of Apple Trees. Master's Thesis, Northwest Agriculture and Forestry University, Xianyang, China, 2018.
45. Hua, J. Research on Tree Canopy Light Distribution Based on 3D Spatial Structure and Ray-Tracing Algorithm. Ph.D. Thesis, Institute of Automation, Chinese Academy of Sciences, Beijing, China, 2018.
46. Li, S.J.; Zhang, H.Q.; Li, Y.L.; Yang, T.; He, J.; Ma, Z.; Shen, K. Visual simulation of Chinese fir stand growth dynamics based on sample library. *For. Sci. Res.* **2019**, *32*, 21–30.
47. Ma, Z.Y.; Zhang, H.C.; Li, Y.L.; Yang, T.; Chen, Z.; Li, S. Visual simulation of fir tree canopy growth based on spatial structure. *For. Sci. Res.* **2018**, *31*, 150–157.
48. Hunan Forestry Bureau: Huangfengqiao State-Owned Forest Farm in Youxian County, Hunan Province. (27 December 2015). Available online: https://lyj.hunan.gov.cn/lyj/ztzl/gdzt/gylc/201512/t20151227_2587491.html (accessed on 11 November 2022).
49. Tang, W.-J. Global long series high resolution photosynthetically active radiation (PAR) (1984–2018). *Natl. Tibet. Plateau Sci. Data Center.* **2021**, *24*, 23–32. [CrossRef]
50. Wu, G.Y.; Wang, S.H. A fast algorithm for inverting the control vertices of cubic B spline curves. *J. Hangzhou Univ. Electron. Sci. Technol. Nat. Sci. Ed.* **2005**, *25*, 64–66.
51. Li, Y.L.; Zhang, H.Q. Research on Interactive Management Visualization Simulation Technology for Plantation Forest Stands. Ph.D. Thesis, China Academy of Forestry Science, Beijing, China, 2014.

forests

MDPI

Article

Visual Simulation Research on Growth Polymorphism of Chinese Fir Stand Based on Different Comprehensive Grade Models of Spatial Structure Parameters

Xingtao Hu [1,2,3,4], Huaiqing Zhang [1,3,4,*], Guangbin Yang [2], Hanqing Qiu [1,3,4], Kexin Lei [1,3,4], Tingdong Yang [1,3,4], Yang Liu [1,3,4], Yuanqing Zuo [1,3,4], Jiansen Wang [1,3,4] and Zeyu Cui [1,3,4]

1 Institute of Forest Resource Information Techniques, Chinese Academy of Forestry, Beijing 100091, China
2 School of Geography and Environmental Sciences, Guizhou Normal University, Guiyang 550025, China
3 Key Laboratory of Forest Management and Growth Modelling, NFGA, Beijing 100091, China
4 National Long Term Scientific Research Base of Huangfengqiao Forest Monitoring and Simulation in Hunan Province, Beijing 100091, China
* Correspondence: zhang@ifrit.ac.cn; Tel.: +86-136-5116-6672

Abstract: Since tree morphological structure is strongly influenced by internal genetic and external environmental factors, accurate simulation of individual morphological–structural changes in trees is the premise of forest management and 3D simulation. However, existing studies have few descriptions, and the research on the impact of growth environments and stand spatial structures on tree morphological structure and growth is still limited. In our study, we constructed a comprehensive grade model of spatial structure (CGMSS) to comprehensively evaluate individual tree growth states of the stands and grade them from 0 to 10 correspondingly. In addition, we developed a Chinese fir morphological structure growth model based on CGMSS, and dynamically simulate the growth variations of Chinese fir stands. The results showed that the overall stand prediction accuracy of CGMSS-based Chinese fir diameter at breast height, tree height, crown width and under-living branch height growth models was more than 94%. According to the analysis of the comprehensive grade of spatial structure (CGSS) of trees in the stand, except for the prediction accuracy and systematic error of the under-living branch height growth model at the CGSS = 3–5 levels, the systematic error of the Chinese fir growth model at each level was lower than 21.2%, and the prediction accuracy was greater than 73%. Compared with the spatial structural unit (SSU)-based Chinese fir growth model proposed by Ma et al., all growth models fit better at all levels, except for the CGMSS-based Chinese fir tree height and under-living branch height growth models that fit significantly lower than the SSU-based Chinese fir growth model at CGSS = 3–5 levels. In this study, the main conclusion is that the simulation results of CGMSS's Chinese fir morphological structure growth model are closer to the real growth state of trees, achieving accurate simulation of differential growth of trees in different growth dominance degrees and spatial structure states in forest stands, making visualized forest management more effective and realistic.

Keywords: tree polymorphism; forest spatial structure grade; growth model; visual simulation; *Cunninghamia lanceolata*

1. Introduction

Forest management practices aim to maintain or improve the forest ecosystem function, product, and service value by changing the spatial and temporal distribution of forest resources, intra- and inter-competition relationships and stand spatial distribution pattern [1]. Compared with traditional 2D descriptions, 3D visual simulation technology can directly reveal differences before and after forest management practices, improve the understanding of forest management benefits [2], and enrich information expression approaches [3]. Therefore, realistic individual tree growth and morphological structure simulation with high complexity and diversity [4] play a key role in the analysis of the relationship between tree growth, climate, and spatial environment [5], and is also the premise of the effective visualization of forest management.

However, forest structure and tree growth have complex and variable characteristics, which make it difficult to visualize the stand growth [6]. In the early stages, research into 3D visual simulation of stands was mainly focused on the plant function–structure [7–9] under a given environment, combined with the algorithm of plant growth rules or fractal geometry [10–12] to simulate and visualize the individual growth process. It can effectively improve the realism of the tree growth process and reflect tree morphological structure in a virtual environment using a refined individual tree growth model [9]. However, the visual simulation of stand growth based on the individual tree model has the characteristics of forest morphological structure similarity and growth environment ideality [13]. Trees are usually faced with complex and diverse environments [14] and different spatial structure of stands during their growth [15]. Therefore, existing research is mainly focused on site conditions, climate factors and combining the biochemical–physics processes to establish a tree growth visual model from the perspective of considering environmental factors, such as using different organ biomass [16–18], and achieve the interactive process between tree growth and environmental factors. However, existing studies lack interactive feedback among the target tree and neighboring trees [17], but forest spatial structure parameters based on the relationship between the target tree and neighboring trees play a key role in the quantitative description of the forest state [15]. Therefore, some scholars have realized a visual simulation of forest stands by optimizing forest stand spatial structure parameters [2,19,20], incorporating one or more independent spatial structure parameters to construct tree growth models [21–24], etc., or using the spatial structure parameters as the constraints on tree growth, combined with the tree shape growth model [25–27]. These studies successfully simulated the growth process of tree morphology structure from different perspectives, which could provide effective technical references for visualizing the process of forest management practices [28].

However, there is a lack of comprehensive analysis of tree morphological structure growth changes under different growth potential and spatial structure states in existing visual simulation studies. The growth models for the visual simulation of forest stands are mostly focused on the plot–stand growth model, or based on the single-tree model to realize the visual growth simulation of forest stands [6], resulting in similar growth of trees [26]. They fail to reflect the morphological and structural differences among trees under different growth states in the stand and lack a true simulation of stand growth [4]. Therefore, it is essential to comprehensively analyze the growth and spatial structure of different trees in a stand and to incorporate the differences in tree morphology and structure under different growth states into the existing growth model to realize the growth polymorphism of the stand. We took the Chinese fir plantation in the Huangfengqiao State-owned Forest Farm in Hunan Province, China, as an example. Our main research work is as follows: (1) From the three dimensions of a single-tree–spatial structure unit–stand, through the comprehensive analysis of the growth difference of single-tree morphology and structure in the stand, the comprehensive grade model of forest spatial structure (CGMSS) was constructed to realize the comprehensive classification of the growth state of forest trees; (2) a growth model was constructed of Chinese fir morphological structure based on CGMSS to provide the theoretical basis for the visual simulation of Chinese fir growth; (3) with the help of three-dimensional dynamic visualization technology, the visual simulation of the polymorphic growth of Chinese fir forest was realized.

2. Materials and Methods

2.1. Study Area and Data Collection

The study area is in Huangfengqiao State-owned Forest Farm, which is distributed in a belt across the east and west of Youxian County, Hunan Province, with latitude and longitude between 113°04′ and 113°43′ E, 26° 43′ and 27°06′ N. The altitude of the forest farm ranges from 115 m to 1270 m, with a humid subtropical monsoon climate, average annual temperature of 17.8 °C, annual precipitation of 1410.8 mm, mid–low mountain landforms, and the percentage of forest cover is over 90%. For six consecutive years (2012–2017), we conducted field investigation and sampling in the Huangfengqiao State-owned Forest Farm, and selected six Chinese fir plantation plots with the same range of forest sites, slopes, aspects, and elevations (site index:18, aspect: south slope, slope: 5–10°, elevation: 270–320 m), and used equipment to collect data such as individual tree coordinates, age, diameter at breast height (DBH), tree height (H), crown width (CW), under-living branch height (UBH), etc. Its attribute distribution is shown in Table 1. In this study, the center tree and its four nearest neighbor trees were used as the spatial structure unit of the tree [29]. To eliminate the influence of the edge effect of the sample plot, a buffer zone of 5 m was set inward of the sample plot, and 1787 central trees were calculated, among which 1008 center trees whose adjacent trees did not change, and 779 center trees whose spatial structure changed due to the death or thinning of adjacent trees. In this study, 1008 center trees whose adjacent trees have not changed were used as the modeling data of the tree morphological structure growth model, of which 70% were used for modeling and 30% were used for testing. The location of the study area is shown in Figure 1.

Table 1. Statistical table of tree factors in the sample plot.

Plot Number	Age Span	Quantity/Plant	Area/hm^2	DBH Range/cm	H Range/m	UBH Range/m	CW Range/m
A	17~22	533	0.36	7.1~16.2~28.6	4.3~10.6~18.0	1.7~5.0~9.3	0.6~2.7~3.8
B	23–28	362	0.4	13.3~22.6~33.4	8.4~15.4~20.2	1.8~8.2~12.2	1.3~2.9~4.2
C	11–16	309	0.25	5.9~14.2~21.2	5.3~9.5~13.5	1.9~4.3~7.5	0.8~2.7~4.0
D	10–15	955	0.48	3.2~11.6~22.4	3.2~8.4~13.7	1.3~4.4~8.0	0.3~2.2~3.8
E	16–21	230	0.36	10.5~20.0~29.8	6.7~12.8~16.8	2.6~6.6~9.2	1.6~3.0~4.5
F	13–18	120	0.16	6.2~16.5~23.2	5.6~11.0~13.8	1.8~6.3~9.5	1.8~3.4~5.6

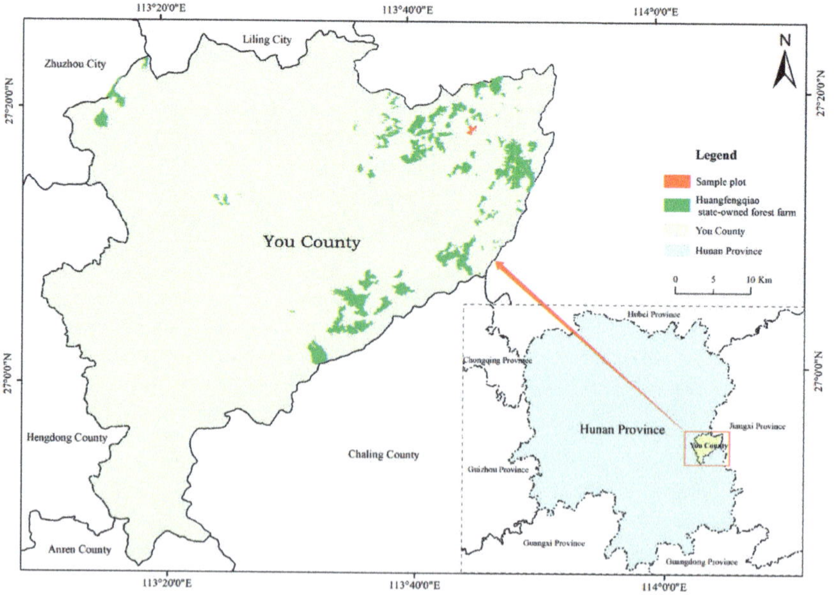

Figure 1. Location of study area.

2.2. Construction of Comprehensive Grade Model of Spatial Structure (CGMSS)

The core of comprehensive, integrated, and three-dimensional reflection of the real growth state of a stand is a selection of factors or parameters that can express the growth and spatial structure of different trees in the stand [30], and to evaluate and quantify them [31]. This paper conducts a comprehensive analysis from the three dimensions of individual tree–spatial structure unit–stand, and selects tree spatial advantage degree (S), forest density of stocking (B), direction proportion (DP), neighborhood comparison (U), crowding (C), and forest competition index based on intersection angle (UCI) as parameters to quantitatively describe the growth state of trees. A comprehensive grade model of spatial structural of trees (CGMSS) was constructed using the multiplication and division method to provide a comprehensive and scientific classification of trees in different growth states. Among them, S and B reflect the current growth state and potential maximum growth value of the tree; DP, U, C, and UCI, respectively, reflect the distribution pattern of adjacent trees, the degree of tree differentiation, the degree of crowding, and the magnitude of the competitive pressure on the top of the center tree and on the flanks. The smaller the value of the final calculated comprehensive grade of spatial structure (CGSS), the more sufficient the space for tree growth and the higher the spatial dominance. The calculation formulas and meanings describing each parameter index are shown in Table 2. What needs to be explained is the calculation index of direction proportion proposed in this paper.

Table 2. Calculation parameters of the comprehensive grade of Chinese fir spatial structure.

Model Name	Model Expression	Explanation
Neighborhood comparison [32]	$U_i = \frac{1}{n}\sum_{j=1}^{n} k_{ij}$	n represents the number of adjacent trees (the same as below), and k_{ij} is a discrete variable, which means that when the diameter of the center tree i is smaller than that of the adjacent tree j, $k_{ij} = 1$, otherwise $k_{ij} = 0$.
Crowding [33]	$C_i = \frac{1}{n}\sum_{j=1}^{n} y_{ij}$	C_i represents the density of the center tree i; y_{ij} is a discrete variable, which takes a value of 1 when the crown of adjacent tree j overlaps with center tree i, otherwise, it is 0.
a forest competition index based on intersection angle [34]	$UCI_i = \frac{1}{n}\sum_{j=1}^{n} \frac{(\alpha_1+\alpha_2\cdot C_{ij})}{180^\circ}\cdot U_i$	When the adjacent tree j is larger than the center tree i, $\alpha_1 = \tan^{-1}\frac{H_i}{d_{ij}}$, otherwise $\alpha_1 = \tan^{-1}\frac{H_j}{d_{ij}}$; $\alpha_2 = \tan^{-1}\frac{H_j-H_i}{d_{ij}}$ When the adjacent tree j is larger than the center tree i, $C_{ij} = 1$, otherwise $C_{ij} = 0$. UCI_i means the intersection angle competition index; α_1, α_2 means the included angle; U_i means the neighborhood comparison, which means the weight; H_i, H_j means the height of the center tree and adjacent trees; C_{ij} means the parameter, value is 0 or 1.
tree spatial advantage degree [35]	$S_d = \sqrt{P_{D_i}\frac{1+G_{max}}{1+G_{max}-\overline{G}}}$	P_{D_i} is the probability that the DBH of other trees in the stand is smaller than the center tree; G_{max} is the potential maximum cross-sectional area of the tree, and the value of the largest cross-sectional area of the tree in the stand is used as the potential size of the tree growth; $\overline{G}$ is the cross-sectional area of the center tree(the same as below).
forest density of stocking [36]	$B_0 = \overline{G}/G_{max}$	

In research, the neighborhood pattern parameter is usually used by judging the relationship between the small angle and the standard angle (72°) between the center tree and any two nearest adjacent trees to obtain the random (0.475 < W< 0.517), uniform (W < 0.475), and cluster (W > 0.517) distribution pattern of trees [15]; however, only the relationship between the small angle and the standard angle cannot determine the influence of adjacent trees on the growth of its morphological structure in the specific orientation to the center tree. When considering the distribution of adjacent trees in the east, south, west, and north directions of the center tree [15], as shown in Figure 2a,b, the angle between any two adjacent trees is smaller than the standard angle, i.e., when the value of the neighborhood pattern is 0, the adjacent trees may be distributed in one or two orientations of the center tree. However, when the angle between any two adjacent trees is larger than the standard angle, and the value of neighborhood pattern is 1, the adjacent trees may be distributed in three or four directions of the center tree (Figure 2c,d). To determine the influence of the nearest four adjacent trees on the morphological structure of the center tree in each direction, we use specific parameter values to characterize the number of positions occupied by the adjacent trees in each direction of the center tree. According to the distribution of adjacent trees in the four directions of the center tree, this study proposes the calculation index of direction proportion (DP). Let c_{ij} be the proportion of the nearest four adjacent trees in the east, south, west and north directions of the center tree. When c_{ij} = 1, it means that the four adjacent trees are in the same direction of the center tree; when c_{ij} = 2, it means that the four adjacent trees are distributed in two different directions of the center tree; when c_{ij} = 3, it means that the four adjacent trees are distributed in three different directions of the center tree; when c_{ij} = 4, it means that the four adjacent trees are evenly distributed in four different directions of the center tree, and finally the calculated expression formula is as follows:

$$DP_i = \frac{1}{n}\sum_{j=1}^{n} c_{ij}, \quad (1)$$

where DP_i represents the direction proportion; n represents the number of adjacent trees; $DP_i \in (0,1]$, DP_i takes four values—0.25, 0.5, 0.75, and 1—respectively indicating that the adjacent tree is distributed in 1, 2, 3, and 4 directions of the center tree.

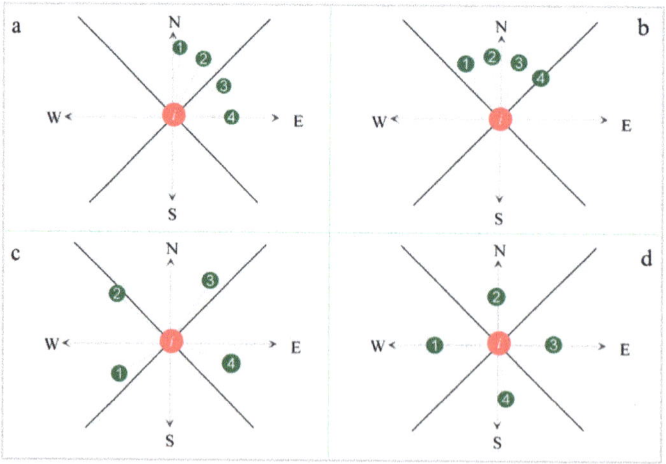

Figure 2. Schematic diagram of Adjacent trees in different orientations of the center tree. (**a**,**b**) Distribution when the angle between any two adjacent trees is smaller than the standard angle. (**c**,**d**) Distribution when the angle between any two adjacent trees is greater than the standard angle. The numbers 1–4 represent the distribution of the four adjacent trees.

Different parameters are of great significance for analyzing the growth changes in tree morphological structure under different conditions; however, due to the interdependence and mutual exclusion of each parameter, each parameter is required to achieve the best judgment on the growth of tree morphology and structure at the same time, and the above parameters need to be weighed comprehensively. The construction of the CGMSS in this study is shown in Figure 3.

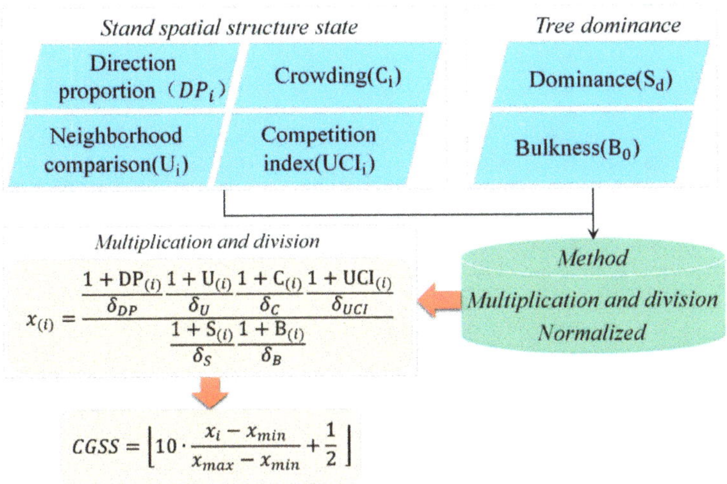

Figure 3. Construction of comprehensive grade model of Chinese fir spatial structure.

The comprehensive grade value of spatial structure is obtained by synthesizing the six sub-objectives that characterize the growth advantage of trees and the state of spatial structure, and the calculation formula is as follows:

$$x_{(i)} = \frac{\frac{1+DP(i)}{\sigma_{DP}} \cdot \frac{1+U(i)}{\sigma_U} \cdot \frac{1+C(i)}{\sigma_C} \cdot \frac{1+UCI(i)}{\sigma_{UCI}}}{[1+S(i)] \cdot \sigma_S \cdot [1+B(i)] \cdot \sigma_B}, \tag{2}$$

where $DP(i)$, $U(i)$, $C(i)$, $UCI(i)$, $S(i)$ and (i), respectively, represent direction proportion, neighborhood comparison, crowding, a forest competition index based on intersection angle, tree spatial advantage degree and forest density of stocking, and σ_{DP}, σ_U , σ_C, σ_{UCI}, σ_S, σ_B are the standard deviation of direction proportion, neighborhood comparison, crowding, a forest competition index based on intersection angle, tree spatial advantage degree and forest density of stocking, respectively.

To make the data indicators comparable, normalization and other standardized processing methods are usually used to transform the data equivalently into the range of [0, 1]. To analyze and compare the comprehensive grades of the spatial structure of the center tree, this study combined qualitative and quantitative methods [30] to divide the continuous data in the range [0, 1], as shown in Equation (3). By rounding the comprehensive grade value of spatial structure and rounding down, the normalized value is rounded to 0~10, and there are 11 comprehensive grades of tree spatial structure.

$$\text{CGSS} = 10 \cdot \frac{x_i - x_{min}}{x_{max} - x_{min}} + \frac{1}{2}, \tag{3}$$

where CGSS represents the comprehensive grade of spatial structure, x_i represents the value of comprehensive grades of the center tree spatial structure before normalization, and x_{max} and x_{min} represent the maximum and minimum values of the normalized values, respectively.

2.3. *Fitting of Growth Model of Chinese Fir's Morphological Structure*

2.3.1. DBH Growth Model

The DBH is the basic and the most important variable for expressing both spatial and non-spatial structural characteristics of a stand [37], and it is also the parameter that is easy to obtain relatively and has the highest accuracy in field measurements (accuracy of 0.1 cm) [38,39]. Therefore, DBH is often used as a key variable in predicting tree growth. In this study, the basic growth equation for DBH was established based on the Richards equation [40], as in Equation (4).

$$D = a\left(1 - e^{-kt}\right)^{c} \qquad a > 0,\ k > 0,\ c > 0, \tag{4}$$

where D represents the diameter at breast height; a represents the maximum value of tree growth; k represents the growth rate of the tree; c is related to the shape of the equation curve and determines the location of the inflection point; and t represents the age of the tree. Using this growth equation to fit the tree diameter for different CGSS, 11 DBH growth equations with CGSS of 0 to 10 levels can be obtained. To study the relationship between different grades and growth equation parameters, the function expression $f(x)$ between the parameters and CGSS was established by drawing scatter diagrams of a, k, c in the 11 DBH growth equations and the CGSS, respectively. By fitting the correlation analysis between each parameter and the corresponding grade (Figure 4), this paper will use the quadratic polynomial to fit the relationship between a and the rank where it is located and obtain the function $f_a(x)$; the Gauss to fit the relationship between k and the corresponding rank to obtain the function $f_k(x)$; and the logistics to fit the relationship between c and the rank where it is located to obtain the function $f_c(x)$. The formula is shown in Table 3. We finally obtain the CGMSS-based DBH growth model, as in Equation (5).

$$CD = f_a(x)\ \left(1 - e^{-f_k(x)t}\right)^{f_c(x)}, \tag{5}$$

where $f_a(x)$ is to replace parameter a, $f_k(x)$ is to replace parameter k, and $f_c(x)$ is to replace parameter c. CD represents the diameter at breast height based on CGMSS, x represents the comprehensive grade value of the spatial structure of the tree, and t represents the current tree age.

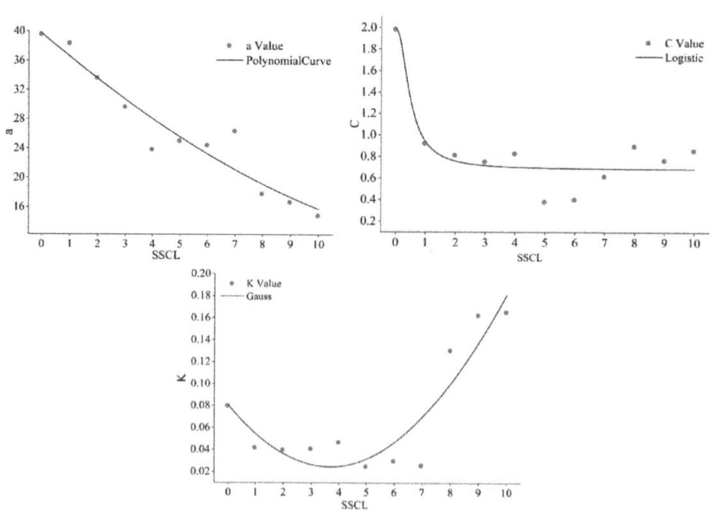

Figure 4. The relationship between the parameters of the DBH growth equation and the CGSS.

Table 3. Relational expression of each parameter and CGSS.

Equation	Parameters	Model Name	Model Expression
(6)	a	Quadratic polynomial	$f_a(x) = bx^2 + cx + d$
(7)	k	Gauss	$f_k(x) = A_1 + \frac{A_2}{w\sqrt{\frac{\pi}{2}}}e^{-2(\frac{x-x_0}{w})^2}$
(8)	c	Logistic	$f_c(x) = A_2 + \frac{A_1-A_2}{1+\left(\frac{x}{x_0}\right)^p}$

In the formula: x represents the CGSS value, and b, c, k_0, w, A represent the fitting parameter values.

2.3.2. H, CW and UBH Growth Model

As basic parameters of tree morphological structure [39], DBH, H, CW and UBH are often used as predictors of forest growth models [41]. We conducted a matrix analysis on 1008 Chinese firs whose spatial structure had not changed (Figure 5) and found that the correlation coefficient between H and DBH was the highest at 0.89, followed by the correlation coefficients between H and UBH and CW at 0.70 and 0.66, respectively. Through nonlinear fitting analysis among various morphological structure factors, this study intends to use the Gauss function describing the relationship between DBH and H and H and UBH, and use the logistic function describing the relationship between H and CW. The expressions are shown in Table 4.

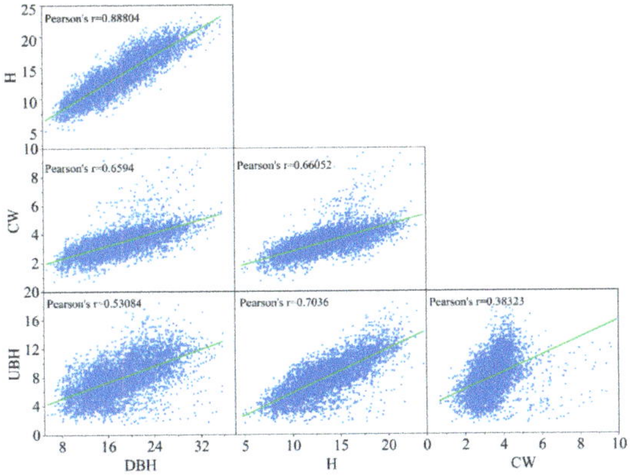

Figure 5. Matrix analysis of morphological structure factors of 1008 center trees.

Table 4. Fitting models of Chinese fir tree height, branch height and crown growth.

Equation	Parameters	Model Expression
(9)	H	$\mathrm{H} = A_1 + \frac{A_2}{w\sqrt{\frac{\pi}{2}}}e^{-2(\frac{DBH-x_0}{w})^2}$
(10)	UBH	$\mathrm{UBH} = A_1 + \frac{A_2}{w\sqrt{\frac{\pi}{2}}}e^{-2(\frac{H-x_0}{w})^2}$
(11)	CW	$\mathrm{CW} = A_2 + \frac{A_1-A_2}{1+\left(\frac{H}{x_0}\right)^p}$

2.4. *Evaluation of Model Accuracy*

To scientifically and objectively evaluate the fitting effect and validity of the model, this study first used the determination adjustment coefficient (R^2, Equation (12)), residual sum of squares (RSS, Equation (13)), root mean square error (RMSE, Equation (14)), and mean absolute error (MAE, Equation (15)) [42] to evaluate the fitting accuracy and applicability of models for DBH, H, CW and UBH, then used the systematic error (Equation (16)) and prediction accuracy (Equation (17)) to test the effectiveness of the Chinese fir growth model based on CGMSS and SSU, respectively, and the expressions are as follows.

$$R_{adj}^2 = 1 - \frac{\sum_{i=1}^{n}(y_i - \hat{y}_i)^2/(n-p-1)}{\sum_{i=1}^{n}(y_i - \overline{y_i})^2/(n-1)}, \tag{12}$$

$$\text{RSS} = \sum_{i=1}^{n}(y_i - \hat{y}_i)^2, \tag{13}$$

$$\text{RMSE} = \sqrt{\frac{1}{n-k}\sum_{i=1}^{n}(y_i - \hat{y}_i)^2}, \tag{14}$$

$$\text{MAE} = \frac{1}{n}\sum_{i=1}^{n}|y_i - \hat{y}_i|, \tag{15}$$

$$\text{Systematic error} = \frac{1}{n}\sum\frac{\hat{y}_i - y_i}{y_i} \times 100\%, \tag{16}$$

$$\text{Accuracy} = 1 - \frac{t_{\partial} \times \sqrt{\Sigma(y_i - \hat{y}_i)^2}}{\hat{\bar{y}} \times \sqrt{n(n-k)}}, \tag{17}$$

where y_i, $\overline{y_i}$, $\hat{y}_i$, $\hat{\bar{y}}$, n, p, k, t_{∂}, respectively, represent the observed value, the average of the observed value, the predicted value, the average of the predicted value, the number of samples, the number of independent variables in the model, the number of parameters in the model, and the t distribution value of the confidence level α.

2.5. Dynamic Visual Simulation of Chinese Fir Morphological Structure

In this study, Unreal Engine 4 (UE4) was used as the visualization platform for the Chinese fir forest [43]. First, using the tree-modeling software combined with the model sample, the Chinese fir foundation subdivision model was made, and it was divided into two parts: the crown and the trunk. The coordinate origin (0, 0, 0) of the trunk model at the root was set, took the tree height as the Z-axis variation value of the crown model was taken, and the coordinate origin (0, 0, H) of the crown model at the top of the crown was set, as shown in Figure 6. Using the UE4 blueprint system, through the development of the growth control blueprint of DBH, H, CW, and UBH of Chinese fir, the subdivision model of Chinese fir was coupled with the growth model of morphological structure based on CGMSS, and used the time axis animation curve technology to realize the visual simulation of the differential growths of Chinese fir forest morphology and structure.

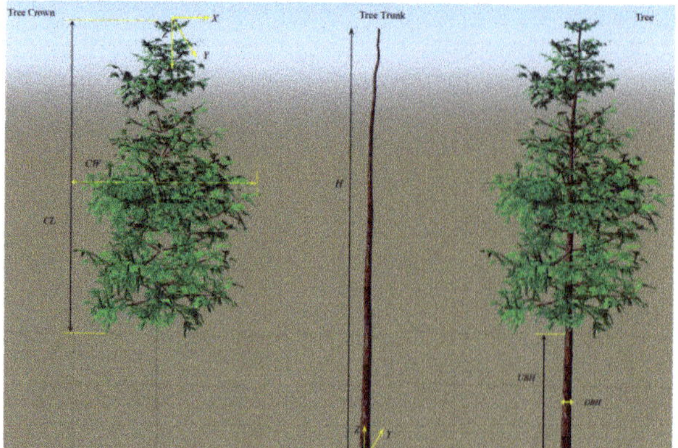

Figure 6. Schematic diagram of fir split model.

3. Results

3.1. Comprehensive Grade Model of Spatial Structure (CGMSS)

It can be seen from Table 5 that, except for some of the Pearson correlation coefficients between the crown width of Chinese fir and other morphological factors, CGSS = 1–4 are slightly lower than those when the correlation analysis is carried out without considering the classification of forest trees, and the correlation coefficients among the other grades of morphological structure factors are all higher than the former. As the CGSS value increases, the correlation among morphological factors becomes higher, demonstrating that the use of CGMSS to classify trees in a stand into classes can improve the correlation among tree morphological structure factors better and achieve a more accurate simulation of tree morphological structure under different growth and spatial structures.

Table 5. Correlation coefficient analysis of Chinese fir morphological structure factors.

Pearson's Correlation	No	CGSS										
		0	1	2	3	4	5	6	7	8	9	10
H:DBH	0.90	0.94	0.91	0.94	0.94	0.95	0.96	0.95	0.95	0.97	0.97	0.98
H:CW	0.84	0.71	0.65	0.78	0.87	0.87	0.86	0.91	0.91	0.94	0.98	0.96
H:UBH	0.84	0.94	0.74	0.85	0.87	0.74	0.91	0.91	0.91	0.95	0.98	0.94
DBH:UBH	0.61	0.89	0.6	0.77	0.76	0.69	0.84	0.86	0.86	0.93	0.96	0.91
DBH:CW	0.79	0.67	0.67	0.76	0.86	0.85	0.85	0.87	0.87	0.92	0.96	0.95
CW:UBH	0.71	0.78	0.37	0.61	0.74	0.63	0.74	0.81	0.81	0.91	0.97	0.9

When CGSS = 0, tree morphological structure parameters (DBH, H, UBH, CW) are at the maximum state (Figure 7). However, with the gradual increase of CGSS, each morphological structure parameter gradually decreases and shows a significant negative correlation with CGSS. According to the CGMSS calculation, when CGSS = 0, it indicates that the trees in the stand have the maximum growth potential and the growth is not constrained by other neighboring trees. However, as the value of CGSS increasing the potential growth of trees will become less and less, and the competition with the nearest neighboring trees will become more intense. Therefore, when CGSS = 10, its morphological parameters are all the minimum, the tree growth space is severely squeezed, and the growing competitive pressure will also reach the maximum. We also analyzed 1008 center trees with unchanged adjacent trees and obtained that the CGSS at different stand ages basically remained the same as the rank classified at the initial stand age (Figure 8), indicating that the growth condition and spatial structure environment of the classified CGSS will remain relatively unchanged in stands where the spatial structure tends to be stable.

The above results indicate that the CGMSS constructed in this study can make comprehensive judgments and give reasonable ranks to trees with different degrees of growth dominance and spatial structure environment in the stand, while the results of center tree classification are relatively stable for neighboring trees that have not changed.

3.2. Morphological–Structural Growth Model of CHINESE Fir

As shown in Table 6, the R^2_{adj} of the three parameter models of $f_a(x)$, $f_k(x)$ and $f_c(x)$ are more than 0.8. Since parameter a represents the maximum value of DBH growth, the residuals of the predicted values are usually greater than 1, so the RSS is relatively large. The RSS value of each model shows that all three models can express the correlation between parameters and CGSS, and have high fitting accuracy. Substituting the three parameter models into the base DBH growth equation (Equation (4)), we obtain that the R^2_{adj} = 0.836 of the CGMSS-based DBH growth model (*CD*) is higher than the R^2_{adj} = 0.5 of the base DBH growth model (D), and the results show that CGMSS improves the fitting accuracy of DBH growth. The R^2_{adj} of the H, UBH and CW growth models are 0.871, 0.703 and 0.658, respectively, and the RMSE and MAE are basically less than 1, implying that the fitted model has good applicability.

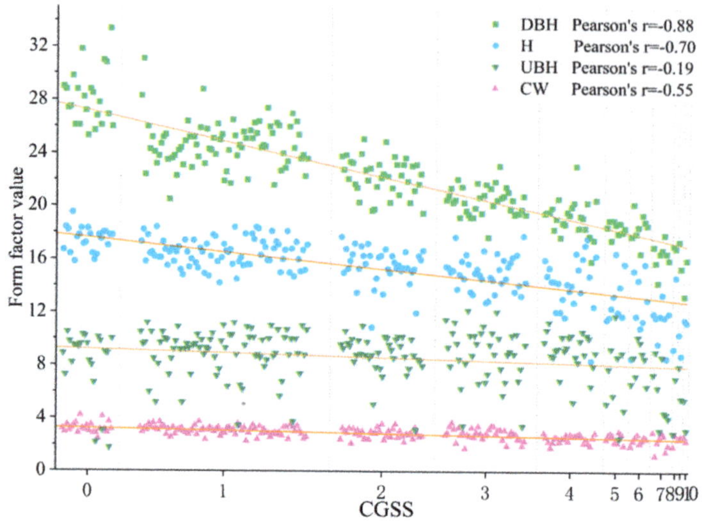

Figure 7. Distribution map of morphological structure factor values of Chinese fir of the same forest age under different comprehensive grades of spatial structure.

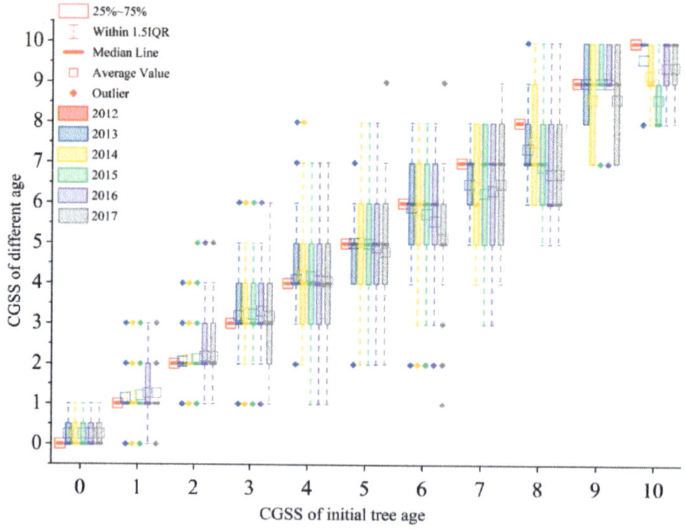

Figure 8. Comparison of CGSS of 1008 center trees at the initial stand age and CGSS at stand age changed.

Table 6. Estimation and testing of parameters of Chinese fir morphological structure growth model.

Parameter Estimates/Test	Model Name							
	D	$f_a(x)$	$f_k(x)$	$f_c(x)$	*CD*	H	UBH	CW
a	29.205							
b		−3.274						
c	1.487	0.088						
d		39.735						
k	0.078							

Table 6. *Cont.*

Parameter Estimates/Test	Model Name							
	D	$f_a(x)$	$f_k(x)$	$f_c(x)$	*CD*	H	UBH	CW
*A*1			1.318	1.987		19.419	12.193	2.209
*A*2			40.207	0.684		102.575	67.988	1114.545
*X*0			3.691	0.495		13.846	9.284	434.601
w			24.800			10.645	8.386	
p				2.008				2.008
R^2_{adj}	0.500	0.913	0.819	0.820	0.836	0.871	0.703	0.658
RSS	1875.627	54.127	0.005	0.290	205.53	60.398	90.909	7.863
RMSE	5.414	2.452	0.023	0.180	1.792	0.971	1.192	0.351
MAE	23.420	1.544	0.017	0.125	20.020	0.745	0.930	0.232

The CD model is used to test the DBH of trees divided into grades in the forest, and the estimated accuracy of DBH of 0–10 grades is at least 79.19% (Table 7). Combined with Figure 9, it can be seen that the prediction accuracy of more than 73.5% of the trees is above 85%. A negative value in the systematic error indicates that the predicted value is lower than the observed value [44]. Among all levels, the minimum prediction accuracies of the H, CW, and UBH growth models are 73%, 80%, and 55%, respectively, while the percentages of trees with an estimated accuracy greater than 85% are 69.6%, 85.2%, and 54.3%, respectively. From the prediction accuracy and error of each grade, except for the fluctuation of grades 3–5, the prediction accuracy of the rest of the grades is basically more than 80%, the systematic error is less than 10%, and their changes show a fluctuating downward and then upward trend with the increase of CGSS.

Table 7. Test indicators of morphological–structural growth model of Chinese fir based on CGMSS.

Model Name	Test Indicators	NO	CGSS										
			0	1	2	3	4	5	6	7	8	9	10
D	systematic error	34.17	0.48	7.51	−1.93	−13.81	−17.53	−10.39	−4.70	9.91	4.82	2.89	6.26
	accuracy	0.98	0.99	0.93	0.98	0.84	0.79	0.89	0.95	0.91	0.95	0.97	0.94
H	systematic error	−2.19	−0.42	4.98	0.95	−15.36	−21.12	−12.61	−7.79	11.8	5.98	11.25	−1.72
	accuracy	0.97	0.94	0.93	0.95	0.82	0.73	0.84	0.89	0.89	0.94	0.88	0.94
UBH	systematic error	−1.40	12.67	9.21	4.44	−21.26	−30.06	−18.22	−15.29	12.8	24.75	0.56	4.95
	accuracy	0.94	0.84	0.87	0.90	0.65	0.55	0.71	0.78	0.85	0.75	0.89	0.74
CW	systematic error	−0.03	1.51	9.77	5.85	−8.96	−9.1	−6.72	−4.93	9.81	11.35	−8.31	−0.56
	accuracy	0.96	0.88	0.88	0.90	0.85	0.82	0.86	0.80	0.88	0.88	0.82	0.91

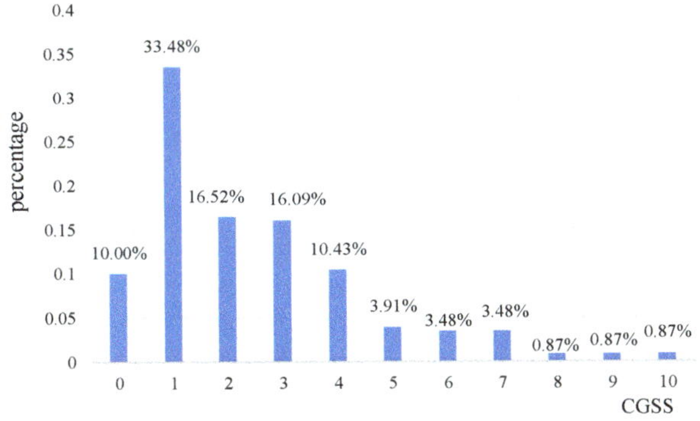

Figure 9. Percentage of trees with different CGSS.

In this paper, to verify the accuracy of the CGMSS-based Chinese fir morphological–structural growth model for simulating Chinese fir in different growth states, it was compared with the Chinese fir growth model [27] constructed based on the spatial structure unit (SSU) and integrated with independent spatial structure parameters in the existing research. Table 8 shows the results of the index test of the SSU-based Chinese fir growth model. The overall prediction accuracy of the CGMSS- and SSU-based Chinese fir DBH, H, CW, and UBH growth models reach more than 92% when the CGMSS classification classes are not applied to the sample data for analysis, indicating that both growth models can achieve accurate simulation of most of the morphological–structural growth changes in Chinese fir. In Table 8, the value of systematic error indicates that the estimated value of DBH, H and UBH is high, and the estimated value of CW is low. However, when applying CGMSS classification to individual trees in a forest and testing for different grades of trees, the lowest values of prediction accuracy of SSU-based Chinese fir DBH, H, CW and UBH growth models are obtained below 55%. The proportions of these models for the number of trees with prediction accuracies above 85% in each grade are 66.1%, 86.5%, 16.5%, and 86.5%, respectively, but as the CGSS gradually increases, the prediction accuracy of each model decreases significantly and has great systematic errors. Among them, the prediction accuracy of the DBH growth model at levels 4–10, the H growth model at levels 8–10, the UBH growth model at levels 7–10 and the CW growth model at levels 5–10 are less than 80%. However, the average prediction accuracy of CGMSS's Chinese fir morphological structure growth model is basically greater than 85% at levels 5–10 except for UBH, and the prediction accuracy of CGMSS's UBH growth model is equally better than that of the SSU-based UBH growth model obtained from Tables 7 and 8. The above results show that the CGMSS-based Chinese fir growth model proposed in this study can achieve a better fit at the stand level and provide a more realistic simulation of the growth of trees in different growth states in the stand, especially for trees with limited growth space and fierce competition (CGSS = 5–10), and the results are better than those of the SSU Chinese fir growth model.

Table 8. Test indicators of morphological−structural growth model of Chinese fir based on SSU.

Model Name	Test Indicators	NO	CGSS										
			0	1	2	3	4	5	6	7	8	9	10
D	systematic error	54.40	−0.08	11.24	7.15	25.54	100.23	143.18	140.73	337.27	335.07	508.27	682.48
	accuracy	0.94	0.82	0.92	0.87	0.85	0.78	0.54	0.57	0.46	0.39	0.65	0.72
H	systematic error	18.94	20.28	18.97	8.66	14.67	18.09	22.36	30.28	31.36	32.61	54.00	19.86
	accuracy	0.98	0.93	0.96	0.95	0.95	0.93	0.84	0.81	0.81	0.46	0.66	0.70
UBH	systematic error	14.89	26.32	−1.32	2.40	9.22	4.29	58.40	20.88	100.20	135.15	78.77	152.06
	accuracy	0.96	0.87	0.95	0.93	0.92	0.91	0.71	0.82	0.61	0.09	0.10	0.15
CW	systematic error	−33.46	−38.77	−40.49	−17.03	−34.56	−26.92	−27.40	−24.17	−13.27	−21.72	−28.81	−27.64
	accuracy	0.92	0.70	0.85	0.80	0.80	0.82	0.65	0.71	0.66	0.28	0.20	0.24

3.3. Dynamic Visual Simulation of the Morphological Structure of Chinese Fir Stand

In this paper, we first imported the tree coordinates and attribute information into the UE4 data structure table, then created a new action class in the blueprint class, organized the two parts into a complete tree model according to the trunk and crown coordinates of the produced Chinese fir split model, and finally developed a Chinese fir morphological structure control blueprint, and used the relative coordinate system to map the tree model into the 3D scene to form the initial plantation scenarios. As shown in Figure 10, the initial fir plantation scenario is the result of a visual simulation using the observed data from study plot F. Figure 10a shows the top view angle of Chinese fir stand with an initial stand age of 23 years old, which can visualize the competition status, CW size and spatial location relationship among trees. Figure 10b clearly shows the differences of the UBH, H, CW and CL among different trees from the side view angle.

Figure 10. Visual simulation results of a 23-year-old Chinese fir plantation. (**a**) Top view angle to observe the crown width extrusion and competition between trees in the stand. (**b**) Side view angle to observe the differences between UBH, H, CW and CL of different trees.

The key to achieving polymorphism in tree morphological structure growth in a stand is to achieve control over the morphological structure of each tree. Therefore, this paper achieved scaling control of trunk height and diameter in the subdivision model using the H and DBH growth models, and scaling control of crown size by the CW and CL growth models, where CL was calculated from the difference between H and UBH. The calculation process of tree morphological structure parameter values is shown in Figure 11. First, the CGSS of trees in the plot is calculated in real time, then the tree growth age is input and combined with the morphological structure growth model to realize the differentiated growth of Chinese fir under different growth dominance conditions and spatial structure environment. The increment of the tree morphological structure parameter is the value of the morphological structure parameter at the current input tree age minus the value of the initial tree age morphological structure parameter, and the scaling value of the morphological structure control blueprint is the parameter increment.

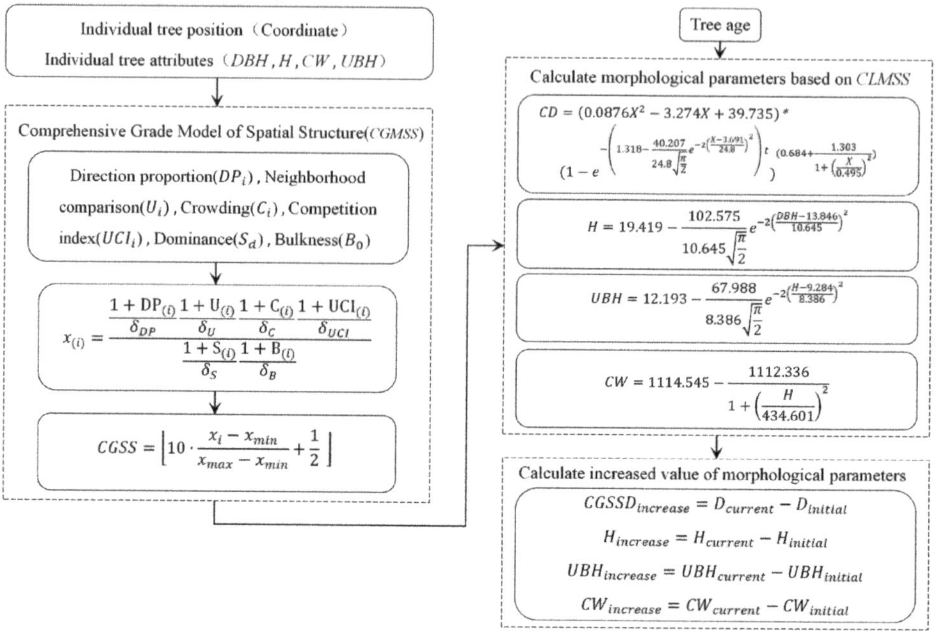

Figure 11. The calculation process of morphological structure parameter values of Chinese fir.

After obtaining the morphological structure parameters and location information of the initial stand age, the user can input the stand age and combine it with the Chinese fir growth model to realize the differentiated growth of tree morphological structure under different growth advantage conditions and spatial structure environment. Figure 12 presents the stand scenario when the Chinese fir stands in plot F grows to 28 years. Figure 12a–c represent the results of the real observation data, the Chinese fir growth model of CGMSS and the visual simulation based on the SSU Chinese fir growth model in the same perspective, respectively. It can be seen from the figure that the simulation results of the Chinese fir growth model of CGMSS and the observation data are highly similar in H, UBH and CL (left panel), but from the right panel, it can be seen that the degree of competition among trees CW such as squeezing and shading is slightly lower than that of the Chinese fir forest in the real growth state. However, the simulation results of UBH, CW and CL for trees visualized based on the SSU Chinese fir growth model were worse than the actual growth state, except for H. In particular, the simulation results for competition among CW differed from the actual growth. The analysis results show that when the tree CGSS > 5 in the Chinese fir forest, the fitting effect of the SSU Chinese fir growth model is gradually lower than that of the fitting model proposed in this paper. Figure 13 shows that the morphological structure of the CGMSS Chinese fir growth model simulating growth-limited trees (Figure 13b) is more visually similar to that of trees in the real growth state (Figure 13a); however, the simulation results of H and UBH of trees under different growth states by the SSU-based Chinese fir growth model (Figure 13c) are relatively consistent, and the tree crown width differs more from that of the real growth trees. The above results show that the visualization of Chinese fir stands based on the CGMSS Chinese fir growth model can reflect the differences in H, CL, UBH, and CW growth under different growth states better, and at the same time, the simulation of the tree morphological structure growth under different growth dominance degrees and spatial structure environments is more realistic.

Figure 12. Results of visual simulation of 28-year-old Chinese fir stands. (**a**) Simulation from observed data; (**b**) Simulation from predicted data of CGMSS Chinese fir growth model; (**c**) Simulation from predicted data of SSU Chinese fir growth model. The left panel shows the side view observation angle of the stand visualization simulation, and the right panel shows the top view observation angle.

Figure 13. Visual simulation results of poor growth and growth space restricted trees (CGSS > 5) in 28-year-old Chinese fir stands. (**a**) Simulation by observed data; (**b**) Simulation by predicted data from CGMSS Chinese fir growth model; (**c**) Simulation by predicted data from SSU Chinese fir growth model. Highlighted trees in the figure are growth-limited Chinese fir trees.

4. Discussion

In this study, tree morphological–structural growth models were represented by establishing H, DBH, CW and UBH growth models. Then CGMSS was coupled with the tree morphological structure growth model to realize the visual simulation of different tree growth polymorphisms in stands with the help of 3D visualization technology. The results show that the morphological structure growth model of CGMSS for Chinese fir can simulate the changes in tree morphological structure growth under different growth dominance conditions and spatial structure environments more accurately, which can provide guidance and reference for simulating the morphological structure polymorphism of trees in different growth states in forest stands. In previous studies, the morphological structure growth model established considering the spatial structure of trees was usually based on single or multiple independent spatial structure parameters to fit the plot data [45], and the results of the studies mostly descried the stand growth in the average state well. However, few studies have considered the differences in morphological structure among trees in the plots and established morphological structure growth models that conform to the growth environment and spatial structure of trees. Therefore, this study achieved the classification of trees in different growth states in the forest by building a comprehensive grade model of spatial structure (CGMSS), and the morphological structure growth model of Chinese fir constructed on this basis successfully simulated the growth polymorphism of Chinese fir forests.

In addition, by comparing with the SSU-based morphological–structural growth model of Chinese fir proposed by other scholars, the prediction accuracy of the two morphological–structural growth models for the stand scale was obtained to be more than 92%, indicating that the two models could provide a better fit for the overall growth variation of trees and the average state of individual tree growth in the stand. However, analysis of the test indicators of the two growth models in different grades showed that the prediction accuracy of the SSU-based morphological–structural growth model of Chinese fir decreased significantly with increasing CGSS and showed a great systematic error, showing completely opposite results compared to the CGMSS-based morphological–structural growth model of Chinese fir for tree prediction with different CGSS. The SSU growth model had a minimum tree prediction accuracy of less than 55% and a maximum of less than 85% at CGSS = 5–10 levels, with the DBH growth model at 5–10 levels, the H growth model at 8–10 levels, the UBH growth model at 5 and 7–10 levels, and the CW growth model at 0 and 5–10 levels all with a prediction accuracy of less than 75%. However, the CGMSS Chinese fir growth model had a minimum prediction accuracy of 71% and a maximum of 97% at CGSS = 5–10 levels, and all models except the UBH growth model had greater than 80% prediction accuracy for CGSS = 5–10 levels. The comparison between Tables 7 and 8 shows that the prediction accuracy of the UBH growth model of CGMSS is equally better than that of the UBH growth model of SSU. The above results show that the CGMSS-based growth model of Chinese fir proposed in this study can achieve a better fit at the stand level, and can simulate the growth of trees in different growth states in the stand, especially for trees with limited growth space and fierce competition (CGSS = 5–10), and the results are better than those of the SSU-based Chinese fir growth model. The analysis of the plot data showed that there were a large number of trees of the same age in the same plot of Chinese fir plantation [40]; however, the reason for the different morphological structures among trees was that trees of the same age had different growth and spatial structure environments [46]. A growth model that considers only spatial structural parameters is only suitable for fitting the average state of tree growth in a stand. The CGMSS-based morphological–structural growth model of Chinese fir proposed in this study allows for better simulation of the variability in morphological structure exhibited by trees of different growth states in stands of the same age.

In the correlation analysis of DBH, H, CW and UBH of 1787 trees with only the influence of the edge effect of the sample plot eliminated and 1008 center trees with no change in adjacent trees in the spatial structure unit, respectively, the results showed that the correlation coefficients among tree morphological structure factors were all higher than those of the former when the spatial structure of the individual tree was not changed, and the relative growth changes among the morphological structure factors were obtained stably. In this study, we used the tree attribute factors with unchanged spatial structure as the fitted data, and then selected DBH as the independent variable for the H growth model and H as the independent variable for the CW and UBH growth models, and constructed a Chinese fir morphological structure growth model based on CGMSS, which better fitted the trees under natural growth conditions. However, in the process of tree growth, especially in plantation, they are often affected by human factors such as thinning [47], resulting in strong changes in the spatial structure of the trees and altering the growth rate of the trees [48,49]. Based on the premise that all trees in the stand are not eliminated and the spatial structure is relatively stable, and the differences in the morphological structure of trees in the two growth states are not considered at the same time, this paper constructs a compatible morphological structure growth model of Chinese fir suitable for both growth states. Therefore, further perfecting the morphological structure growth model to improve the applicability and fitting accuracy of the model under dynamic conditions will be the core and difficulty of subsequent research.

As previously mentioned, the data of the sample plots in this paper mainly came from the six plots of the Huangfengqiao State-owned Forest Farm in Hunan Province with the age range of 10–28 years. Therefore, the morphological structure growth model of Chinese fir we constructed was validated well for the morphological structure changes in Chinese fir from 10–28 years, but in terms of time scale, there was a lack of verification of the morphological structure of Chinese fir in other forest age stages. There are great differences in the growth rate and spatial structure environment of trees in different forest age stages [50]. Therefore, to describe the tree morphological and structural changes accurately in all growth stages, it is not sufficient to use measurements from certain ages or individual plots only. At the spatial scale, the stand conditions of the six sample plots in this paper are basically the same, and the trees are in a similar environment. However, the forest area for management planning is usually huge, which means that the environmental conditions of forest areas for management planning will have huge differences [51], which will make the site conditions for tree growth change, which in turn will affect the accuracy of the tree growth simulation by this study method, and this will lead to uncertainty in the results of this study. Additionally, the form construction methods of different tree species vary greatly [52], and different methods need to be used when dissecting the basic 3D model of trees to achieve an accurate simulation of the tree morphological structure. Therefore, the application of the tree 3D subdivision model and CGMSS-based tree growth modeling methods proposed in this study needs to be adapted to the location to realize the visual simulation of forest stand growth polymorphism better.

5. Conclusions

In this study, trees in different growth states were analyzed comprehensively from three dimensions: individual tree–spatial structure unit–stand, and combined with quantitative indicators representing tree growth and spatial structure states, a comprehensive grade model of spatial structure (CGMSS) was constructed and based on this. A morphological structure growth model of Chinese fir was developed for simulating tree morphological structure in different growth dominance degrees and spatial structure environments, to achieve stand growth polymorphism. The results showed that the overall prediction accuracy of morphological–structural growth models of Chinese fir DBH, H, CW and UBH based on CGMSS exceeded 90%. The trees in the stand were divided into comprehensive spatial structure classes (CGSS) for analysis, and among the Chinese fir growth models in CGSS, the systematic error of all growth models in other classes was less than 19%, and the prediction accuracy was greater than 80%, except for the UBH growth model in CGSS = 3–5 class, which had a smaller prediction accuracy and larger systematic error. The prediction accuracy of the SSU-based Chinese fir growth model constructed by Ma et al. for the stand was also more than 90%, but the systematic errors of the model fits were all greater than 20% when the CGSS was >5, and the prediction accuracy was mostly below 80%. The results of visual simulation showed that the simulation of the Chinese fir growth model of CGMSS was better than the Chinese fir growth model of SSU, and it was closer to the real tree growth state, and realized the dynamic visual simulation and growth polymorphism of Chinese fir stands better. However, forest management is huge in terms of time and spatial scales relative to the study area, and there is also a greater diversity of tree species. To make the simulation more accurate, we will consider the limitations of tree species, time, and spatial scales for in-depth study in the future.

Author Contributions: Conceptualization, H.Z. and X.H.; methodology, X.H. and H.Q.; software, X.H. and K.L.; formal analysis, G.Y., H.Q. and Z.C.; resources, T.Y., Y.Z., Y.L. and J.W.; writing—original draft preparation, X.H.; writing—review and editing, X.H., H.Z. and H.Q.; visualization, K.L.; supervision, H.Z. and G.Y. All authors have read and agreed to the published version of the manuscript.

Funding: This research was funded by the National Natural Science Foundation of China, grant number 32071681, 32271877 and the Foundation Research Funds of IFRIT, grant number CAFYBB2021ZE005, CAFYBB2019SZ004.

Data Availability Statement: Not applicable.

Conflicts of Interest: The authors declare that there are no conflicts of interest.

References

1. Meyer, P.; Ammer, C. Forest Management. In *Disturbance Ecology*; Landscape Series; Springer: Berlin, Germany, 2022; pp. 315–347. [CrossRef]
2. Li, Y.; Zhang, H.; Ju, H.; Jiang, X. Visual Simulation Technique of Decision Making of Interactive Stand Management Methods. In Proceedings of the 2014 International Conference on Virtual Reality and Visualization, Shenyang, China, 30–31 August 2014; pp. 459–466. [CrossRef]
3. Yu, H.; Wu, M.M.; He, H.S. Developing platform of 3-d visualization of forest landscape. *Environ. Modell. Softw.* **2022**, *157*, 105524. [CrossRef]
4. Bowman, D.M.; Brienen, R.J.; Gloor, E.; Phillips, O.L.; Prior, L.D. Detecting trends in tree growth: Not so simple. *Trends Plant Sci* **2013**, *18*, 11–17. [CrossRef]
5. Jin, S.; Zhang, W.; Shao, J.; Wan, P.; Cheng, S.; Cai, S.; Yan, G.; Li, A. Estimation of Larch Growth at the Stem, Crown, and Branch Levels Using Ground-Based LiDAR Point Cloud. *J. Remote Sens.* **2022**, *2022*, 12. [CrossRef]
6. Han, Y.; Wu, B.; Wang, K.; Guo, E.; Dong, C.; Wang, Z. Individual-tree form growth models of visualization simulation for managed larix principis-rupprechtii plantation. *Comput. Electron. Agric.* **2016**, *123*, 341–350. [CrossRef]
7. de Reffye, P.; Hu, B.; Kang, M.; Letort, V.; Jaeger, M. Two decades of research with the GreenLab model in agronomy. *Ann Bot* **2021**, *127*, 281–295. [CrossRef] [PubMed]
8. Griffon, S.; De Coligny, F. AMAPstudio: A software suite for plants architecture modelling. In Proceedings of the IEEE 4th International Symposium on Plant Growth Modeling, Simulation, Visualization and Applications (PMA12), Shanghai, China, 31 October–3 November 2012; pp. 141–147.
9. Kang, M.; Wang, X.; Hua, J.; Hu, B.; Wang, F.; De Reffye, P. Over two decades of research with Greenlab model. *J. Agric. Big Data* **2021**, *3*, 3–12. [CrossRef]
10. Peng, F.; Zheng, H.; Lu, S.; Shi, Z.; Liu, X.; Li, L. Growth model and visualization of a virtual jujube tree. *Comput. Electron. Agric.* **2019**, *157*, 146–153. [CrossRef]
11. Söderbergh, I.; Ledermann, T. Algorithms for simulating thinning and harvesting in five european individual-tree growth simulators: A review. *Comput. Electron. Agric.* **2003**, *39*, 115–140. [CrossRef]
12. Chen, K.; Zhang, H.-Q.; Liu, M.; Lu, K.-N. Cunninghamia Lanceolata Three-Dimensional Modeling Technology Based on IFS Fractal Algorithms. In Proceedings of the 2010 International Conference on Computational Intelligence and Software Engineering, Wuhan, China, 10–12 December 2010; pp. 1–4. [CrossRef]
13. Lei, X.; Chang, M.; Lu, Y.; Zhao, T. A review on growth modelling and visualization for virtual trees. *Sci. Silvae Sin.* **2006**, *42*, 123.
14. Looney, C.E.; Zhang, J. Site quality and intensive early stand management practices affect growth dominance, structural complexity, and tree growth in ponderosa pine plantations. *For. Ecol. Manag.* **2022**, *519*, 120318. [CrossRef]
15. Hui, G.; Hu, Y.; Zhao, Z. Research progress of structure-based forest management. *For. Res.* **2018**, *31*, 85–93. [CrossRef]
16. Ma, Z.; Zhang, H.; Li, Y.; Yang, T.; Huang, R.; Li, S. 3D visual simulation of Chinese fir based on the influence of different stand spatial structures. In Proceedings of the 2017 2nd International Conference on Image, Vision and Computing (ICIVC), Chengdu, China, 2–4 June 2017; pp. 559–565. [CrossRef]

17. Kang-ning, L.; Huai-qing, Z.; Min, L.; Kang, C. Research on Tree Growth Models and Visual Simulation Technology. In Proceedings of the 2010 International Conference on Computational Intelligence and Software Engineering, Wuhan, China, 10–12 December 2010. [CrossRef]
18. Falcão, A.O.; Santos, M.P.D.; Borges, J.G. A real-time visualization tool for forest ecosystem management decision support. *Comput. Electron. Agric.* **2006**, *53*, 3–12. [CrossRef]
19. Kuang, Y.; Zhang, H.; Tan, J.; Jiang, X.; Lu, K.; Zhang, N. The study of the stand simulation based on the spatial structure. In Proceedings of the 2012 IEEE 4th International Symposium on Plant Growth Modeling, Simulation, Visualization and Applications, Shanghai, China, 31 October–3 November 2012; pp. 200–203. [CrossRef]
20. Dong, L.; Liu, Z. Visual management simulation for *Pinus svlvestris* var. *mongolica* plantation based on optimized spatial structure. *Sci. Silvae Sin.* **2012**, *48*, 77–85.
21. Li, S.; Zhang, H.; Li, Y.; Yang, T.; Shen, K. Three-Dimensional Visualization Simulation of Chinese Fir Stand Growth Based on Unity3D. In Proceedings of the 2018 3rd International Conference on Mechanical, Control and Computer Engineering (ICMCCE), Huhhot, China, 14–16 September 2018; pp. 530–534. [CrossRef]
22. Wang, H. Dynamic Simulating System for Stand Growth of Forests in Northeast China. Ph.D. Thesis, Northeast Forestry University, Harbin, China, 2012.
23. Gao, G.; Ding, G.; Xiao, M.; Zhang, L.; Lv, R.; Zhang, X.; Li, W.; Liu, Y. A Study on Stand Visualization of Artificial Mixed Forests. *Bull. Soil Water Conserv.* **2012**, *32*, 158–162. [CrossRef]
24. Kershaw, J.A.; Richards, E.W.; Mccarter, J.B.; Oborn, S. Spatially correlated forest stand structures: A simulation approach using copulas. *Comput. Electron. Agric.* **2010**, *74*, 120–128. [CrossRef]
25. Zhu, N.; Zhang, H.; Cui, Z.; Yang, T.; Li, Y.; Liu, H. Visual simulation of Chinese fir under branch height in consideration of spatial structure. *J. Nanjing For. Univ.* **2022**, *46*, 51–57.
26. Li, S.; Zhang, H.; Li, Y.; Yang, T.; He, J.; Ma, Z.; Shen, K. Dynamic Visual Simulation of Chinese Fir Stand Growth Based on Sample Library. *For. Res.* **2019**, *32*, 21–30. [CrossRef]
27. Ma, Z.; Zhang, H.; Li, Y.; Yang, T.; Chen, Z.; Li, S. Visual Simulation of Chinese Fir Crown Growth Based on Spatial Structure. *For. Res.* **2018**, *31*, 150–157. [CrossRef]
28. Haara, A.; Pykäläinen, J.; Tolvanen, A.; Kurttila, M. Use of interactive data visualization in multi-objective forest planning. *J. Environ. Manag.* **2018**, *210*, 71–86. [CrossRef]
29. Hui, G.; Gadow, K.V. *Principles on Structure-Based*; China Forestry Publishing House: Beijing, China, 2016; p. 360.
30. Cao, X.; Li, J.; Feng, Y.; Hu, Y.; Zhang, C.; Fang, X.; Deng, C. Analysis and evaluation of the stand spatial structure of *Cunninghamia lanceolata* ecological Forest. *Sci. Silvae Sin.* **2015**, *51*, 37–48.
31. Põldveer, E.; Korjus, H.; Kiviste, A.; Kangur, A.; Paluots, T.; Laarmann, D. Assessment of spatial stand structure of hemiboreal conifer dominated forests according to different levels of naturalness. *Ecol. Indic.* **2020**, *110*, 105944. [CrossRef]
32. Hui, G.; Gadow, K.V.; Albert, M. A new parameter for stand spatial structure neighborhood comparison. *For. Res.* **1999**, *12*, 4–9.
33. Hu, Y.; Hui, G. How to describe the crowding degree of trees based on the relationship of neighboring trees. *J. Beijing For. Univ.* **2015**, *37*, 1–8. [CrossRef]
34. Hui, G.; Hu, Y.; Zhao, Z.; Yuan, S.; Liu, W. A Forest Competition Index Based on Intersection Angle. *Sci. Silvae Sin.* **2013**, *49*, 68–73.
35. Zhao, Z.; Hui, G.; Hu, Y.; Li, Y.; Wang, H. Method and application of stand spatial advantage degree based on the neighborhood comparison. *J. Beijing For. Univ.* **2014**, *36*, 78–82. [CrossRef]
36. Hui, G.; Zhao, Z.; Zhang, G. Priority of management measures for natural forests based on the stand state. *J. Beijing For. Univ.* **2016**, *38*, 1–10. [CrossRef]
37. Xie, Y.; Yang, H. Relationship between stand spatial structure and DBH increment of principal species in natural spruce-fir mixed forest in Changbai Mountains of northeastern China. *J. Beijing For. Univ.* **2022**, *44*, 1–11.
38. Cortini, F.; Filipescu, C.N.; Groot, A.; Macisaac, D.A.; Nunifu, T. Regional models of diameter as a function of individual tree attributes, climate and site characteristics for six major tree species in Alberta, Canada. *Forests* **2011**, *2*, 814–831. [CrossRef]
39. Zhou, Z.; Fu, L.; Zhou, C.; Sharma, R.P.; Zhang, H. Simultaneous compatible system of models of height, crown length, and height to crown base for natural secondary forests of Northeast China. *Forests* **2022**, *13*, 148. [CrossRef]
40. Wei, X.; Sun, Y.; Ma, W. A height growth model for *Cunninghamia lanceolata* based on Richards' equation. *J. Zhejiang A&F Univ.* **2012**, *29*, 661–666.
41. Sánchez-González, M.; Cañellas, I.; Montero, G. Generalized height-diameter and crown diameter prediction models for cork oak forests in Spain. *Forest Syst.* **2007**, *16*, 76–88. [CrossRef]
42. Dong, C.; Wu, B.; Zhang, H. Parametric prediction models of DBH and height for *Cunninghamia lanceolata* plantation based on crown width. *J. Beijing For. Univ.* **2016**, *38*, 55–63. [CrossRef]
43. Mayerich, D.; Chen, G.; Childs, H. Unreal Chimera: An Unreal Engine Platform for Scientific Visualization of Next-Generation Data. Available online: https://stim.ee.uh.edu/resources/software/unreal-microscopy/ (accessed on 20 February 2022).
44. Zhang, H. Researchs on Single Wood Growth Model of *Fokienia hodginsii*. *Cent. South For. Inventory Plan.* **2006**, *25*, 1–4.
45. Pretzsch, H. Tree growth as affected by stem and crown structure. *Trees* **2021**, *35*, 947–960. [CrossRef]
46. González-Moreno, P.; Quero, J.L.; Poorter, L.; Bonet, F.J.; Zamora, R. Is spatial structure the key to promote plant diversity in mediterranean forest plantations? *Basic Appl. Ecol.* **2011**, *12*, 251–259. [CrossRef]

47. Duchateau, E.; Schneider, R.; Tremblay, S.; Dupont-Leduc, L.; Pretzsch, H. Modelling the spatial structure of white spruce plantations and their changes after various thinning treatments. *Forests* **2021**, *12*, 740. [CrossRef]
48. Robert, A. Simulation of the effect of topography and tree falls on stand dynamics and stand structure of tropical forests. *Ecol. Model.* **2003**, *167*, 287–303. [CrossRef]
49. Gonzalez-Benecke, C.A.; Gezan, S.A.; Leduc, D.J.; Martin, T.A.; Cropper, W.P., Jr.; Samuelson, L.J. Modeling survival, yield, volume partitioning and their response to thinning for longleaf pine plantations. *Forests* **2012**, *3*, 1104–1132. [CrossRef]
50. Qiu, H.; Liu, S.; Zhang, Y.; Li, J. Variation in height-diameter allometry of ponderosa pine along competition, climate, and species diversity gradients in the western united states. *For. Ecol. Manag.* **2021**, *497*, 119477. [CrossRef]
51. Zhang, M.; Liu, N.; Harper, R.; Li, Q.; Liu, K.; Wei, X.; Ning, D.; Hou, Y.; Liu, S. A global review on hydrological responses to forest change across multiple spatial scales: Importance of scale, climate, forest type and hydrological regime. *J. Hydrol.* **2017**, *546*, 44–59. [CrossRef]
52. Lu, K.; Zhang, H. A Review of Plant Architecture. *World For. Res.* **2010**, *23*, 17–20. [CrossRef]

forests

MDPI

Article

A Novel 3D Tree-Modeling Method of Incorporating Small-Scale Spatial Structure Parameters in a Heterogeneous Forest Environment

Linlong Wang [1,2,3], Huaiqing Zhang [1,3,*], Huacong Zhang [1,4], Tingdong Yang [1,3], Jing Zhang [1,3] and Yang Liu [1]

1 Institute of Forest Resource Information Techniques, Chinese Academy of Forestry, Beijing 100091, China; wangll@ifrit.ac.cn (L.W.); zhanghc@ifrit.ac.cn (H.Z.); yangtd@ifrit.ac.cn (T.Y.); zhangjing@ifrit.ac.cn (J.Z.); liuyang@ifrit.ac.cn (Y.L.)

2 Research Institute of Forest Policy and Information, Chinese Academy of Forestry, Beijing 100091, China

3 Key Laboratory of Forest Management and Growth Modeling, NFGA, Beijing 100091, China

4 Experimental Center of Subtropical Forestry, Chinese Academy of Forestry, Fenyi 336600, China

* Correspondence: zhang@ifrit.ac.cn; Tel.: +86-136-5116-6672

Abstract: Currently, 3D tree modeling in a highly heterogeneous forest environment remains a significant challenge for the modeler. Previous research has only focused on morphological characteristics and parameters, overlooking the impact of micro-environmental factors (e.g., spatial-structural diversification and habitat heterogeneity) and providing less structural information about the individual tree and decreasing the applicability and authenticity of 3D tree models in a virtual forest. In this paper, we chose a mixed-forest conversion of Chinese fir (*Cunninghamia lanceolata*) plantations in a subtropical region of China as our study subject and proposed a novel 3D tree-modeling method based on a structural unit (TMSU). Our approach modified traditional rule-based tree modeling (RTM) by introducing a nonlinear mixed-effect model (NLME) to study the coupling response between the spatial structures and morphological characteristics (e.g., tree height (H), height-to-crown base (H_{CB}), and crown width (CW)) of three dominant trees (e.g., *Cunninghamia lanceolata* (SM), *Machilus pauhoi* (BHN), and *Schima superba* (MH)) and develop a prediction model of the morphological characteristic by incorporating forest-based structural parameters. The results showed that: (1) The NLME model in TMSU was found to better fit the data and predict the morphological characteristics than the OLS model in RTM. As compared to the RTM morphological model, the prediction accuracy of the TMSU model of morphological features was improved by 10.4%, 3.02%, and 17.8%, for SM's H, H_{CB}, and CW, respectively; 6.5%, 7.6%, and 8.9% for BHN's H, H_{CB}, and CW, respectively; and 13.3%, 15.7%, and 13.4% for MH's H, H_{CB}, and CW, respectively. (2) The spatial-structural parameters of crowding (C_i), mingling (M_i), and dominance (U_i) had a significant impact on the morphological characteristics of SM, BHN, and MH in TMSU. The degree of crowding, for example, had a positive relationship with tree height, height-to-crown base, and crown width in SM, BHN, and MH; under the same crowding conditions, mingling was positively correlated with tree crown width in SM, and dominance was positively correlated with tree height but negatively correlated with height-to-crown base in BHN; under the same crowding and mingling, dominance was positively correlated with height-to-crown base in MH. (3) Using 25 scenes based on the value class of C_i, M_i for SM, 25 scenes based on the value class of C_i, U_i for BHN, and 125 scenes based on the value class of C_i, M_i, U_i for MH, we generated the model libraries for the three dominating species based on TMSU. As a result, our TSMU method outperformed the traditional 3D tree-modeling method RTM in a complex and highly heterogeneous spatial structure of a forest stand, and it provided more information concerning the spatial structure based on the neighborhood relationships than the simple morphological characteristics; a higher morphological prediction accuracy with fewer parameters; and the relationship between the spatial-structural parameters and the morphological characteristics of a reference tree.

Keywords: tree modeling; spatial structure; heterogeneous conversion forest; morphological characteristics model

1. Introduction

The current 3D tree-modeling methods, such as those based on LIDAR point-cloud data and RGB-D images, have high precision but are costly and computationally intensive [1–3]. Furthermore, it is not appropriate to reconstruct 3D tree models in a dense and highly heterogeneous forest stand because there is insufficient space to obtain sufficient data. Therefore, 3D tree modeling, based on forestry rules, provides a set of generative rules to create plant organs iteratively, though it requires expertise in botany to manually adjust the many parameters in order to model a specific tree [4]. This method has proven to be an effective and popular method for virtual forest reconstruction [5,6]. For example, functional–structural plant models (FSPMs) require specific parameters to model a tree, such as tree height, diameter at breast height, crown width, and height-to-crown base [7]. Therefore, rule-based tree modeling can be easily controlled by modelers without a computer science background [8]. However, the applicability of rule-based modeling is limited because it is based on only a few parameters and overlooks the influence of micro-environmental factors (e.g., spatial-structure diversification and habitat heterogeneity) in a complex and changing forest ecosystem.

Exploring the coupling response between variables related to the spatial structure and morphological characteristics is vital in 3D tree modeling based on a structural unit in a highly heterogeneous forest environment. However, the current research has mostly focused on the coupling responses between unilateral or incomplete characteristics of the spatial structure and tree-crown attributes. For example, researchers suggested that when mixed with other species, the reference tree changed the typical structural attributes of the crown [9–12], such as branching patterns [10,13] and the general shape of the crown [11–14]. These changes could be attributed to the amount of light available, which is the most important resource for which trees compete in a forest canopy [12–15]. Researchers discovered that trees adjusted the shapes and sizes of their crowns in response to competition with their neighbors under different spatial-structure conditions [16], such as in crowded conditions [17,18]. With the rapid application of airborne laser-scanning techniques, more studies are being conducted at the individual tree level and in a more intelligent direction. For example, Pommerening et al. (2021) introduced the CanopyShotNoise model to analyze the response of the canopy pattern under different forest conditions [19]. Fischer et al. (2020) proposed a new simulation-based approach, the canopy constructor, to quantify the forest structure and infer the allometric-scaling relationship in a crowded environment [20].

Morphological models, such as height–diameter, crown width–diameter, and height-to-crown base height–diameter, are critical components of rule-based 3D tree modeling. The prediction accuracy of a model significantly influences the applicability and authenticity of 3D tree models in a virtual forest. Traditional morphological models have adopted the ordinary least squares (OLS) regression for the simple form and its variables. However, the morphological data from many sample plots have hierarchical structures and may be correlated, which has resulted in biased estimations of the variance of the morphological parameters and invalidated the hypothesis when the parameter estimates were analyzed [21,22]. Lately, the popularity and use of mixed-effect models have increased in forestry and morphological modeling practices [23–25]. Researchers have suggested that, as compared to OLS, the nonlinear mixed-effect model (NLME) was more reliable due to its assumption of random and independent observations, as well as the presence of autocorrelations [26–29].

Therefore, in order to model 3D trees in a spatially diverse and heterogeneous forest environment, we chose a mixed forest of Chinese fir (*Cunninghamia lanceolata*) as our subject, which was based on a typical mixed forest stand in a subtropical region of China. The following two aspects were explored in this study: (1) to develop the morphological characteristics (e.g., tree height, height-to-crown base, and crown width) of the model at the individual tree level and with highly heterogeneous spatial structures by introducing a NLME model and (2) to optimize the traditional rule-based 3D tree-modeling method by incorporating individual parameters of spatial structures based on a small-scale forest stand in order to develop a new dominant species' model library under heterogeneous forest conditions.

2. Materials and Methods

2.1. Study Areas

Our study site was located in the ShanXia Forest Farm in Jiangxi Province in Southern China (27°44′45″ N, 114°39′15″ E) (see Figure 1). The site has a mid-subtropical climate and a mean annual temperature of 19.15 °C, with the highest temperature of 40.3 °C occurring in June and July and the lowest temperature of −2.1 °C occurring in February. Furthermore, annual precipitation totaled 1722.3 mm, with the majority of the precipitation falling in April, May, and June. The annual photoperiod was 1657.4 h, with the longest photoperiod occurring in July, August, September, and October and the shortest occurring in January and February. The Chinese fir mixed plantations selected in this study were established in 1986 with an initial planting spacing of 1m × 1m. Thinning measures were taken when the plantations were 13 years and 20 years old. The first thinning measure was to reduce the intraspecific competition for resource by increasing the growth space and the second thinning measure was to fell the low-quality trees and replant broadleaved trees, such as *Machilus pauhoi* and *Schima superba*, which are high-value, fire-resistant trees. During the development of the mixed plantations, some Chinese firs died, allowing companion species or generations to develop in natural succession, such as *Vernicia fordii, Liquidambar formosana Hance, Paulowinia fortune, Sapium discolorr, Castanopsis fargesii*, etc.

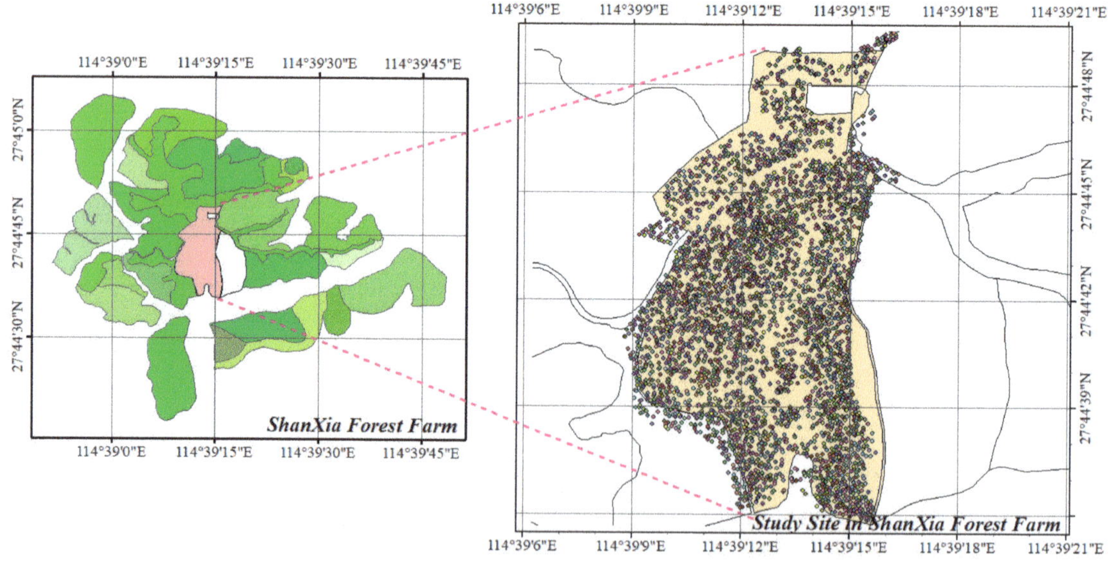

Figure 1. Location of our study site in a subtropical area of China. Notes: the pink area in the left map represents the study region at ShanXia Forest Farm; colorful points in the right map represent different tree species in the study site.

2.2. Data Collection

The data was acquired from 5 ha sample plots of a planted mixed forest dominated by *Cunninghamia lanceolata, Machilus pauhoi*, and *Schima superba* in March and April 2022 (Figure 2). To accurately locate the trees, the plot was divided into 10 subplots of 100 m

× 50 m each (Figure 2). Tree height (H), height-to-crown base (H_{CB}), diameter at breast height (DBH), and crown width (CW) were measured for all trees with a diameter at breast height of (DBH) $\geq$ 5 cm, and their locations were recorded using a total station (TOPCON-GTS-602AF). Crown width was computed as the arithmetic mean of two crown widths obtained from measurements of four crown radii, representing two azimuths [30]. The first azimuth was defined as the direction from the subject tree to the center of the measurement plot; the second azimuth was perpendicular to the first. In each quadrant, the crown radii were measured as the horizontal distance from the center of the tree bole to the greatest extent of the crown from the bole. The branch tip was located by vertical sighting with a clinometer [31]. Figure 2 depicts a distribution map of the tree species, and each tree population was assigned three layer classes, based on tree height, to select the dominant trees: upper layer trees were H $\geq$ 20 m; middle layer trees were 10 m $\leq$ H < 20 m; and lower layer trees were 1.3 m $\leq$ H < 10 m.

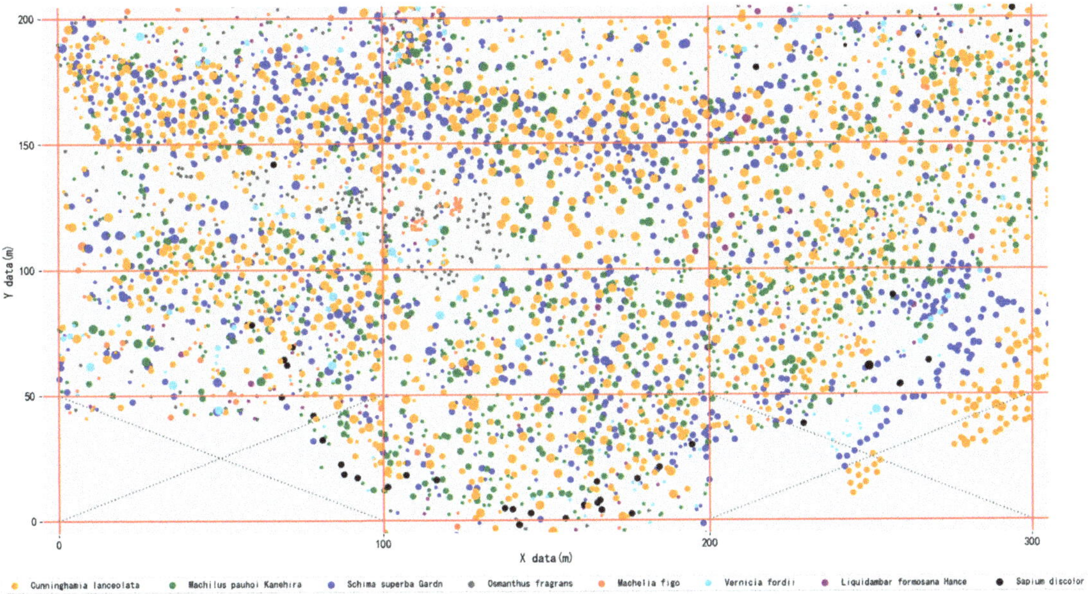

Figure 2. Spatial distribution of eight dominant tree populations at different layers in 5 ha (50 m × 100 m) planted mixed forest in Shanxia Forest Farm. The upper layer tree was tree height (H) > 20 m; the middle layer tree was 10 m $\leq$ H <20 m; and the lower layer tree was 1.3 m $\leq$ H < 10 m. Different sizes of circles represent the layers in the forest, where the large-size circle is the upper layer and the middle-size circle is the middle layer, and the small-size circle is the lower layer in the mixed forest.

To eliminate edge effects and improve the accuracy of the spatial-structure parameters (SSP), we established a 5 m buffer zone around the plot. In the statistical analysis, only the trees in the reduced window (X, Y) = (300 m × 200 m, except (0 m–100 m, 0–50), (200 m–300 m, 0 m–50 m)) were used as reference trees, and the individual trees in the buffer zone were only considered to be the nearest neighbors of the trees in the reduced window [32]. This edge correction enabled the individual evaluation of each tree to determine whether all n-nearest-neighbors were actually located within the plot.

2.3. Data Analysis

2.3.1. Identification of Dominant Populations for 3D Tree Modeling by the Important Value Index

The importance value (IV) of a species was defined as the average of its relative density (RD), relative frequency (RF), and relative dominance (Rd) and was calculated using the quadratic method [33] for each layer of the planted mixed forest. A total of 32 tree species belonging to 30 genera and 21 families were identified in the planted mixed forest. The total stand density (DBH $\geq$ 5 cm) was nearly 996.8 trees/ha, with a total basal area of 161.2 m^2/ha. Table 1 shows the IVs of the forest's eight dominant tree populations. *Cunninghamia lanceolata, Machilus pauhoi,* and *Schima superba* accounted for 88.24% of the total stand density, and their IVs were clearly superior to those of the other tree species in different stand layers. As a result, they were identified as the dominant populations in the forest for 3D tree modeling (Table 1).

Table 1. *IV* and characteristics of eight dominant tree populations in the planted mixed forest.

Species	Density	*IV* (%)				Stand Characteristics			
	(Trees ha^{-1})	Upper Layer	Median Layer	Lower Layer	All	DBH (cm)	Height (m)	Basal Area (m^2)	Dominant Height (m)
Cunninghamia lanceolata (Lamb.) Hook.	273.0	14.57 ± 7.14	15.13 ± 5.19	3.28 ± 1.86	17.18 ± 3.74	24.86 ± 8.91	16.50 ± 5.40	80.83	26.8
Machilus pauhoi Kanehira	252.6	2.53 ± 1.68	10.38 ± 2.73	7.04 ± 3.13	11.13 ± 3.44	13.82 ± 8.25	10.63 ± 3.83	25.72	23.5
Schima superba Gardn. et Champ.	259.0	1.21 ± 1.25	12.78 ± 4.21	4.32 ± 1.88	12.39 ± 4.39	15.32 ± 7.02	11.85 ± 3.55	27.67	23.4
Osmanthus fragrans (Thunb.) Lour.	62.0	0.00 ± 0.00	0.00 ± 0.00	1.82 ± 3.56	1.65 ± 2.94	10.37 ± 3.31	5.74 ± 1.12	2.88	8.2
Machelia figo (Lour.) Spreng.	48.0	0.07 ± 0.21	0.81 ± 0.90	1.12 ± 1.35	1.59 ± 1.60	13.69 ± 8.03	9.30 ± 3.76	4.36	18.8
Vernicia fordii (Hemsl.) Airy Shaw	35.6	0.21 ± 0.55	1.34 ± 1.23	0.41 ± 0.74	1.52 ± 1.65	22.60 ± 9.47	14.21 ± 3.40	7.54	21.1
Liquidambar formosana Hance	27.0	0.03 ± 0.09	0.63 ± 0.90	0.74 ± 0.86	1.24 ± 1.43	16.26 ± 7.63	12.19 ± 3.45	3.01	18.2
Sapium discolor (Champ. ex Benth.) Muell. Arg.	21.8	0.00 ± 0.00	0.91 ± 1.76	0.13 ± 0.40	0.74 ± 1.21	22.22 ± 7.83	13.33 ± 1.92	4.21	15.4

Notes: *IVs*, DBH, and Height are mean ± SD; the upper layer was tree height (H) $\geq$ 20 m; the middle layer was 10 m $\leq$ tree height (H) < 20 m; and the lower layer was 1.3 m tree height (H) < 10 m.

2.3.2. Nonlinear Mixed-Effect Morphological Model of Three Predominant Trees

(1) Base morphological model selection.

In this paper, we chose 16 H~DBH models, 5 HCB~H and DBH models, and 11 CW~DBH models from the literature for appraisal as the base models [27,28,34,35]. These candidate base models were fitted to the 10 plot data (5 ha) using nonlinear least squares regression. The three statistical variables (e.g., coefficient of determination (R^2), mean error (E), root-mean-squared error (RMSE)) were used to evaluate the performance of the base models (see Figure 3). The H, H_{CB}, and CW base models evaluated in this study are shown in Table 2.

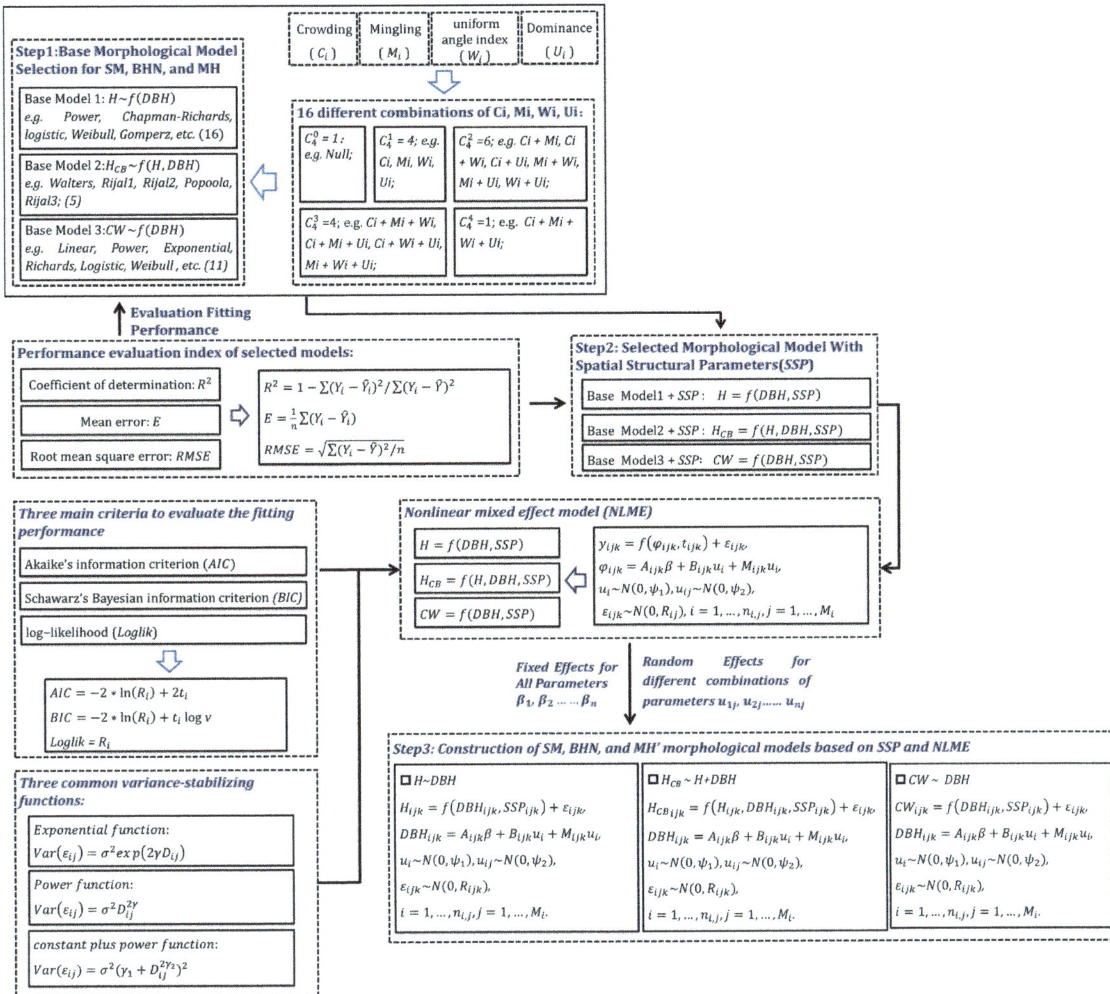

Figure 3. Schematic diagram showing the nonlinear mixed-effect morphological model with spatial-structure parameters (SSP). It includes three steps of morphological construction based on SSP and NLME, evaluation index, and variance-stabilizing functions. The first step is the selection of the base morphological models of SM, BHN, and MH; the second step is the selection of the morphological model with spatial-structure parameters (SSP); and the third step is the construction of the nonlinear mixed-effect of SM, BHN, and MH. SM, BHN, and MH represent *Cunninghamia lanceolata*, *Machilus pauhoi*, and *Schima superba*, respectively.

Table 2. Base model selection: H is the total tree height (m), DBH is the diameter at breast height (cm), H_{CB} is the height-to-crown base (m), and CW is the crown width (m); a, b, and c are the parameters of the model. M1~M32 represents the formulae of the morphological characteristic models.

Model	Formula	Model Form	Model	Formula	Model Form
M1	$H \sim 1.3 + DBH^2/(a*DBH + b)^2$	Naslund	M17	$H_{CB} \sim H/(1 + exp(a + b*DBH))$	Walters
M2	$H \sim 1.3 + a*DBH/(1 + DBH)^b$	Curtis	M18	$H_{CB} \sim H/(1 + exp(a + b*DBH))^{1/2}$	Rijal 1
M3	$H \sim 1.3 + a*exp(-b/DBH)$	Schumacher	M19	$H_{CB} \sim H/(1 + exp(a + b*DBH))^{1/6}$	Rijal 2
M4	$H \sim 1.3 + a*(1 - exp(-b*DBH))$	Meyer	M20	$H_{CB} \sim H/(a + exp(-b*DBH))$	Popoola
M5	$H \sim 1.3 + a*DBH^b$	Power	M21	$H_{CB} \sim H/(1 - a*exp(-b*DBH))$	Rijal 3
M6	$H \sim 1.3 + a*DBH/(b + DBH)$	Michaelis-Menten	M22	$CW \sim a + b*DBH$	Linear
M7	$H \sim 1.3 + exp(a - b/(DBH + 1))$	Wykoff	M23	$CW \sim a + b*DBH + c*DBH^2$	Quadratic
M8	$H \sim 1.3 + a*((1 - exp(-b*DBH)))$	Chapman-Richards	M24	$CW \sim a*DBH^b$	Power
M9	$H \sim 1.3 + DBH^2/(a*DBH^2 + b*DBH + c)$	Prodan	M25	$CW \sim a*(1-exp(-b*DBH))$	Monomolecular
M10	$H \sim 1.3 + a/(1 + b*exp(-c*DBH))$	logistic	M26	$CW \sim (DBH/(a + b*DBH))^2$	Hossfeld I
M11	$H \sim 1.3 + a*(1-exp(-b*DBH^c))$	Weibull	M27	$CW \sim a*(b)^{DBH}$	Compound
M12	$H \sim 1.3 + a*exp(-b*exp(-c*DBH))$	Gomperz	M28	$CW \sim exp(a + b*DBH)$	Growth
M13	$H \sim 1.3 + a*DBH^{(b*DBH-c)}$	Sibbesen	M29	$CW \sim a*exp(b*DBH)$	Exponential
M14	$H \sim 1.3 + a*exp(-b*DBH^{-c})$	Korf	M30	$CW \sim a*(1 - exp(-b*DBH))$	Richards
M15	$H \sim 1.3 + a*exp(-b/(DBH + c))$	Ratkowsky	M31	$CW \sim a/(1 + b*exp(-c*DBH))$	Logistic
M16	$H \sim 1.3 + a/(1 + 1/(b*DBH^c))$	Hossfeld IV	M32	$CW \sim a*(1 - exp(-b*DBH))$	Weibull

(2) Additional predictor variables.

A structural unit was defined as a neighborhood consisting of a focal tree and its four nearest neighbors [36]. In a mixed forest, tree size (DBH or crown), tree crowding, tree species, and tree distribution in space were used to describe any structural unit synchronously. These variables were easily expressed by a set of structural stand parameters: W_i, M_i, C_i, and U_i [37,38]. Therefore, there were 16 different combinations of W_i, M_i, C_i, and U_i for selecting the best performing model according to two rules: higher prediction accuracy and lower computational cost (see Figure 3).

$$W_i = \frac{1}{n}\sum\nolimits_{j=1}^{n} Z_{ij}, Z_{ij} = \begin{cases} 1, & if\ \alpha - angle\ is\ smaller\ than\ \alpha_0 \\ 0, & otherwise \end{cases} \tag{1}$$

where W_i is defined as the proportion of angles α smaller than the standard angle α_0 (72°) and reflects the degree of uniformity of tree distribution.

$$M_i = \frac{1}{n}\sum\nolimits_{j=1}^{n} V_{ij}, V_{ij} = \begin{cases} 1, & if\ neighbor\ j\ is\ not\ the\ same\ species\ as\ reference\ tree\ i \\ 0, & otherwise \end{cases} \tag{2}$$

where M_i is defined as the proportion of the four nearest neighbors that are of a different species than reference tree i, and it reflects the spatial segregation of different species in a multispecies forest.

$$U_i = \frac{1}{n}\sum\nolimits_{j=1}^{n} K_{ij}, K_{ij} = \begin{cases} 0, & if\ neighbor\ j\ is\ smaller\ than\ reference\ tree\ i \\ 1, & otherwise \end{cases} \tag{3}$$

where U_i is defined as the degree of DBH differentiation in this paper and represents the relationship between the size of the reference tree i and its four nearest neighbors.

$$C_i = \frac{1}{n}\sum\nolimits_{j=1}^{n} P_{ij}, P_{ij} = \begin{cases} 1, & if\ average\ crowns\ of\ neighbor\ j\ and\ reference\ tree\ i\ is\ smaller\ than\ that\ of\ reference\ i \\ 0, & otherwise \end{cases} \tag{4}$$

where C_i reflects the relationship between the canopy of the reference tree i and its four nearest neighbors and can reveal the degree of crowding of the reference tree i.

(3) Construction of mixed-effect morphological model.

We have incorporated the fixed effects for all parameters in the morphological characteristics models and compared different combinations of all parameters with random effects in the model to identify the parameters that needed to be included, according to the value of the Akaike information criteria (AIC) and Schwarz's Bayesian information criterion (BIC), which were the main criteria for evaluating the fitting performance. In order to solve the problem of heteroscedasticity in the morphological models, we used three typical variance-stabilizing functions (e.g., exponential function, power function, and constant-plus-power function) (4–6) (Figure 3) and selected the best-performing function based on the AIC and the log-likelihood (Loglik) values (see Figure 3).

The graphic exploration and correlational analysis were used for evaluation of the contributions of the covariate predictors to the H, H_{CB}, and CW models. All analyses were conducted using R, version 4.1.2. The best-performing model was selected to include additional covariate predictors for formulations of the H, H_{CB}, and CW mixed-effect models.

(4) Assessment and analysis of morphological models

The main differences between rule-based tree modeling (RTM) and 3D tree modeling based on a structural unit (TMSU) were the morphological models' parameter-estimation model and whether the tree modeling considered the spatial structure. In general, the parameter estimation in a traditional morphological model in RTM was based on the ordinary least squares (OLS) model, whereas in TMSU, it was based on a nonlinear mixed-effect model (NLME). In order to validate the feasibility and superiority of the morphological models in TMSU for 3D tree modeling, statistics (R^2, E, RMSE) and a residual plot of three dominant tree-based morphological models were applied to assess and analyze the morphological models based on OLS and NLME.

2.3.2.1. 3D Tree Modeling Based on a Structural Unit

Rule-based tree modeling (RTM) employs morphological parameters, such as H, H_{CB}, DBH, and CW, to reconstruct a reference tree *i*. However, ignoring the influence of spatial structure on reference tree *i* at a small scale would result in a decrease in the quality of 3D tree modeling and the reality of a virtual forest in forest visualization applications. Our research proposed a new method of 3D tree modeling based on a structural unit (TMSU) that incorporated structural forest theory into 3D tree modeling and achieved 3D tree modeling at the individual tree level with few parameters (see Figure 4). A nonlinear mixed-effect model was introduced in this study to improve the prediction accuracy of the morphological characteristics and parameters (e.g., H, H_{CB}, CW) by avoiding the random effect from the plot and increasing the applicability of the method by reducing the time and computational demand. Based on the predicted morphological data, we reconstructed a 3D tree model with our Creating Tree System (CTS) [39–42], which was developed by the Institute of Forest Resource Information Techniques of the Chinese Academy of Forestry (CAF). The CTS used in this study was based on an improved IFS algorithm [43], and it provided a graphical user interface, 3D visualization tools, and a python scripting interface, as well as the ability to create 3D tree models by selecting the tree species and inputting the H, H_{CB}, DBH, and CW parameters.

Figure 4. Three-dimensional tree modeling based on RTM and TMSU.

3. Results

3.1. Three-Dimensional Tree Morphological Model Development at the Individual Tree Level

(1) Function selection.

Table 2 displays the fit statistics of the functions. Although the evaluation indices for all the functions were nearly identical, M6(Chapman − Richard for SM and BHN's H − D), M5 (Power for MH's H − D), M17 (Walters for SM, BHN, and MH's H_{CB} − D), M31(Logistic for SM and BHN's CW − D), and M24 (Power for MH's CW − D) performed slightly better with the fewest parameters. As a result, these functions were chosen as the basic nonlinear model for building the morphological model.

Morphological models of the three predominant trees were the following:

$$\text{SM: } H_{ij} = 1.3 + a * (1 - exp(-b * D_{ij})) + \varepsilon_{ij} \quad (5)$$

$$H_{CB\,ij} = H_{ij}/(1 + exp(a + b * D_{ij})) + \varepsilon_{ij} \quad (6)$$

$$CW_{ij} = a/(1 + b * exp(-c * D_{ij})) + \varepsilon_{ij} \quad (7)$$

$$\text{BHN: } H_{ij} = 1.3 + a * (1 - exp(-b * D_{ij})) + \varepsilon_{ij} \quad (8)$$

$$H_{CB\,ij} = H_{ij}/(1 + exp(a + b * D_{ij})) + \varepsilon_{ij} \quad (9)$$

$$CW_{ij} = a/(1 + b * exp(-c * D_{ij})) + \varepsilon_{ij} \quad (10)$$

$$\text{MH: } H_{ij} = 1.3 + a * {D_{ij}}^{b} + \varepsilon_{ij} \quad (11)$$

$$H_{CB\,ij} = H_{ij}/(1 + exp(a + (b * M_i + c * C_i + d * U_i) * D_{ij})) + \varepsilon_{ij} \quad (12)$$

$$CW_{ij} = a * {D_{ij}}^{b} + \varepsilon_{ij} \quad (13)$$

where H_{ij}, D_{ij}, H_{CBij}, CW_{ij} are the height (m), the diameter at breast height (cm), the height-to-crown base (m), and the crown width (m) of tree *i* in the plot *j*; *a*, *b*, and *c* are the formal parameters; and ε_{ij} is an error term.

(2) Incorporating additional predictor variables.

We tested 16 different combinations of W_i, M_i, C_i, and U_i to select the variables that made a significant contribution to the variation of the morphological characteristic (H, H_{CB}, CW) in order to determine the influence of the spatial structure of individual trees. The statistical results showed that C_i, M_i, and U_i were strongly correlated with the morphological variation and, thus, were included in Equations (5)–(13). The following was the final base model used to build the morphological variation mixed-effect model:

$$\text{SM: } H_{ij} = 1.3 + a * (1 - exp((-b + c * C_i) * D_{ij})) + \varepsilon_{ij} \quad (14)$$

$$H_{CB_{ij}} = H_{ij}/(1 + exp(a + (b + c * C_i) * D_{ij})) + \varepsilon_{ij} \quad (15)$$

$$CW_{ij} = a/(1 + b * exp((-c + d * C_i + e * M_i) * D_{ij})) + \varepsilon_{ij} \quad (16)$$

$$\text{BHN: } H_{ij} = 1.3 + a * (1 - exp((-b + c * C_i + d * U_i) * D_{ij})) + \varepsilon_{ij} \quad (17)$$

$$H_{CB_{ij}} = H_{ij}/(1 + exp(a + (b * C_i + c * U_i) * D_{ij})) + \varepsilon_{ij} \quad (18)$$

$$CW_{ij} = a/(1 + b * exp((-c + d * C_i) * D_{ij})) + \varepsilon_{ij} \quad (19)$$

$$\text{MH: } H_{ij} = 1.3 + (a + c * M_i + d * C_i) * {D_{ij}}^b + \varepsilon_{ij} \quad (20)$$

$$H_{CB_{ij}} = H_{ij}/(1 + exp(a + (b * M_i + c * C_i + d * U_i) * D_{ij})) + \varepsilon_{ij} \quad (21)$$

$$CW_{ij} = (a + c * C_i + d * M_i) * {D_{ij}}^b + \varepsilon_{ij} \quad (22)$$

where H_{ij}, D_{ij}, H_{CBij}, CW_{ij} are the height (m), the diameter at breast height (cm), the height-to-crown base (m), and the crown width (m) of tree i in the plot j; a, b, and c are the formal parameters; C_i, M_i, and U_i are the degree of crowding, mingling, and dominance of the reference tree i base on structural unit; and ε_{ij} is an error term.

(3) Constructing mixed-effect model with parameters for the spatial structure.

After considering the model parameters (a, b, c, d, e) involved, there were seven different combinations of random effects at the sample plot level for Equations (14), (15) and (18), 15 different combinations for Equations (17), (19) and (20)–(22), and 31 different combinations for Equation (16). When fitted to the data, only model alternatives 5, 6, and 10 converged in Equations (14)–(16), respectively; whereas 12, 6, and 1 converged in Equations (17)–(19), respectively. Model alternatives 5, 9, and 6 converged in Equations (20)–(22), respectively. Among the models that converged, the NLME model with random parameters on $a/b/c$ in Equation (14), c in Equation (15), a/b in Equation (16), b/c in Equation (17), b/c in Equation (18), $a/b/d$ in Equation (19), a in Equation (20), a in Equation (21), and c/d in Equation (22) showed the smallest AIC and BIC. Three dominant trees' morphological models with mixed effects became the following:

$$\text{SM: } H_{ij} = 1.3 + (\beta_1 + u_{1j}) * (1 - exp((-\beta_2 - u_{2j} + (\beta_3 + u_{3j}) * C_i) * D_{ij})) + \varepsilon_{ij} \quad (23)$$

$$H_{CB_{ij}} = H_{ij}/(1 + exp(\beta_1 + (\beta_2 + (\beta_3 + u_{1j}) * C_i) * D_{ij})) + \varepsilon_{ij} \quad (24)$$

$$CW_{ij} = (\beta_1 + u_{1j})/(1 + (\beta_2 + u_{2j}) * exp((-\beta_3 + \beta_4 * C_i + \beta_5 * M_i) * D_{ij})) + \varepsilon_{ij} \quad (25)$$

$$\text{BHN: } H_{ij} = 1.3 + \beta_1 * (1 - exp((-\beta_2 - u_{1j} + (\beta_3 + u_{2j}) * C_i + \beta_4 * U_i) * D_{ij})) + \varepsilon_{ij} \quad (26)$$

$$H_{CB_{ij}} = H_{ij}/(1 + exp(\beta_1 + (\beta_2 + u_{1j}) * C_i + (\beta_3 + u_{2j}) * U_i) * D_{ij})) + \varepsilon_{ij} \quad (27)$$

$$CW_{ij} = (\beta_1 + u_{1j})/(1 + (\beta_2 + u_{2j}) * exp((-\beta_3 + (\beta_4 + u_{3j}) * C_i) * D_{ij})) + \varepsilon_{ij} \quad (28)$$

$$\text{MH: } H_{ij} = 1.3 + (\beta_1 + u_{1j} + \beta_3 * M_i + \beta_4 * C_i) * {D_{ij}}^{\beta_2} + \varepsilon_{ij} \quad (29)$$

$$H_{CB_{ij}} = H_{ij}/(1 + exp(\beta_1 + u_{1j} + (\beta_2 * M_i + \beta_3 * C_i + \beta_4 * U_i) * D_{ij})) + \varepsilon_{ij} \quad (30)$$

$$CW_{ij} = (\beta_1 + (\beta_3 + u_{1j}) * C_i + (\beta_4 + u_{2j}) * M_i) * {D_{ij}}^{\beta_2} + \varepsilon_{ij} \quad (31)$$

where $\beta_1 - \beta_5$ are mixed-effect parameters; u_{1j}, u_{2j}, u_{3j}, and u_{4j} are random-effect parameters generated by the plot on $\beta_1, \beta_2, \beta_3, \beta_4$, respectively; and ε_{ij} is an error term.

(4) Within-plot variance–covariance (R) structure.

Because our data did not show spatial autocorrelations, we assumed the matrix of the within-plot error autocorrelation was an identity matrix. However, we found heteroscedasticity in our preliminary analysis even after the random effects were included in the morphological model with SSP. We then evaluated three variance functions for their heteroscedasticity-reducing performances. The results showed that Equations (24), (27)–(30) and (31) produced significantly different results than the models without variance functions, whereas Equations (23), (25) and (26) did not. According to the AIC and Loglik values, the power variance function and constant power variance function performed the best among the tested variance functions (Table 3). Therefore, the final modes were the following: Equation (23) + null, Equation (24) + power function, Equation (25) + power function, Equation (26) + power function, Equation (27) + power function, Equation (28) + constant plus power function, Equation (29) + power function, Equation (30) + power function, and Equation (31) + constant plus power function. By including the variance function, the model's AIC decreased from 3438.982 to 3421.136 in Equation (24), from 1935.082 to 1934.881 in Equation (25), from 3825.354 to 1934.881 in Equation (26), from 3322.544 to 3231.132 in Equation (27), from 3160.599 to 2342.501 in Equation (28), from 4316.121 to 4310.291 in Equation (29), from 3798.104 to 3699.936 in Equation (30), and from 2462.686 to 2416.827 in Equation (31).

Table 3. Performance of mixed-effect model Equations (23)–(31) using morphological data with different variance functions. SM, BHN, and MH represent *Cunninghamia lanceolata*, *Machilus pauhoi*, and *Schima superba*, respectively.

Species	Model	Evaluation Indices	Residual Variance Function			
			Null	Exponential Function	Power Function	Constant Plus Power Function
SM	Equation (23)	AIC	4336.224	4337.334	4337.458	4339.477
		Loglik	−2158.11	−2157.667	−2157.729	−2157.739
		p Value		$p > 0.05$	$p > 0.05$	$p > 0.05$
	Equation (24)	AIC	3438.982	3421.818	3421.136	3423.136
		Loglik	−1712.491	−1702.909	−1702.568	−1702.568
		p Value		$p < 0.0001$	$p < 0.0001$	$p < 0.0001$
	Equation (25)	AIC	1935.082	1935.461	1934.881	1936.881
		Loglik	−960.5411	−959.7304	−959.441	−959.4404
		p Value		$p > 0.05$	$p > 0.05$	$p > 0.05$
BHN	Equation (26)	AIC	3825.354	3827.209	3825.14	3827.14
		Loglik	−1906.677	−1906.604	−1905.57	−1905.57
		p Value		$p > 0.05$	$p > 0.05$	$p > 0.05$
	Equation (27)	AIC	3322.544	3249.631	3231.132	3235.042
		Loglik	−1906.677	−1616.815	−1607.57	−1608.521
		p Value		$p < 0.0001$	$p < 0.0001$	$p < 0.0001$
	Equation (28)	AIC	3160.599	2365.384	2449.84	2342.501
		Loglik	−1574.299	−1175.692	−1217.92	−1163.251
		p Value	$p < 0.0001$	$p < 0.0001$	$p < 0.0001$	$p < 0.0001$
MH	Equation (29)	AIC	4316.121	4313.26	4310.291	4312.291
		Loglik	−2150.061	−2147.63	−2146.145	−2146.145
		p Value		$p < 0.05$	$p < 0.05$	$p < 0.05$
	Equation (30)	AIC	3798.104	3701.172	3699.936	3701.532
		Loglik	−1888.052	−1838.586	−1837.968	−1837.766
		p Value		$p < 0.0001$	$p < 0.0001$	$p < 0.0001$
	Equation (31)	AIC	2462.686	2418.486	2427.477	2416.827
		Loglik	−1223.343	−1200.243	−1204.739	−1198.413
		p Value		$p < 0.0001$	$p < 0.0001$	$p < 0.0001$

(5) Estimation of parameters.

All parameter estimates of the OLS and NLME models at the sample plot level with the power variance function or constant plus power variance function were significantly different from zero ($p < 0.05$), and the parameters are listed in Table 4. When the fixed parameters were substituted into Equations (23)–(31), the morphological model for SM, BHN, and MH, respectively, became the following:

$$M:\ H_{ij} = 1.3 + (36.7543 + u_{1j}) * (1 - exp((-0.0236 - u_{2j} + (-0.0045 + u_{3j}) * C_i) * D_{ij})) + \varepsilon_{ij} \quad (32)$$

$$H_{CB\,ij} = H_{ij}/(1 + exp(\,0.3896 + (0.0073 + (-0.0096 + u_{1j}) * C_i) * D_{ij})) + \varepsilon_{ij} \quad (33)$$

$$CW_{ij} = (\,5.5212 + u_{1j})/(1 + (2.4130 + u_{2j}) * exp((-0.0520 + (-0.0310 * C_i) + 0.0162 * M_i) * D_{ij})) + \varepsilon_{ij} \quad (34)$$

$$N:\ H_{ij} = 1.3 + 20.0990 * (1 - \exp((-0.0399 - u_{1j} + (-0.0133 + u_{2j}) * C_i + (-0.0078 * U_i)) * D_{ij})) + \varepsilon_{ij} \quad (35)$$

$$H_{CB\,ij} = H_{ij}/(1 + exp(0.0408 + (-0.0308 + u_{1j}) * C_i + (0.0133 + u_{2j}) * U_i) * D_{ij})) + \varepsilon_{ij} \quad (36)$$

$$CW_{ij} = (6.7421 + u_{1j})/(1 + (3.2953 + u_{2j}) * exp((-0.0721 + (-0.0455 + u_{3j}) * C_i) * D_{ij})) + \varepsilon_{ij} \quad (37)$$

$$H:\ H_{ij} = 1.3 + (2.3741 + u_{1j} - 0.2014 * M_i + 0.0622 * C_i) * {D_{ij}}^{0.5686} + \varepsilon_{ij} \quad (38)$$

$$H_{CB\,ij} = H_{ij}/(1 + exp(0.5479 + u_{1j} + (0.0277 * M_i + 0.0132 * C_i - 0.0026 * U_i) * D_{ij})) + \varepsilon_{ij} \quad (39)$$

$$CW_{ij} = (0.6514 + (0.2820 + u_{1j}) * C_i + (0.0519 + u_{2j}) * M_i) * {D_{ij}}^{0.5591} + \varepsilon_{ij} \quad (40)$$

Table 4. Parameter estimates and evaluation indices of each model. SM, BHN, and MH represent *Cunninghamia lanceolata*, *Machilus pauhoi*, and *Schima superba*, respectively. If the letters a, b, and c, are the same, it represents no significant difference between models; otherwise, there are significant differences between models. ** and *** represent significant and extremely significant, respectively.

Species	Formula	Model	Fixed Parameters					Evaluation Indexes		
			β_1	β_2	β_3	β_4	β_5	AIC	Loglik	p Value
SM	H-D	Equation (5)	33.4665	0.0266	——	——	——	4686.538	−2340.27	a
		Equation (14)	32.5791	0.0252	−0.0037	——	——	4350.543	−2168.27	b ***
		Equation (23)	36.7543	0.0236	−0.0045	——	——	4336.224	−2158.11	c ***
	H_{CB}-H	Equation (6)	0.4501	−0.0012	——	——	——	3488.336	−1741.17	a
		Equation (15)	0.4230	0.0061	−0.0087	——	——	3441.651	−1716.83	b ***
		Equation (24)	0.3896	0.0073	−0.0096	——	——	3421.136	−1702.57	c **
	CW-D	Equation (7)	5.9591	2.4952	0.0538	——	——	2224.718	−1108.36	a
		Equation (16)	5.5826	2.3971	0.0461	−0.0367	0.0153	2030.871	−1009.44	b ***
		Equation (25)	5.5212	2.4130	0.0520	−0.0310	0.0162	1934.881	−959.441	c ***
BHN	H-D	Equation (8)	19.3411	0.0540	——	——	——	3970.13	−1982.07	a
		Equation (17)	19.6529	0.0399	−0.0123	−0.0072	——	3935.849	−1962.93	b ***
		Equation (26)	20.0990	0.0399	−0.0133	−0.0078	——	3825.14	−1905.57	c ***
	H_{CB}-H	Equation (9)	0.2933	0.0043	——	——	——	3404.81	−1699.41	a
		Equation (18)	0.0384	−0.0310	0.0202	——	——	3335.732	−1663.87	b ***
		Equation (27)	0.0408	−0.0308	0.0133	——	——	3231.132	−1607.57	c ***
	CW-D	Equation (10)	8.3246	4.1547	0.0885	——	——	3218.003	−1605	a
		Equation (19)	8.4723	3.7601	0.0501	−0.0386	——	3188.058	−1589.03	b ***
		Equation (28)	6.7421	3.2953	0.0721	−0.0455	——	2342.501	−1163.25	c ***
MH	H-D	Equation (11)	2.5166	0.5378	——	——	——	4577.814	−2285.91	a
		Equation (20)	2.6616	0.5135	−0.3857	0.3370	——	4526.287	−2258.14	b ***
		Equation (29)	2.3741	0.5686	−0.2014	0.0622	——	4310.291	−2146.15	c ***
	H_{CB}-H	Equation (12)	0.8568	−0.0036	——	——	——	3980.823	−1987.41	a
		Equation (21)	0.5636	0.0253	0.0185	−0.0103	——	3919.156	−1954.58	b ***
		Equation (30)	0.5479	0.0277	0.0132	−0.0026	——	3699.936	−1837.97	c ***
	CW-D	Equation (13)	0.8478	0.5819	——	——	——	2749.965	−1371.98	a
		Equation (22)	0.5918	0.5606	0.3178	0.1041	——	2554.176	−1272.09	b ***
		Equation (31)	0.6514	0.5591	0.2820	0.0519	——	2416.827	−1198.41	c ***

In order to validate the effect of the plot level and SSP on the model parameters, we recorded the performance statistics and parameters of the final NLME and OLS models. According to the AIC, Loglik, and p-value in Table 4, As shown, Equations (23)–(31) had the smallest AIC and the largest Loglik values, and there was a significant difference between Equations (5) and (14); Equations (14) and (23); Equations (6) and (15); Equations (15) and (24); Equations (7) and (16); Equations (16) and (25); Equations (8) and (17); Equations (17) and (26); Equations (9) and (18); Equations (18) and (27); Equations (10) and (19); Equations (19) and (28); Equations (11) and (20); Equations (20) and (29); Equations (12) and (21); Equations (21) and (30); Equations (13) and (22); and, finally, between Equations (22) and (31), which indicated that the plot and SSP had significant effects on the morphological models of SM, BHN, and MH.

(6) Assessment and analysis of morphological models.

Most models of the morphological characteristics in TMSU differed significantly from the models in RTM, according to Table 5. For example, as compared to the morphological models of Equations (5)–(13) and (23)–(31), which were based on NLME and incorporated variables of the spatial structurers, we were able to significantly increase the prediction accuracy of the morphological characteristics by 10.4%, 3.02%, and 17.8% for SM's H, HCB, and CW, respectively; 6.5%, 7.6%, and 8.9% for BHN's H, HCB, and CW, respectively; and 13.3%, 15.7%, and 13.4% for MH's H, HCB, and CW, respectively.

Table 5. Parameter estimates and evaluation indices of each model. SM, BHN, and MH represent *Cunninghamia lanceolata*, *Machilus pauhoi*, and *Schima superba*, respectively; u_{1j}, u_{2j},u_{3j} are the random effect parameters caused by the jth sample plot on β_1, β_2, β_3, respectively.

Species	Formula	Model	Variance Components						Fit Statistics		
			$\sigma^2_{u_{1j}}$	$\sigma^2_{u_{2j}}$	$\sigma^2_{u_{3j}}$	$\sigma_{u_{1j}}\sigma_{u_{2j}}$	$\sigma_{u_{1j}}\sigma_{u_{3j}}$	$\sigma_{u_{2j}}\sigma_{u_{3j}}$	R^2	E	RMSE
SM	H-D	Equation (5)	——	——	——	——	——	——	0.7155	−0.0201	2.7644
		Equation (14)	——	——	——	——	——	——	0.7369	−0.0199	2.6584
		Equation (23)	4.99×10^0	7.51×10^{-5}	2.90×10^{-5}	-9.86×10^{-1}	8.83×10^{-1}	-9.47×10^{-1}	0.8203	−0.0190	2.1971
	H_{CB}-H	Equation (6)	——	——	——	——	——	——	0.6121	0.0635	1.4820
		Equation (15)	——	——	——	——	——	——	0.6313	0.0739	1.4450
		Equation (24)	3.78×10^{-6}	——	——	——	——	——	0.6406	0.0670	1.4265
	CW-D	Equation (7)	——	——	——	——	——	——	0.4276	−0.0004	0.7671
		Equation (16)	——	——	——	——	——	——	0.5341	−0.0004	0.6921
		Equation (25)	1.24×10^{-1}	2.95×10^{-2}	——	-1.81×10^{-1}	——	——	0.5965	−0.0013	0.6441
BHN	H-D	Equation (8)	——	——	——	——	——	——	0.6752	0.0250	2.1741
		Equation (17)	——	——	——	——	——	——	0.6887	0.0201	2.1285
		Equation (26)	3.17×10^{-6}	9.95×10^{-5}	——	-8.83×10^{-1}	——	——	0.7376	0.0205	1.9542
	H_{CB}-H	Equation (9)	——	——	——	——	——	——	0.4095	0.0482	1.5897
		Equation (18)	——	——	——	——	——	——	0.4542	0.0383	1.5284
		Equation (27)	2.49×10^{-5}	1.61×10^{-4}	——	-4.30×10^{-1}	——	——	0.4860	0.0467	1.4832
	CW-D	Equation (10)	——	——	——	——	——	——	0.4594	−0.0012	1.4319
		Equation (19)	——	——	——	——	——	——	0.4782	−0.0015	1.4068
		Equation (28)	3.59×10^0	3.41×10^0	3.47×10^{-13}	1.00×10^0	-3.20×10^{-2}	-3.20×10^{-2}	0.5271	0.0032	1.3393
MH	H-D	Equation (11)	——	——	——	——	——	——	0.5161	0.0008	2.4594
		Equation (20)	——	——	——	——	——	——	0.5426	−0.0021	2.3912
		Equation (29)	7.43×10^{-2}	——	——	——	——	——	0.6473	−0.0029	2.0999
	H_{CB}-H	Equation (12)	——	——	——	——	——	——	0.3018	−0.0148	1.8170
		Equation (21)	——	——	——	——	——	——	0.3468	−0.0177	1.7575
		Equation (30)	3.77×10^{-2}	——	——	——	——	——	0.4282	0.0078	1.6443
	CW-D	Equation (13)	——	——	——	——	——	——	0.5322	0.0020	0.9734
		Equation (22)	——	——	——	——	——	——	0.6180	0.0054	0.8796
		Equation (31)	1.10×10^{-2}	2.76×10^{-3}	——	-9.95×10^{-1}	——	——	0.6658	0.0077	0.8228

A box plot of the residuals of the dependent variable classes of three dominant trees was examined to further assess the feasibility of the fitted models based on OLS and NLME (see Figure 5). In each plot, nearly all absolute values of the standard residuals based on OLS and NLME models were less than two. Furthermore, the residuals from the NLME model showed no significant trends, whereas some residual plots from the OLS model did, such as the HCB model of *Cunninghamia lanceolata*, *Machilus pauhoi*, and *Schima superba*, as well as the CW model of *Machilus pauhoi* and *Schima superba*. Only a few larger deviations were observed for trees that were unusually large or small. The residual from the NLME model exhibited no variance in heteroskedasticity. As a result, the NLME models Equations (32)–(40) had superior statistical performance and could be used for SM, BHN, and MH's 3D tree modeling in TMSU, respectively.

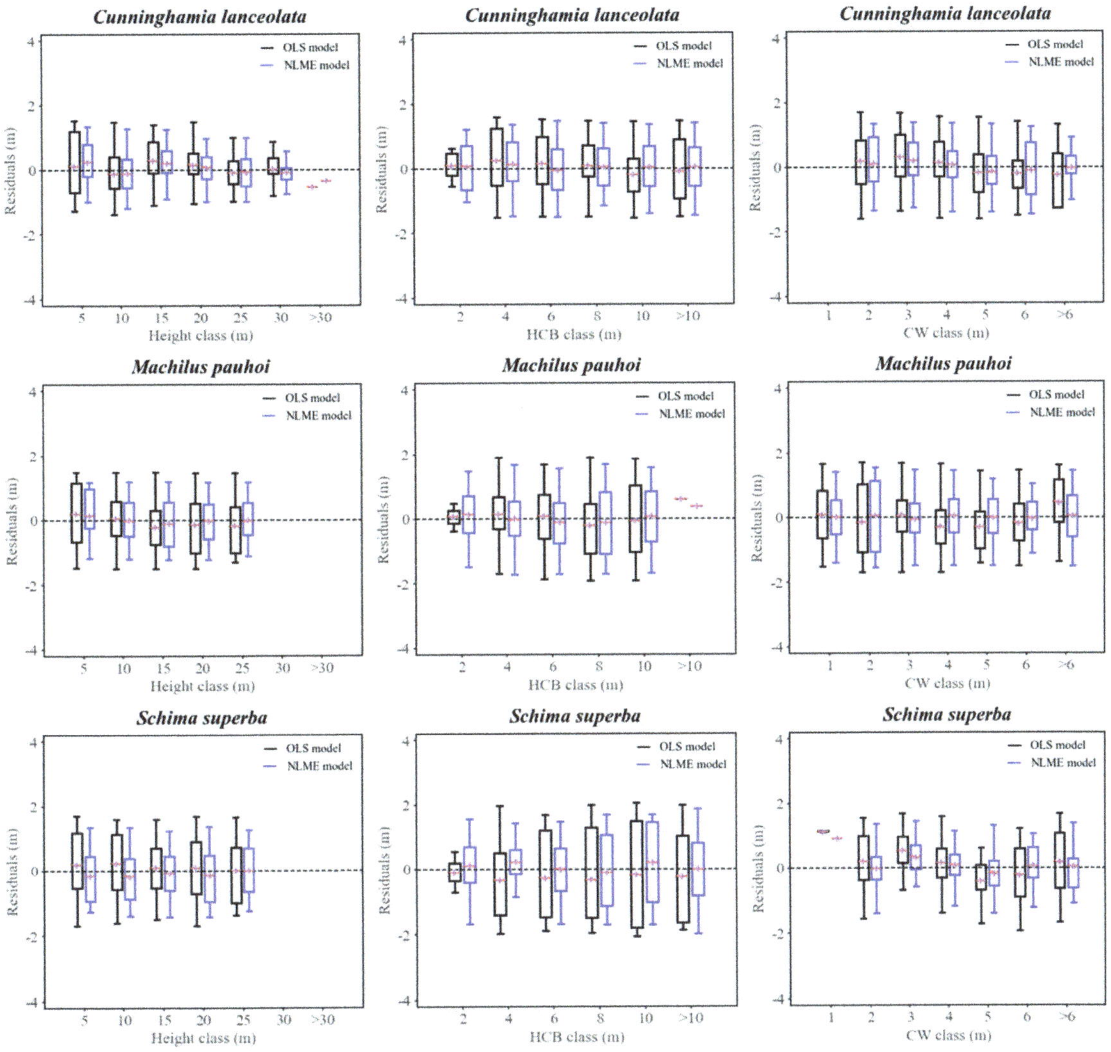

Figure 5. Box plots of standard residuals of three dominant tree morphological models. The box length, whisker length, horizontal lines in red, and plus signs in a box represent the interquartile range (IQR); the class minimum and maximum values in the IQR; and the class median and mean values, respectively. *Cunninghamia lanceolata*, *Machilus pauhoi*, and *Schima superba* are represented by SM, BHN, and MH, respectively.

3.2. Analysis of 3D Tree Modeling Based on a Structural Unit

We discovered that the reference tree's C_i, M_i, and U_i had a substantial influence on the morphological features of morphological characteristic of SM, BHN, and MH, according to the analytical results of the nonlinear mixed-effect morphological models of three dominant species. For example, (i) crowding and mingling had a significant impact on SM's morphological characteristics: the higher the degree of crowding, the higher the tree height and the height-to-crown base, the smaller the crown width, while under the same crowding, the higher the mingling, the bigger the crown width. (ii) Crowding and dominance had an impact on BHN's morphological characteristics: the bigger the crowding, the higher the tree height and the height-to-crown base, the smaller the crown width, while under the same crowding, the higher the dominance, the bigger the tree height and the lower the height-to-crown base. (iii) Crowding, dominance, and mingling all had an impact on the morphological characteristics of MH: the bigger the crowding, the higher the tree height and the smaller the crown width, while under the same crowding and mingling, the higher the dominance, the higher the height-to-crown base. (see Figure 6).

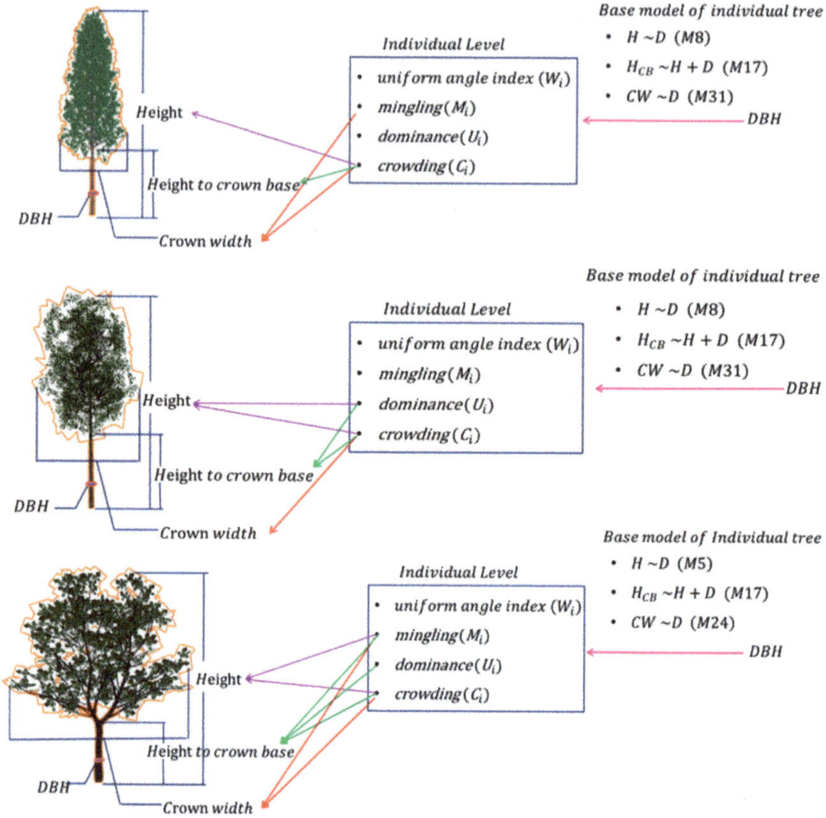

Figure 6. Three-dimensional tree modeling of three predominant trees based on a structural unit.

In this study, we have selected three very different scenes for SM, BHN, and MH as examples to reconstruct 3D tree modeling with a structural unit (see Figure 7). Based on the value classes of C_i, M_i, and U_i, we constructed the SM and BHN model libraries with 25 scenes each (see Appendix A Figures A1 and A2) and an MH model library with 125.

Figure 7. Three-dimensional modeling of *Cunninghamia lanceolata*, *Machilus pauhoi*, and *Schima superba* under different C_i, U_i, and M_i. The values of C_i, U_i, and M_i were 0.00, 0.25, 0.50, 0.75, and 1.00, respectively. *Cunninghamia lanceolata*, *Machilus pauhoi*, and *Schima superba* are represented by SM, BHN, and MH, respectively. Since *Cunninghamia lanceolata* (DBH = 25.9 cm), *Machilus pauhoi* (DBH = 14.4 cm), and *Schima superba* (DBH = 15.1 cm) occurred the most frequently in the study region, we chose them as our examples.

4. Discussion

4.1. The Development of Morphological Characteristic Models Involving Forest Spatial Structure in TMSU

A model of morphological characteristics involving forest spatial structure is an essential component of 3D tree modeling based on a structural unit. By acquiring several morphological parameters for 3D tree reconstruction, such as tree height, height-to-crown base, diameter at breast height, and crown width, RTM used fewer micro-characteristics of the individual trees [4–7]. However, forest environments are not homogeneous even at very small scales. The spatial structures must be incorporated into the morphological characteristics in order to accurately model 3D trees in an environment with a high degree of structural diversity and heterogeneity [12–14]. In our study, TMSU used four spatial-structure variables (e.g., W_i, M_i, C_i, and U_i) to describe the spatial structures and their effects on morphological traits. According to Table 4, we discovered that these spatial structures could increase the precision with which morphological characteristics were predicted due to the spatial heterogeneity of a reference tree. For example, as compared to the RTM model Equations (5)–(13), the TMSU model Equations (14)–(22) incorporated variables of the spatial structures and then improved its prediction accuracy of the morphological characteristics by 2.1%, 1.9%, and 10.7% for SM's H, H_{CB}, and CW, respectively; 1.4%, 4.5%, and 1.9% for BHN's H, H_{CB}, and CW, respectively; and finally, 2.7%, 4.5%, and 8.9% for MH's H, H_{CB}, and CW, respectively. However, when developing morphological characteristic models, researchers have suggested that adding excessive parameters affected model convergence and lengthened the time to compute parameter estimates, especially for the nonlinear mixed-effect modeling strategy used in this study [27,28,44]. Furthermore, adding a number of tree or spatial-structure variables to the inventory increased the cost and time required to obtain the measurements for those variables [29,44]. Therefore, for the final morphological prediction model of SM in TMSU, we selected one tree variable (D) and two spatial-structure variables (C_i, M_i); one tree variable (D) and two spatial-structure variables (C_i, U_i) for the final morphological prediction model of BHN; and one tree variable (D) and three spatial-structure variables (C_i, U_i, M_i) for the final morphological prediction model of MH.

Nonlinear mixed-effect models are appealing for the analysis of hierarchically structured data that are auto-correlated, as compared to the OLS model because of their flexibility in accounting for the covariate structure that is overlooked in conventional regression procedures. Therefore, in this study, we selected the Chapman–Richards, Walters, logistic, and power models as the base morphological models based on the statistical variables R^2, E, and RMSE, and then we integrated them with a nonlinear mixed-effect model approach at the plot level to construct morphological models of three dominant trees involving spatial-structure parameters. As shown in Table 5, a random effect model in TMSU Equations (23)–(31) outperformed a model without a random effect in RTM Equations (14)–(22) in terms of prediction accuracy for the full dataset. The result was consistent with most research conclusions [24–28]. As a result, morphological models of three dominant trees in TMSU were chosen for 3D tree modeling due to their higher precision and spatial capabilities. In addition, the morphological models in TMSU were limited in their capacity to handle changes in spatial structures (e.g., those caused by human activity or natural succession) [45] and complete data acquisition that could reflect different spatial-structure conditions. Generally, as the spatial structure of the planted mixed forest would not be significantly altered in a short time, the models in TMSU could be utilized for producing a short-term forecast (less than 5 years). Therefore, in order to improve the applicability of the morphological model in TMSU, we will consider climate variables, in addition to natural succession and human activities, and devise more plots with different spatial-structure conditions in our next project [46,47].

4.2. Three-Dimensional Tree Modeling in a Highly Heterogeneous Forest Environment

In forest visualization research, accurate 3D tree modeling with heterogeneous spatial forest structures is required (e.g., forest quality assessment, health supervision, and forest management planning and assessment) [48]. Given the complexity of forest structures and heterogeneous habitats, describing the relationship between any individual tree and its neighbors in a 3D tree model is a significant issue. In this study, we addressed the challenges of 3D tree modeling in a complex and highly heterogeneous forest environment in order to determine the spatial-structure variables, explore the relationships between spatial structures and morphological characteristics, and achieve 3D tree modeling based on a structural unit.

Most researchers have developed various qualitative and quantitative indices to describe the reference tree and its neighbor trees in a forest stand, such as the index of Clark and Evans [49], the diffusion index [50], Ripley's *K*-function [51], the *K*-function [52], the *O*-ring statistic [53], the Gini coefficient [54], Pielou's isolation index [55], the mean directional index [56], Gadow's species mingling [57], the mixed ratio [58], and the uniform angle index [59,60]. However, some of these indices have been predominantly used in forestry and ecology, and they do not meet the prerequisites for visually and intuitively portraying 3D tree modeling based on a structural unit [19,20]. In addition, certain approaches for analyzing spatial structures consider the neighboring tree species to some extent, although they can only describe the overall or unilateral structure of the reference tree population. For example, the size differentiation, mixed conditions, and the degree of crowding of the reference tree at different distance scales remained unknown when using W_i to analyze the distribution patterns. Therefore, these methods solely considered the distribution pattern and overlooked factors such as the tree size, the degree of crowding, and the degree of species aggregation. Furthermore, an entirely different structure could yield the same outcome. Obviously, it has been challenging to accurately depict the spatial structure of a tree population. The forest is a complex and dynamic ecosystem, particularly when it contains many different species and has a highly heterogeneous spatial structure. The four most significant characteristics of the population structure for any species in a mixed forest may be interspecific and intraspecific variances in tree sizes, species mingling, crowding, and distribution patterns [60]. The spatial structural parameters W_i, M_i, C_i, and U_i were used in TMSU to describe the relationships between any reference tree and its four adjacent neighboring trees. It uncovered reference tree population characteristics and modeled a 3D tree in a more accurate and thorough manner by simultaneously employing two or more pieces of structural information.

According to TMSU-based morphological models, we found that C_i, M_i, and U_i had a strong influence on the morphological characteristic of SM, BHN, and MH. For example, in SM, BHN, and MH, there was an identical phenomenon where the degree of crowding had a positive relationship with tree height, height-to-crown base, and crown width, which could be related to a growth response strategy of trees in a crowded and highly competitive environment to increase the tree's height to occupy the main layer of the forest and obtain more resources, such as light, moisture, and nutrition [17,60]. This could also promote the self-thinning of the branches, which increases the height-to-crown base for the limited availability of the light resource [10], as well as retaining a narrow crown width due to competition with conspecific and nonspecific species [16]. Furthermore, under the same degree of crowding, mingling was positively correlated with tree crown width in SM, and dominance was positively correlated with tree height but negatively correlated with the height-to-crown base in BHN; under the same crowding and mingling conditions, the dominance was positively correlated with the height-to-crown base in MH. The results could be linked to the competitive response mechanism of various tree species, such as light transmittance, shade tolerance, and biological characteristics. Therefore, considering the unique coupling responses between the spatial-structure parameters and the morphological characteristics, we created model libraries for the three dominant species with 25 scenes based on the value classes of C_i and M_i for SM; 25 scenes based on the value classes of

C_i and U_i for BHN; and 125 scenes based on the value classes of C_i, M_i, and U_i for MH, respectively.

5. Conclusions

This study provided a new approach for 3D tree modeling TMSU. The differences between RTM and TMSU were focused on: (i) Four spatial structural parameters (W_i, M_i, C_i, and U_i) were incorporated into the morphological characteristic models of the reference trees and revealed the relationship between the spatial-structure variables and the morphological features. (ii) The nonlinear mixed-effect model was introduced into the parameter estimation of the TMSU-based morphological model for higher precision and applicability. By comparing the two approaches, RTM and TMSU, we found that (i) the prediction accuracy of the TMSU-based morphological model increased by 10.4%, 3.02%, and 17.8% for SM's H, HCB, and CW, respectively; 6.5%, 7.6%, and 8.9% for BHN's H, HCB, and CW, respectively; and 13.3%, 15.7%, and 13.4% for MH's H, HCB, and CW, respectively, as compared to the RTM morphological models. (ii) TMSU provided more spatial-structure information with fewer parameters in the morphological models and revealed the relationships between the spatial structures and the morphological traits of the reference tree.

Author Contributions: L.W. and H.Z. (Huaiqing Zhang) proposed the novel 3D tree modeling method by incorporating spatial structural parameters under highly heterogeneous forest environment. L.W. and H.Z. (Huacong Zhang), T.Y., J.Z. and Y.L. collected and processed the morphological and spatial structural data of study site. L.W. and H.Z. (Huaiqing Zhang) contributed to writing the paper. All authors have read and agreed to the published version of the manuscript.

Funding: This research was funded by the National Natural Science Foundation of China, grant number 32071681, 32271877. The sponsor of the research is Huaiqing Zhang.

Data Availability Statement: Not applicable.

Acknowledgments: This work is supported by VISLAB Team in IRIFIT, CAF. Thank you to the teammates who help collect field data and assist with its processing.

Conflicts of Interest: The authors declare that there are no conflict of interest.

Appendix A

Figure A1. Three-dimensional modeling of SM under different C_i and M_i. The values of C_i and M_i were 0.00, 0.25, 0.50, 0.75, and 1.00. Since *Cunninghamia lanceolata* (SM) (DBH = 25.9 cm) occurred most frequently in the study region, we chose it as an example.

Figure A2. Three-dimensional modeling of BHN under different C_i and U_i. Notes: $D_i < D_a, D_b, D_c, D_d$ indicated that the diameter of tree *i* was smaller than tree *a*, *b*, *c*, and *d*; $T_i \cap T_a, T_b, T_c, T_d$ indicated that the canopy of tree *i* and tree *a*, *b*, *c*, and *d* overlapped; C_i and U_i had values of 0.00, 0.25, 0.50, 0.75, and 1.00. Since *Machilus pauhoi* (BHN) (DBH = 14.4 cm) occurred most frequently in the study region, we chose it as an example.

References

1. Paulus, S.; Dupuis, J.; Mahlein, A.K.; Kuhlmann, H. Surface feature based classification of plant organs from 3D laserscanned point clouds for plant phenotyping. *BMC Bioinform.* **2013**, *14*, 238. [CrossRef]
2. Thapa, S.; Zhu, F.; Walia, H.; Yu, H.; Ge, Y. A novel LiDAR-based instrument for high-throughput, 3D measurement of morphological traits in maize and sorghum. *Sensors* **2018**, *18*, 1187. [CrossRef]
3. Indirabai, I.; Nair, M.V.H.; Jaishanker, R.N.; Nidamanuri, R.R. Terrestrial laser scanner based 3D reconstruction of trees and retrieval of leaf area index in a forest environment. *Ecol. Inform.* **2019**, *53*, 100986. [CrossRef]
4. Sun, R.; Jia, J.; Jaeger, M. Intelligent tree modeling based on L-system. In Proceedings of the 2009 IEEE 10th International Conference on Computer-Aided Industrial Design & Conceptual Design, Wenzhou, China, 26–29 November 2009; pp. 1096–1100. [CrossRef]
5. Wang, G.; Zhang, D.; Zhou, K.; Jia, J. Rule and reuse based lightweight modeling and real time web3D rendering of forest scenes. In Proceedings of the 23rd International ACM Conference on 3D Web Technology, Poznań, Poland, 20–22 June 2018; pp. 1–8. [CrossRef]
6. Tang, L.; Peng, X.; Chen, C.; Huang, H.; Lin, D. Three-dimensional Forest growth simulation in virtual geographic environments. *Earth Sci. Inform.* **2019**, *12*, 31–41. [CrossRef]
7. Kim, D.; Kang, W.H.; Hwang, I.; Kim, J.; Kim, J.H.; Park, K.S.; Son, J.E. Use of structurally-accurate 3D plant models for estimating light interception and photosynthesis of sweet pepper (*Capsicum annuum*) plants. *Comput. Electron. Agric.* **2020**, *177*, 105689. [CrossRef]
8. Zhang, X.; Li, H.; Dai, M.; Ma, W.; Quan, L. Data-driven synthetic modeling of trees. *IEEE Trans. Vis. Comput. Graph.* **2014**, *20*, 1214–1226. [CrossRef]
9. Seidel, D.; Hoffmann, N.; Ehbrecht, M.; Juchheim, J.; Ammer, C. How neighborhood affects tree diameter increment–new insights from terrestrial laser scanning and some methodical considerations. *For. Ecol. Manag.* **2015**, *336*, 119–128. [CrossRef]
10. Bayer, D.; Seifert, S.; Pretzsch, H. Structural crown properties of Norway spruce (*Picea abies* [L.] Karst.) and European beech (*Fagus sylvatica* [L.]) in mixed versus pure stands revealed by terrestrial laser scanning. *Trees* **2013**, *27*, 1035–1047. [CrossRef]
11. Seidel, D.; Leuschner, C.; Müller, A.; Krause, B. Crown plasticity in mixed forests—Quantifying asymmetry as a measure of competition using terrestrial laser scanning. *For. Ecol. Manag.* **2011**, *261*, 2123–2132. [CrossRef]
12. Forrester, D.I.; Benneter, A.; Bouriaud, O.; Bauhus, J. Diversity and competition influence tree allometric relationships–developing functions for mixed-species forests. *J. Ecol.* **2017**, *105*, 761–774. [CrossRef]
13. Vincent, G.; Harja, D. Exploring ecological significance of tree crown plasticity through three-dimensional modelling. *Ann. Bot.* **2008**, *101*, 1221–1231. [CrossRef]
14. Purves, D.W.; Lichstein, J.W.; Pacala, S.W. Crown plasticity and competition for canopy space: A new spatially implicit model parameterized for 250 North American tree species. *PLoS ONE* **2007**, *2*, e870. [CrossRef]
15. Valladares, F.; Niinemets, Ü. Shade tolerance, a key plant feature of complex nature and consequences. *Annu. Rev. Ecol. Evol. Syst.* **2008**, *39*, 237–257. [CrossRef]
16. Jucker, T.; Bouriaud, O.; Coomes, D.A. Crown plasticity enables trees to optimize canopy packing in mixed-species forests. *Funct. Ecol.* **2015**, *29*, 1078–1086. [CrossRef]
17. Lines, E.R.; Zavala, M.A.; Purves, D.W.; Coomes, D.A. Predictable changes in aboveground allometry of trees along gradients of temperature, aridity and competition. *Glob. Ecol. Biogeogr.* **2012**, *21*, 1017–1028. [CrossRef]
18. Grueters, U.; Seltmann, T.; Schmidt, H.; Horn, H.; Pranchai, A.; Vovides, A.G.; Peters, R.; Vogt, J.; Dahdouh-Guebas, F.; Berger, U. The mangrove forest dynamics model mesoFON. *Ecol. Model.* **2014**, *291*, 28–41. [CrossRef]
19. Pommerening, A.; Gaulton, R.; Magdon, P.; Myllymäki, M. CanopyShotNoise–An individual-based tree canopy modelling framework for projecting remote-sensing data and ecological sensitivity analysis. *Int. J. Remote Sens.* **2021**, *42*, 6837–6865. [CrossRef]
20. Fischer, F.J.; Labriere, N.; Vincent, G.; Hérault, B.; Alonso, A.; Memiaghe, H.; Bissiengou, P.; Kenfack, D.; Saatchi, S.; Chaveet, J. A simulation method to infer tree allometry and forest structure from airborne laser scanning and forest inventories. *Remote Sens. Environ.* **2020**, *251*, 112056. [CrossRef]
21. Yang, Z.; Liu, Q.; Luo, P.; Ye, Q.; Sharma, R.P.; Duan, G.; Zhang, H.; Fu, L. Nonlinear mixed-effects height to crown base model based on both airborne LiDAR and field datasets for Picea crassifolia Kom trees in northwest China. *For. Ecol. Manag.* **2020**, *474*, 118323. [CrossRef]
22. Ma, A.; Miao, Z.; Xie, L.; Dong, L.; Li, F. Crown width prediction for *Larix olgensis* plantations in Northeast China based on nonlinear mixed-effects model and quantile regression. *Trees* **2022**, *36*, 1761–1776. [CrossRef]
23. Sharma, R.P.; Štefančík, I.; Vacek, Z.; Vacek, S. Generalized nonlinear mixed-effects individual tree diameter increment models for beech forests in Slovakia. *Forests* **2019**, *10*, 451. [CrossRef]
24. Sharma, R.P.; Vacek, Z.; Vacek, S.; Kučera, M. A nonlinear mixed-effects height-to-diameter ratio model for several tree species based on Czech national forest inventory data. *Forests* **2019**, *10*, 70. [CrossRef]
25. Bronisz, K.; Mehtätalo, L. Mixed-effects generalized height–diameter model for young silver birch stands on post-agricultural lands. *For. Ecol. Manag.* **2020**, *460*, 117901. [CrossRef]
26. Ciceu, A.; Garcia-Duro, J.; Seceleanu, I.; Badea, O. A generalized nonlinear mixed-effects height–diameter model for Norway spruce in mixed-uneven aged stands. *For. Ecol. Manag.* **2020**, *477*, 118507. [CrossRef]

27. Fu, L.; Zhang, H.; Sharma, R.P.; Pang, L.; Wang, G. A generalized nonlinear mixed-effects height to crown base model for Mongolian oak in northeast China. *For. Ecol. Manag.* **2017**, *384*, 34–43. [CrossRef]
28. Fu, L.; Sharma, R.P.; Hao, K.; Tang, S. A generalized interregional nonlinear mixed-effects crown width model for Prince Rupprecht larch in northern China. *For. Ecol. Manag.* **2017**, *389*, 364–373. [CrossRef]
29. Zhang, S.; Sun, J.; Duan, A.; Zhang, J. Variable-exponent taper equation based on multilevel nonlinear mixed effect for Chinese fir in China. *Forests* **2021**, *12*, 126. [CrossRef]
30. Bragg, D.C. A local basal area adjustment for crown width prediction. *North. J. Appl. For.* **2001**, *18*, 22–28. [CrossRef]
31. Marshall, D.D.; Johnson, G.P.; Hann, D.W. Crown profile equations for stand-grown western hemlock trees in northwestern Oregon. *Can. J. For. Res.* **2003**, *33*, 2059–2066. [CrossRef]
32. Li, Y.; Hui, G.; Zhao, Z.; Hu, Y.; Ye, S. Spatial structural characteristics of three hardwood species in Korean pine broad-leaved forest—Validating the bivariate distribution of structural parameters from the point of tree population. *For. Ecol. Manag.* **2014**, *314*, 17–25. [CrossRef]
33. Mueller-Dombois, D.; Ellenberg, H. *Aims and Methods of Vegetation Ecology*; Wiley: New York, NY, USA, 1974.
34. Mehtätalo, L.; de-Miguel, S.; Gregoire, T.G. Modeling height-diameter curves for prediction. *Can. J. For. Res.* **2015**, *45*, 826–837. [CrossRef]
35. Leduc, D.; Goelz, J. A height–diameter curve for longleaf pine plantations in the Gulf Coastal Plain. *South. J. Appl. For.* **2009**, *33*, 164–170. [CrossRef]
36. Von Gadow, K.; Hui, G.Y. Characterizing forest spatial structure and diversity. In *Sustainable Forestry in Temperate Regions*; Björk, L., Ed.; SUFOR, University of Lund: Lund, Sweden, 2002; pp. 20–30.
37. Hui, G.; Zhang, G.; Zhao, Z.; Yang, A. Methods of forest structure research: A review. *Curr. For. Rep.* **2019**, *5*, 142–154. [CrossRef]
38. Gadow, K.; Zhang, G.Q.; Durrheim, G.; Drew, D.; Seydack, A. Diversity and production in an Afromontane Forest. *For. Ecosyst.* **2016**, *3*, 15. [CrossRef]
39. Li, S.; Zhang, H.; Li, Y.; Yang, T.; Shen, K. Three-Dimensional Visualization Simulation of Chinese Fir Stand Growth Based on Unity3D. In Proceedings of the 2018 3rd International Conference on Mechanical, Control and Computer Engineering (ICMCCE), Huhhot, China, 14–16 September 2018; pp. 530–534. [CrossRef]
40. Liu, H.; Zhang, H.Q.; Ju, H.B.; Tang, X.; Hu, B. 3-D tree model and its visualization in Guozigou forest farm. *For. Res. Beijing* **2016**, *29*, 74–79. [CrossRef]
41. Lu, K.N.; Zhang, H.Q.; Liu, M.; Ouyang, G. Design and implementation of individual tree growth visualization system of *Cunninghamia lanceolata*. *For. Res. Beijing* **2012**, *25*, 207–211. [CrossRef]
42. Wu, Q.; Zhang, H.; Chen, Y.F.; Liu, M. Study on visual simulation technology of *Cunninghamia lanceolata* morphological characters. *For. Res. Beijing* **2010**, *23*, 59–64. [CrossRef]
43. Zhang, H.; Liu, M. Tree growth simulation method based on improved IFS algorithm. In Proceedings of the 2009 International Conference on Computational Intelligence and Software Engineering, Wuhan, China, 11–13 December 2009; pp. 1–5. [CrossRef]
44. Fu, L.; Sun, H.; Sharma, R.P.; Lei, Y.; Zhang, H.; Tang, S. Nonlinear mixed-effects crown width models for individual trees of Chinese fir (*Cunninghamia lanceolata*) in south-central China. *For. Ecol. Manag.* **2013**, *302*, 210–220. [CrossRef]
45. Knutti, R. Should we believe model predictions of future climate change? *Philos. Trans. R. Soc. A Math. Phys. Eng. Sci.* **2008**, *366*, 4647–4664. [CrossRef]
46. Leng, W.; He, H.S.; Bu, R.; Dai, L.; Hu, Y.; Wang, X. Predicting the distributions of suitable habitat for three larch species under climate warming in Northeastern China. *For. Ecol. Manag.* **2008**, *254*, 420–428. [CrossRef]
47. Lei, X.; Yu, L.; Hong, L. Climate-sensitive integrated stand growth model (CS-ISGM) of *Changbai larch* (Larix olgensis) plantations. *For. Ecol. Manag.* **2016**, *376*, 265–275. [CrossRef]
48. Tasoulas, E.; Varras, G.; Tsirogiannis, I.; Myriounis, C. Development of a GIS application for urban forestry management planning. *Procedia Technol.* **2013**, *8*, 70–80. [CrossRef]
49. Clark, P.J.; Evans, F.C. Distance to nearest neighbor as a measure of spatial relationships in populations. *Ecology* **1954**, *35*, 445–453. [CrossRef]
50. Morisita, M. Estimation of population density by spacing method. *Mem. Fac. Sci. Kyushu Univ. Ser. E Biol.* **1954**, *1*, 187–197. [CrossRef]
51. Ripley, B.D. Modelling spatial patterns. *J. R. Stat. Soc. Ser. B Methodol.* **1977**, *39*, 172–192. [CrossRef]
52. Upton, G.; Fingleton, B. *Spatial Data Analysis by Example. Volume 1: Point Pattern and Quantitative Data*; John Wiley & Sons Ltd.: Chichester, UK, 1985.
53. Wiegand, T.; AMoloney, K. Rings, circles, and nullal. Rule and reuse based lightweight mecology. *Oikos* **2004**, *104*, 209–229. [CrossRef]
54. Lexerød, N.L.; Eid, T. An evaluation of different diameter diversity indices based on criteria related to forest management planning. *For. Ecol. Manag.* **2006**, *222*, 17–28. [CrossRef]
55. Getzin, S.; Dean, C.; He, F.; Trofymow, J.A.; Wiegand, K.; Wiegand, T. Spatial patterns and competition of tree species in a Douglastems, chronosequence on Vancouver Island. *Ecography* **2006**, *29*, 671–682. [CrossRef]
56. Corral-Rivas, J.J.; Wehenkel, C.; Castellanos-Bocaz, H.A.; Vargas-Larreta, B.; Diéguez-Aranda, U. A permutation test of spatial randomness: Application to nearest neighbour indices in forest stands. *J. For. Res.* **2010**, *15*, 218–225. [CrossRef]
57. Gadow, K.; Füldner, K. Zur bestandesbeschreibung in der forsteinrichtung. *Forst Und Holz* **1993**, *48*, 602–606.

58. Gadow, V.K.; Hui, G.Y. *Modelling Forest Development*; Springer Science & Business Media: Göttingen, Germany, 1999.
59. Aguirre, O.; Hui, G.; von Gadow, K.; Jiménez, J. An analysis of spatial forest structure using neighbourhood-based variables. *For. Ecol. Manag.* **2003**, *183*, 137–145. [CrossRef]
60. Hui, G.; Wang, Y.; Zhang, G.; Zhao, Z.; Bai, C.; Liu, W. A novel approach for assessing the neighborhood competition in two different aged forests. *For. Ecol. Manag.* **2018**, *422*, 49–58. [CrossRef]

forests

MDPI

Article

Research on the Identification of Particleboard Surface Defects Based on Improved Capsule Network Model

Chengcheng Wang [1], Yaqiu Liu [1,*], Peiyu Wang [2] and Yunlei Lv [1]

[1] College of Information and Computer Engineering, Northeast Forestry University, Harbin 150040, China; chengchengcauc@163.com (C.W.); yunleilv@nefu.edu.cn (Y.L.)
[2] College of Electrical Mechanical Engineering, Northeast Forestry University, Harbin 150040, China; pywang1_11@163.com
* Correspondence: yaqiuliu@126.com; Tel.: +86-0451-82191528

Abstract: Aiming at the problems of low classification accuracy and overfitting caused by the limited number of particleboard image samples, a Capsule Network algorithm based on the improved CBAM (Convolutional Block Attention Module) attention model is proposed. The improved algorithm utilizes the GELU equation to improve the CBAM attention model and incorporates it into the convolutional layer of the Capsule Network. In this way, the improved algorithm optimizes the feature maps of surface defects and, meanwhile, improves the training efficiency and stability of the model. The improved algorithm alleviates the overfitting problem by adding a dropout layer, which makes the model more suitable for small sample classification. The effectiveness of the method proposed in this paper is verified by classification experiments on the dataset of particleboard surface defect images.

Keywords: particleboard defects detection; image classification; CBAM attention model; capsule network

1. Introduction

Particleboard, also known as bagasse board, is one of the three main products of man-made board at present. Particleboard is a kind of wood-based panel made of wood or other lignocellulosic materials, which is glued under the action of heat and pressure after applying adhesive. The raw materials of particleboard come from a wide range of sources, and the board has high strength. The sales market of particleboard-related products is also broad [1].

At present, there are many kinds of automatic production equipment for particleboard, among which the continuous flat press is the leading technology. It has the advantages of high automation and outstanding production capacity. However, due to the particularity of the raw material source of particleboard, there are some problems in the production line of the continuous flat press. Due to factors such as uneven size and type of raw materials, machine errors in production equipment, and workshop environmental interference, individual products may have surface defects. These defects can impact the subsequent veneer process, the appearance and quality of the finished board, and make secondary processing more challenging. With the refinement and high-end development of the furniture manufacturing industry, the demand for high-quality particleboard is rising. Consequently, particleboard manufacturers pay more attention to product surface quality testing than ever before, making it one of the core measurement factors for board grading. Among the available quality testing methods, surface defect image classification is a critical component for improving the quality of particleboard products.

During the particleboard production process, surface quality inspection is a crucial step that greatly impacts the subsequent veneer and edge bonding processes. The influence of surface defects on the board quality can vary widely depending on factors such as the damage degree, the defect type, and the defect area [2]. Currently, the treatment schemes usually involve either leaving the defect untreated or directly cutting the defective parts of the board. The former approach often leads to problems such as blistering and layering of the board after veneering, thereby affecting the quality of the final product and subsequent processing. The latter approach can result in the wastage of raw materials and production resources, which is not conducive to the effective utilization of wood resources. Detecting and identifying surface defects on the board and implementing different treatment methods based on the detection and identification results can address the aforementioned issues.

With the development of industrial technology, non-destructive testing methods are emerging as a replacement for manual operations in detecting and identifying defects in industrial products. In the wood-based panel industry, researchers have been exploring defect detection technology, resulting in progress in non-destructive testing and identification methods. These methods include ultrasonic detection [3], X-ray detection [4], and machine vision detection [5]. Ultrasonic testing uses ultrasonic propagation attenuation law to judge the defect position in the panel, but it has drawbacks such as long detection time, the need for coupling agents, and operator experience. X-ray testing targets internal defects in the panel, utilizing ray beam attenuation imaging to display the defect size and position. However, protective equipment is required due to the harmful effects of radiation, limiting its applicability in open production environments.

Machine vision technology emerged in the 1970s and 1980s, owing to the progress of digital image processing and computer pattern recognition technologies. In the realm of product surface defect detection, this technology acquires product surface images through detection and perception and subsequently produces results of diverse projects such as defect measurement, classification, and quality evaluation based on algorithms. It is capable of multi-mode detection, such as quantitative, qualitative, and semi-quantitative. This method is highly suitable for industrial production fields associated with panel defect detection due to its fast detection speed, flexible processing, high accuracy, and easy equipment configuration. However, the industrial production line's production environment is intricate, and the requirements for surface defect detection differ for different boards. Thus, surface defect detection and recognition technology based on machine vision still present extensive research prospects.

In the problem of surface defect detection and recognition based on machine vision, extracting the defect features to form a sample set, building a defect-type classifier, training the classifier, and classifying the defects according to their feature values are necessary. Currently, the main classifiers are the Bayesian Classifier, K-Nearest Neighbor Classifier, Decision Tree Classifier, Random Forest Classifier, Support Vector Machine Classifier, and Neural Network Classifier. Scholars have conducted much in-depth research and innovations on relevant issues in recent years [6–11]. He, T and Liu, Y used the improved AP (Affinity Propagation) clustering algorithm and SOM (Self-Organizing Maps) neural network to classify wood defects. The research achieves recall rates of 85.5% and 83.6% and precision rates of 90.3% and 92.7% [6]. Zhang, H improved the classification speed by using compressed Random Forest to classify the defects on the surface of solid wood flooring, achieving a one-time prediction time of 3.44 ms [7]. Liu, S classified wood defects using a BP (Back Propagation) Neural Network and Support Vector Machine [9]. The experimental result showed that the Support Vector Machine is more accurate in classification, with a recognition accuracy of 92% of live joints, dead joints, and cracks. To address the problem of high labor cost and low efficiency in wood defect detection, Ding, F and Zhuang, Z used a color charge-coupled equipment camera to collect surface images of red wood and Pinus sylvestris [11]. They applied the transfer learning method to a Single-Shot Multi-Box Detector (SSD) and proposed a target detection algorithm. The average accuracy of detecting a live joint, dead joint, and three types of defects reached 96.1%.

Based on recent research, it is evident that Neural Networks perform particularly well in related problems and yield satisfactory results for defect detection. Although many scholars have improved and innovated Neural Networks, current algorithms such as Convolutional Neural Networks and BP Neural Networks used in defect classification have their own limitations. In particular, trained network models may recognize targets as other objects when interference such as translation, rotation, or scale transformation occurs. Moreover, many Neural Network algorithms are greatly affected by the number of training samples, requiring a large number of training samples to achieve better recognition results, which involves significant computational cost and a long training time.

In this article, we address the challenge of identifying surface defects in particleboards, which is limited by the availability of defect samples. To overcome this challenge, we propose using the Capsule Network model, which has been shown to perform well in small sample classification problems. We improve the model to achieve a better recognition effect.

The Capsule Network (CapsNet) proposes replacing neurons in a Convolutional Neural Network with capsules to preserve detailed pose information and spatial hierarchical relationships between objects [12–14]. A capsule, in this network, refers to a small group of neurons that can identify a specific object in a defined region. The output of a capsule is a vector with a specific length, representing the probability of object existence, and the vector direction records the object's attitude parameters. If there is any slight change in the object's position, such as rotation or movement, the capsule would output a vector with the same length but a slightly varied direction.

In recent years, a Capsule Network has been increasingly utilized in image classification and recognition and has shown promising results [15–19]. Shakhnoza, M applied a Capsule Network to fire smoke image recognition, which demonstrated high recognition accuracy and robustness for outdoor camera image classification in the presence of smoke and fire [15]. Wang, W combined a Convolutional Neural Network and Capsule Network to classify gastrointestinal endoscopic images with a classification accuracy of over 85% for multiple datasets [16]. Sreekala, K developed a deep transfer learning model for face recognition using the Capsule Network and utilized a Grey Wolf Optimization (GWO) and Stacked Autoencoder (SAE) model to classify faces in images [18].

In our study of particleboard surface defect recognition, the recognition process of the defect type is greatly interfered with by the special characteristic of the surface texture of the board. To address this, we introduce an improved CBAM attention model into the Capsule Network. The CBAM (Convolutional Block Attention Module) model is a channel space mixed attention model [20–22]. It generates attention feature map information in both channel and space dimensions and multiplies the two feature maps with the original input feature map to perform adaptive feature correction and obtain an optimized feature map. By using the CBAM attention model in the particleboard surface defect classification and recognition problem, important feature information is strengthened, unnecessary features are suppressed, and feature optimization is achieved, resulting in improved recognition accuracy with reduced calculation parameters and computing power being saved.

In the particleboard production line, surface defects must be detected, extracted, classified, and identified to assess the degree of damage to the board. Subsequently, the boards that do not meet surface quality requirements must be removed from production. To achieve this objective, we use the CBAM attention model to improve the Capsule Network and use the improved method to identify five common surface defects (bare shavings, oil spots, glue spots, sand lines, and holes) on the production line. The study has two main aspects. Firstly, we propose an algorithm for the Capsule Network based on the improved CBAM attention model, making it more suitable for identifying particleboard surface defect types. Secondly, we conduct recognition experiments using defect samples from the particleboard surface defect dataset and compare the effectiveness of our model with that of other common recognition methods.

2. Materials and Methods

2.1. The Capsule Network

The network structure of the Capsule Network can be divided into two main parts: the primary part and the reconstruction part [23,24]. The primary part comprises convolution layers, convolution capsule layers, and fully connected capsule layers. On the other hand, the reconstruction part decodes the characteristic vector to assist in reconstructing the input image in cooperation with the Capsule Network. The main part structure of the Capsule Network is illustrated in Figure 1.

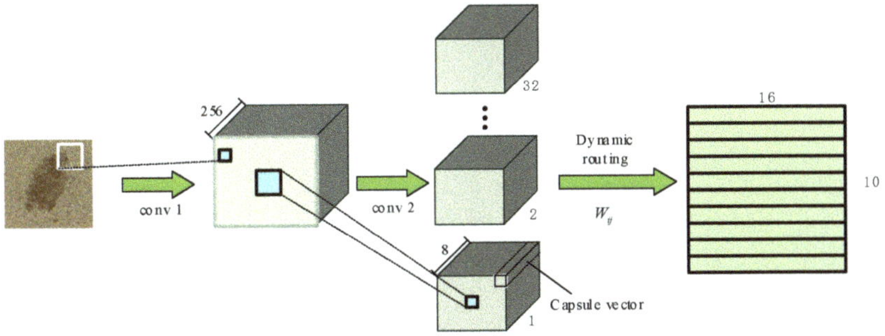

Figure 1. The main structure diagram of the Capsule Network.

The Capsule Network has a network structure that includes two convolution layers and a classified capsule layer based on the dynamic routing algorithm, as shown in Figure 1. The first two layers are similar to those of a Convolutional Neural Network. Notably, the feature map output at the second layer is used to construct the capsule vector. In the third layer, each capsule vector serves as the input and output of the dynamic routing algorithm, and each capsule vector has its corresponding weight matrix. The dynamic routing algorithm is applied to transform the capsules. After the third layer of conversion, the capsule vector is represented by two equations: one represents the current data sample, and the other represents the probability of the corresponding category of the input sample.

In the second convolution layer, the primary capsule layer follows a typical Convolution Network structure. However, it differs from traditional Convolutional Neural Networks in that its role is to process the feature map. As shown in Figure 1, the feature map output from the first convolutional layer is presented on the left side of the primary capsule layer. Unlike traditional Convolutional Neural Networks, which stretch multiple feature map matrices and concatenate them into a vector, the Capsule Network groups all the feature maps according to a specified number, as demonstrated in Figure 2. Capsule vectors are extracted by grouping feature maps based on the channel direction, allowing the convolutional kernel to be separated into multiple groups.

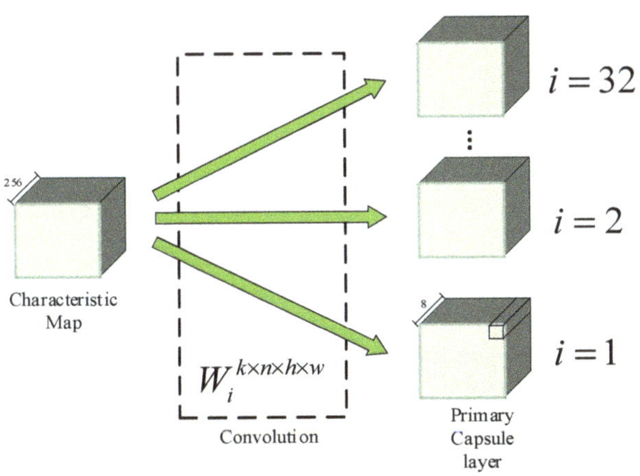

Figure 2. Calculation principle of the capsule layer.

The Capsule Network's dynamic routing capsule layer performs the conversion of capsule vectors using the principle depicted in Figure 3, as mentioned earlier. The dynamic routing algorithm is a crucial component of the Capsule Network and not only facilitates the implementation of the capsule mechanism but also embodies its ideas. To analyze the principle of the Capsule Network more easily, the lower-level capsule is transformed into a high-level capsule, which serves as the target of the conversion.

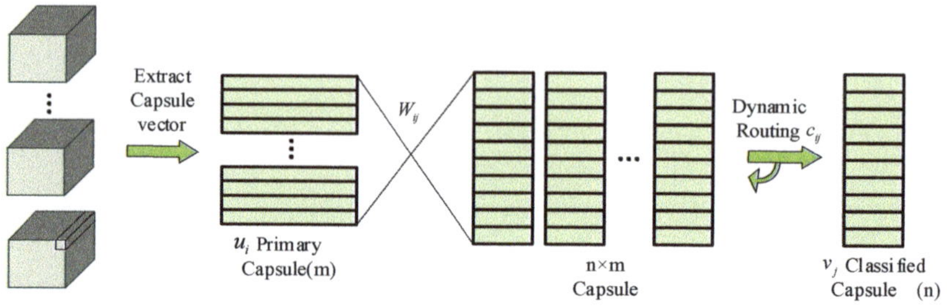

Figure 3. Schematic diagram of dynamic routing.

Consider u_i as the capsule vector of layer l, v_j as the capsule vector of layer $l+1$, which represents the weighted sum of all prediction vectors $\hat{u}_{j|i}$. The change formula can be expressed as:

$$\hat{u}_{j|i} = W_{ij} \cdot u_i \tag{1}$$

where, W_{ij} is the weight matrix that can be learned. Consider c_{ij} is the coupling coefficient corresponding to each prediction vector $\hat{u}_{j|i}$ mapped to v_i, which is the weight coefficient. c_{ij} determines the weight required to convert the prediction vector of layer l to the corresponding v_i of layer $l+1$. Since $\sum_j c_{ij} = 1$ can be calculated by dynamic routing algorithm, there are:

$$s_j = \sum_i c_{ij} \hat{u}_{j|i} \tag{2}$$

Each capsule vector represents the probability of the entity's occurrence, and the entity can be either an object in the picture or a part of the object. So, the s_j obtained through the calculation of the above formula cannot be used as the vector expression of the high-level capsule j. In order to solve this problem, a nonlinear extrusion equation is introduced into the Capsule Network to scale the length of the capsule. By scaling the length of the capsule in this way, the length of the short capsule approaches 0 and the length of the long capsule approaches 1. The length of the capsule represents the probability of the existence of the entity, and its length is between $(0,1)$. The extrusion equation can be expressed as:

$$v_j = \frac{||s_j||^2}{1+||s_j||^2} \cdot \frac{s_j}{||s_j||} \tag{3}$$

According to Formulas (2) and (3), we can know how to calculate the prediction vector of u_i to obtain the vector v_i of high-level capsules. In this process, the coupling coefficient c_{ij} is the amount that determines that u_i is "routed" to v_i. When the similarity between u_i and v_i is high, the contribution of u_i to v_i is greater, and the value of c_{ij} is greater. Similarly, when the similarity between u_i and v_i is low, the value of c_{ij} is smaller. In the above dynamic routing process, c_{ij} is not calculated by gradient training in a Convolutional Neural Network, but by an iterative process:

$$c_{ij} = \frac{\exp(b_{ij})}{\sum_k \exp(b_{ik})} \tag{4}$$

where, b_{ij} is the logarithmic prior probability of the coupling relationship between u_i and v_j, and k is the number of iterations. The iterative process described in Formula (4) is the process of dynamic routing, which is the core part of the Capsule Network. Before the first iteration calculation, the initial value of b_{ij} is 0, and the update process can be expressed as:

$$b_{ij} = b_{ij} + \hat{u}_{j|i} \cdot v_j \tag{5}$$

According to the above principles, the calculation process of dynamic routing can be summarized as follows:

1. Calculate the predictive value $\hat{u}_{j|i}$ of u_i from Formula (1), and initialize b_{ij} to 0;
2. The coupling coefficient c_{ij} is calculated from Formula (4) and substituted into Formulas (2) and (3) to obtain the intermediate value s_j and the vector v_j of the upper capsule;
3. Update b_{ij} by Formula (5);
4. Judge whether the iteration is completed. If it is not completed, repeat steps 1–3. If it is completed, output v_j.

Dynamic routing routes the original capsule to the category capsule through the above process. The length of the category capsule describes the probability of the existence of the represented category. The final classification category of the image is the category corresponding to the longest capsule. To make the capsules representing the image category as long as possible, a margin loss function is introduced in the Capsule Network. For the category capsule v_k, the loss function can be described as:

$$L_k = T_k \max\left(0, m^+ - \|v_k\|\right)^2 + \lambda(1 - T_k)\max\left(0, \|v_k\| - m^-\right)^2 \tag{6}$$

where m^+ and m^- are the capsule length thresholds. λ is the loss adjustment parameter. T_k is the category parameter. If the image belongs to category k, then $T_k = 1$, otherwise $T_k = 0$. Calculate the loss function values for all categories of capsules, and the sum of which is the loss value for the Capsule Network.

The reconstruction part of the Capsule Network is used to decode the category capsule and reconstruct the image. The reconstructed image and the input image use the sum of pixel differences squared as the image reconstruction loss value L_I. The total loss of the reconstruction part of the Capsule Network can be expressed as:

$$L = \sum_k L_k + \mu L_I \tag{7}$$

where μ is the reconstruction adjustment parameter. Based on the resulting loss L, the training of the Capsule Network can be realized.

2.2. Capsule Network Based on Improved CBAM Attention Model (CBAM-CN)

The basic structural model of CBAM is shown in Figure 4.

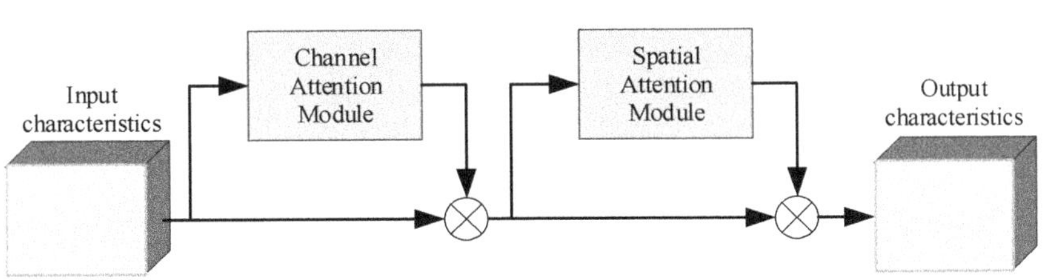

Figure 4. Basic structure of CBAM.

For any input feature map $X \in R^{C\times H\times W}$, the feature map obtained after passing the channel attention module is considered as X'. The feature map obtained after passing the spatial attention module is considered as X'', then:

$$X' = M_C(X) \otimes X \tag{8}$$

$$X'' = M_S(X') \otimes X' \tag{9}$$

where, M_C represent the calculation equations of the channel attention module, and M_S represent the spatial attention module.

In the problem of the board surface defect recognition, due to the limited number of image samples available for training, the training results are prone to overfitting. A good classification effect can be obtained for the test set, but the accuracy in the actual detection is very low. To solve this problem, we introduce a GMLP based on a Gaussian error linear unit, and construct a new attention model of an anti-overfitting channel.

To calculate channel features more efficiently, the channel attention model needs to compress the spatial dimensions of the input features. First, the input feature graph is passed through the parallel maximum pooling layer and the average pooling layer to obtain the maximum pooling feature $F^C_{\max}$ and the average pooling feature F^C_{avg}. Then, input the two features to GMLP, and the hidden layer output H_0 can be expressed as:

$$H_0 = f_{GELU}(XW_h + b_h) \tag{10}$$

where, W_h is the hidden layer weight, b_h is the hidden layer threshold, and f_{GELU} is the Gaussian Error Linear Unit activation equation (GELU). It can be described as follows:

$$f_{GELU}(x) = xP(X \le x) = \frac{x}{2}\left[1 + e_{rf}\left(\frac{x}{\sqrt{2}}\right)\right] \tag{11}$$

where e_{rf} can be expressed as:

$$e_{rf}(x) = \frac{1}{\sqrt{\pi}}\int_{-x}^{x} e^{-t^2}dt = \frac{2}{\sqrt{\pi}}\int_{0}^{x} e^{-t^2}dt \tag{12}$$

Formula (11) can be approximately described as:

$$f_{GELU}(x) = \frac{x}{2}\left\{1 + \tanh\left[\sqrt{\frac{2}{\pi}}\left(x + 0.045x^3\right)\right]\right\} \tag{13}$$

It can be seen from Equation (13) that the multilayer perceptron model using the GELU introduces random regularization while maintaining the nonlinear mapping ability, which can better avoid the generation of the overfitting phenomenon and the disappearance of the gradient. In order to further accelerate the convergence speed of the hidden layers and suppress the gradient disappearance problem in the back-propagation process, each hidden layer is batch normalized, and the process can be expressed as:

$$H_n = f_{BN}(H_0) = \frac{\gamma(H_0 - \mu)}{\sqrt{\sigma^2 + \varepsilon}} + \beta \tag{14}$$

where, f_{BN} is the batch normalization equation, γ is the scale factor, and β is the translation factor.

In order to prevent overfitting, the output characteristics of the hidden layer after batch normalization are randomly deleted by using the Dropout equation, and the final output characteristics of the hidden layer are as follows:

$$H = dropout(H_n) = \begin{cases} 0\ ,\ r < p \\ \frac{H_n}{(1-p)},\ r \ge p \end{cases} \tag{15}$$

where, r is a random value, of which the value range is $[0, 1)$, and p is the dropout coefficient.

The output feature O is obtained by passing the hidden layer output feature H. O can be expressed as:

$$O = f_S(HW_O + b_O) \tag{16}$$

where, W_O is the hidden layer weight, b_O is the hidden layer threshold, and f_S is the Sigmoid equation.

To sum up, the output characteristics of the channel attention model can be expressed as:

$$M_C(F) = f_S(F_{\max}^{CH} + F_{avg}^{CH}) \tag{17}$$

where, $F_{\max}^{CH}$ is the hidden layer output feature of the maximum pooling feature $F_{\max}^{C}$, and F_{avg}^{CH} is the hidden layer output feature of the average pooling feature F_{avg}^{C}. $M_C \in R^{C/r \times 1 \times 1}$, and r is the dimension reduction coefficient. The basic structure of the anti-overfitting channel attention model is shown in Figure 5.

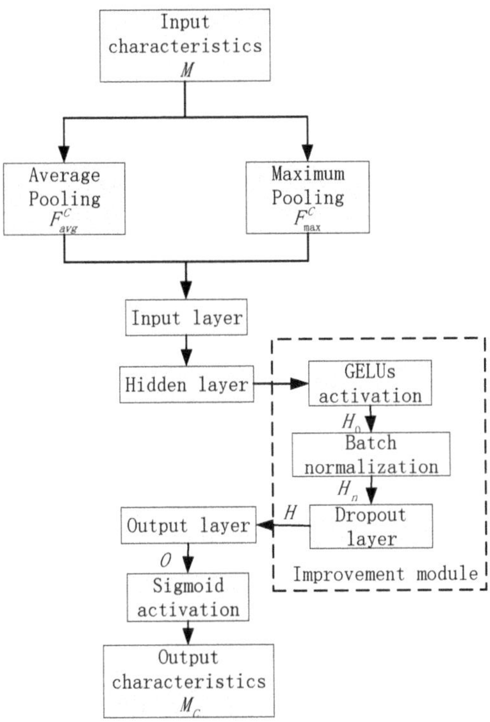

Figure 5. Basic structure of the channel attention model.

After obtaining the channel attention feature M_C, the CBAM model continues to calculate the spatial attention feature M_S. Average pooling and maximum pooling are performed in the channel dimension to obtain the maximum pooling feature $F_{\max}^{S}$ and the average pooling feature F_{avg}^{S}. The two are superposed and the dimension is reduced by convolution. In the same way as the channel attention model, the final spatial attention feature map can be obtained after using the Sigmoid equation. M_S can be expressed as:

$$M_S(F) = f_S(f(F_{\max}^{S}, F_{avg}^{S})) \tag{18}$$

where, $F_{\max}^{S} \in R^{1 \times H \times W}$, $F_{avg}^{S} \in R^{1 \times H \times W}$, and $M_S(F) = R^{H,W}$. The basic structure of the spatial attention model is shown in Figure 6.

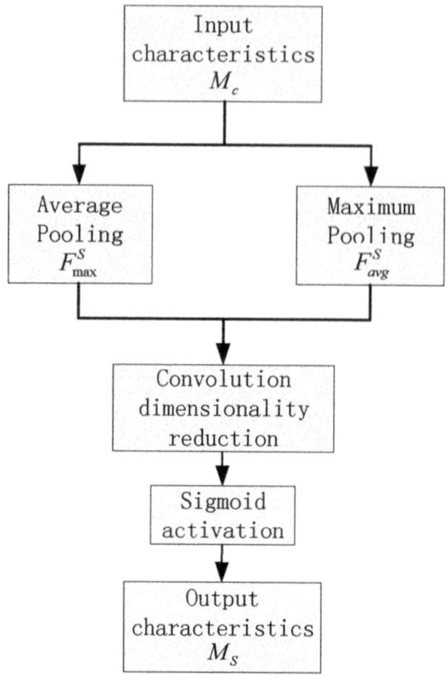

Figure 6. Basic structure of the spatial attention model.

In summary, the improved CBAM attention model is applied to the convolution layer of the Capsule Network, resulting in reduced interference from the particleboard surface textures and improved recognition accuracy through optimized feature maps. Additionally, the improved Capsule Network model has strong anti-overfitting characteristics due to the adoption of the GELU and the addition of the Dropout layer, which can be more suitable for feature recognition with limited sample sizes. The improved model has good applicability to the problem of particleboard surface defect recognition that this article focuses on.

2.3. Experiment Equipment

The research system comprises an image acquisition system and a detection and recognition system. The image acquisition system should capture particleboard surface images clearly, facilitate further processing by the software environment, and transmit data quickly to the computer software system to ensure the overall detection efficiency of the system. The system should also be economically durable and offer good technical support, taking into account the practicality of the production line.

The image acquisition system in this study meets the aforementioned requirements by selecting the GigE color industrial camera model DFK 23GP031 from the imaging source with an acquisition rate of 15 fps. The camera is equipped with an ON Semiconductor CMOS sensor (MT9P031) that ensures high-quality images while also reducing equipment cost. Additionally, the camera uses a surface scanning mode which enables a quicker collection of the particleboard surface image in the field of vision compared to line-scanning cameras. This ensures real-time detection at the image acquisition level while maintaining high definition.

The image acquisition system used in this study employs two industrial cameras positioned side by side, as depicted in Figure 7. By adopting a dual-camera parallel acquisition mode, the system can cover the entire surface of the particleboard, ensuring a comprehensive acquisition range. Moreover, using two cameras enhances the accuracy of the acquisition, which contributes to the precision and reliability of the recognition and analysis of particleboard surface defect images.

(a) (b)

Figure 7. The image acquisition system: (**a**) DFK 23GP031industrial camera; (**b**) image acquisition system architecture.

The detection and identification system comprises computer hardware and internal software. The computer hardware system employed in this research includes Microsoft Windows 10 (64-bit), Inter Xeon Silver 4216 CPU, RAM Tesla V100 GPU, 512 GB memory, and 16GB video memory. IC Capture 2.4 serves as the display software of the camera in the image acquisition system, while the deep learning framework used in the experiment is PyTorch. The programming language used is Python 3.7, and the programming environment is PyCharm 20.1.3.

2.4. Sample Acquisition

At the initial stage, over 2000 particleboard surface images with varying types and degrees of surface defects were collected. Parts of the collected particleboard surface images are shown in Figure 8.

The constructed dataset of the particleboard surface defect images has significant application value in verifying the overall effectiveness of the algorithm, as well as the universality and scope of the real reaction algorithm in the process of simulation experiments. Therefore, it is important that the images in the dataset are typical, inclusive, and authentic. To achieve this, the collected images are further screened, sorted, and classified. Atypical images with incomplete defects, blurred images, excessive interference, and other issues are removed, resulting in a set of 2000 standardized images, each sized 120 × 120 pixels. The number and percentage of images of each defect type included in the dataset are shown in Table 1.

Table 1. Statistical data of the number and percentage of images of defect types.

Category	Number	Percentage
Shavings	895	44.73
Sand grain	292	14.62
Hole	196	9.80
Glue spot	167	8.37
Oil stain	450	22.48

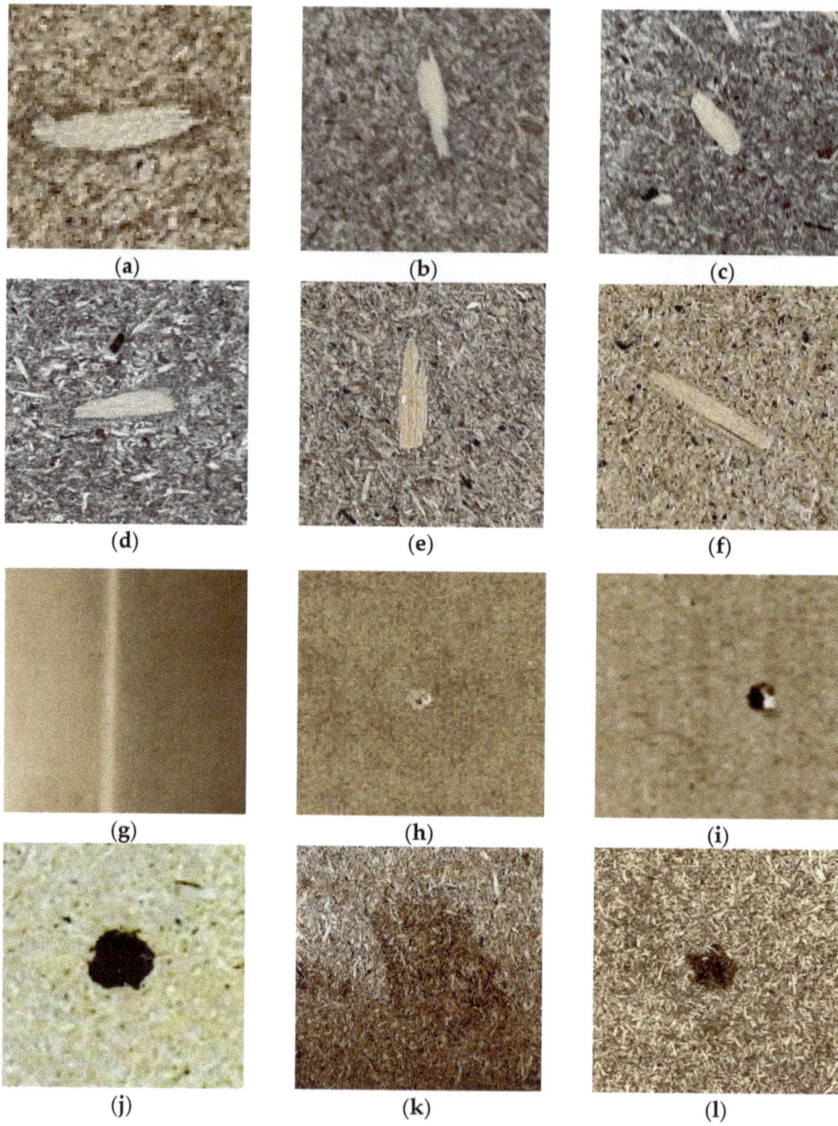

Figure 8. Part of the sample images contains surface defect features (figure (a–f) are shavings; (g) is sand grain; (h,i) are holes; (j) is glue spot; (k,l) are oil stains).

In this study, we collected particleboard surface defect images that encompassed five different types of surface defects: bare particles, sand grains, oil stains, dust stains, and holes. The percentage of frames containing each type of defect in the total sample size of the dataset simulates the frequency difference of these defects in the actual production line. By utilizing these image samples as research objects, we can conduct simulation experiments to ensure the reliability of data analysis and the effectiveness of the results obtained in this article.

3. Results

The sample set of defect features to be tested is divided into a training set and a test set in a 7:3 proportion. The training set comprises 1400 defect feature maps, while the test set contains 600 defect feature maps.

Common classification recognition methods such as an AP Clustering algorithm, SOM Neural Network, Capsule Network (CapsNet), Fast Capsule Network (Fast CapsNet), and the proposed CBAM-Capsule Network (CBAM CapsNet) were selected to identify and compare the types of surface defects. The size of the sliding pane in the AP clustering algorithm is 40×40, and the reference threshold is $P = 20$. The SOM Neural Network takes the number of competing layer nodes as 200. CapsNet, Fast CapsNet, and CBAM-CN employ the loss function shown in Formula (6), wherein the loss regulation parameter $\lambda = 0.5$, the capsule length threshold $m^+ = 0.9$, $m^- = 0.1$, and the reconstruction loss as shown in Formula (7), where the reconstruction regulation parameter $\mu = 0.05$. Table 2 show the other training parameter settings.

Table 2. Training parameter settings.

Parameter	Value
Initial learning rate	0.01
Momentum	0.99
Weight decay rate	0.001
Epoch	100
Batch size	10

Each recognition algorithm is utilized to conduct recognition experiments on the five types of surface defect images. The accuracy rates and F1 score evaluation index results obtained from the recognition experiments on the training set and test set are presented in Tables 3 and 4.

Table 3. Classification results of feature images in the training set.

Experimental Methods	Accuracy	F1-Score
AP	0.9158	0.9153
SOM	0.9315	0.9316
CapsNet	0.9629	0.9625
Fast CapsNet	0.9543	0.9542
CBAM-CN ($p = 1$)	0.9853	0.9851
CBAM-CN ($p = 0.5$)	0.9861	0.9862

Table 4. Classification results of feature images in the test set.

Experimental Methods	Accuracy	F1-Score
AP	0.8315	0.8323
SOM	0.8294	0.8312
CapsNet	0.8536	0.8532
Fast CapsNet	0.9025	0.9029
CBAM-CN ($p = 1$)	0.9459	0.9453
CBAM-CN ($p = 0.5$)	0.9560	0.9556

The data show that the CBAM-CN achieves a higher classification accuracy and F1 score than other algorithms in both the training and test sets when recognizing particleboard surface defect feature images with small sample sizes. This indicates that the CBAM-CN proposed in this paper is effective in eliminating background texture interference and improving classification accuracy. Moreover, by comparing the recognition accuracy of the CBAM-CN at dropout probabilities of 0.5 and 1, it is observed that the introduction of a dropout layer into the improved CBAM attention model results in a relatively small gap between the test set and the training set. These results suggest that the CBAM-CN with a dropout layer can better alleviate overfitting problems caused by small sample sizes.

To verify the effectiveness of the CBAM-CN on the particleboard surface defect image sample set, we analyzed the relationship between recognition accuracy, loss rate, and epochs for CapsNet, Fast CapsNet, and the CBAM-CN. The corresponding curves are presented in Figures 9 and 10.

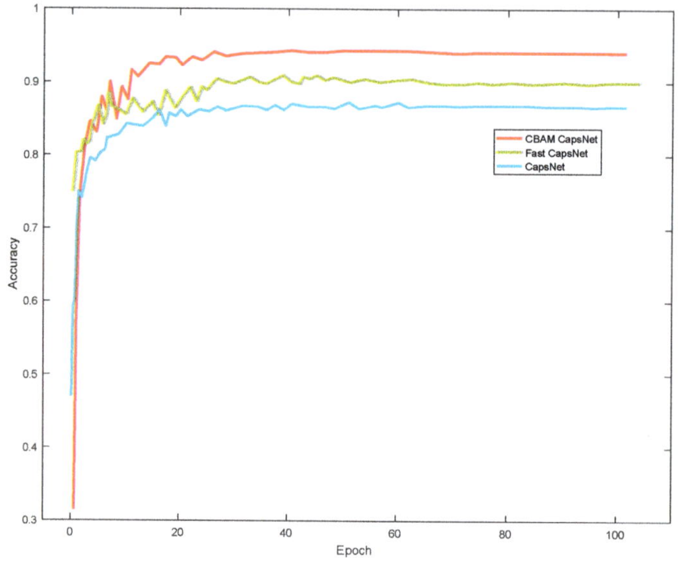

Figure 9. The relationship between the accuracy and epochs.

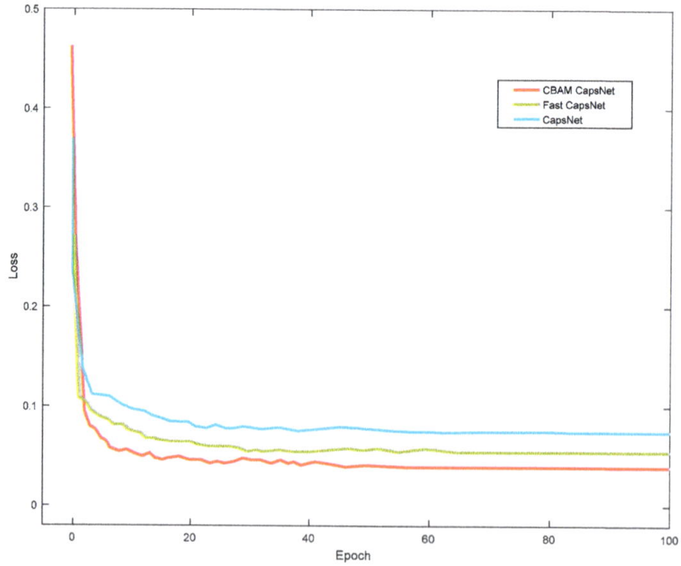

Figure 10. The relationship between the loss and epochs.

The graph above demonstrates that the CBAM-CN achieves a stable training effect at 40 epochs, while the curves of CapsNet and Fast CapsNet tend to stabilize at 60 epochs. This indicates that the CBAM-CN can optimize the feature graph and improve the training efficiency of the capsule network model. Moreover, it is evident from the curves that the CBAM-CN has higher recognition accuracy and lower loss rate when the training curve tends to be stable, indicating that it can achieve a stable and accurate recognition effect in identifying surface defects of particleboard under the same epoch conditions.

In general, image recognition problems require a large number of training samples for Convolutional Neural Networks to achieve optimal performance. However, in the case of particleboard surface defect recognition, the number of available samples is limited. Therefore, we compare and analyze the recognition accuracy and loss rate of the AP Clustering algorithm, SOM Neural Network, and the CBAM-CN, with varying sample sizes. The resulting curves are shown in Figures 11 and 12. To compare the effects of these recognition algorithms on the problem of overfitting, we include the recognition results of the training set as a comparison in Figure 12. The solid line represents the experimental data of the test set, while the dotted line represents the experimental data of the training set.

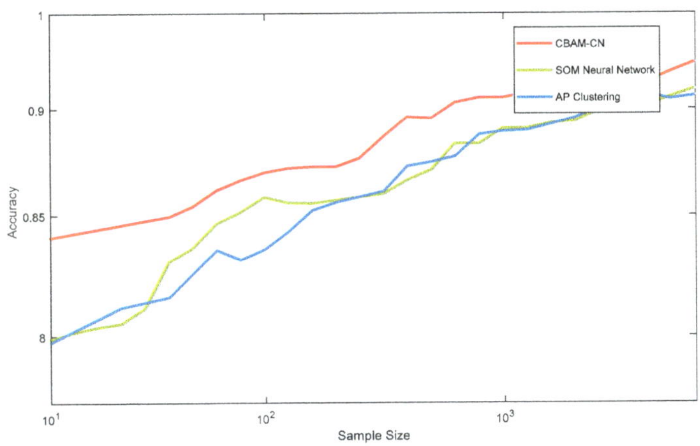

Figure 11. The relationship between the accuracy and the number of samples.

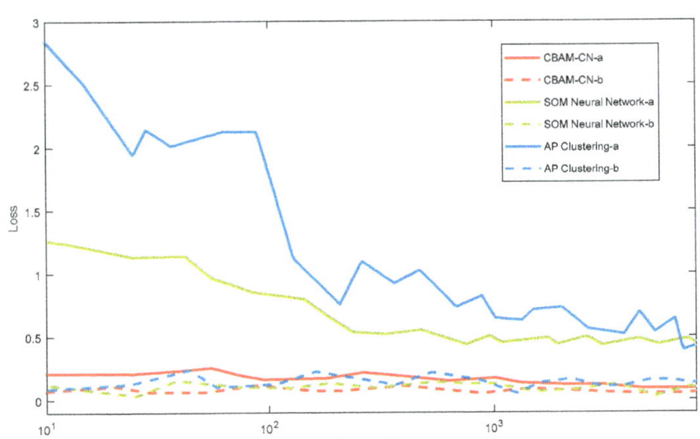

Figure 12. The relationship between the loss and the number of samples.

The curves in Figures 11 and 12 reveal that when the sample size is small, the training performance of the AP Clustering algorithm and SOM Neural Network will be greatly affected, and a satisfactory recognition effect cannot be achieved. In contrast, the CBAM-CN can achieve an accuracy rate of nearly 85% and a loss rate of less than 15% with just 100 training set samples. When the training set of samples is more than 500, the CBAM-CN can maintain an accuracy rate of more than 90% and a loss rate of less than 5%. On the other hand, other neural network models require over 1000 training samples to attain an accuracy rate of more than 90%. Moreover, under the same number of training samples, the loss rate is also higher than the CBAM-CN.

It is evident from Figure 12 that when the number of samples is small, there is a significant difference between the recognition loss rate of the test set and the training set for the AP Clustering algorithm and SOM Neural Network, indicating the presence of an overfitting problem. However, the two experimental data curves of the CBAM-CN are similar, and it can maintain a low loss rate for the test set even when the number of training samples is small. This result highlights that the CBAM-CN can effectively address the overfitting problem caused by the insufficient number of training samples.

4. Discussion

According to the experimental results and analysis presented above, the CBAM-CN proposed in this article outperforms other effective recognition methods in terms of faster training efficiency and stability for the recognition of particleboard surface defects. Additionally, with a small number of samples in the training set, the CBAM-CN can effectively suppress overfitting and maintain high recognition accuracy. These results also indicate that the CBAM-CN requires fewer training samples to achieve satisfactory recognition results, thereby meeting the recognition requirements for a greater variety of challenging scenes.

5. Conclusions

In this article, we propose a Capsule Network algorithm based on the improved CBAM attention model (CBAM-CN) to tackle the challenges faced by particleboard surface defect recognition technology, including the interference of background texture and the limited availability of image samples for training. We introduce the GELU-enhanced CBAM attention model into the convolution layer of the Capsule Network to optimize the feature map of the surface defects to be detected, thereby improving the model's training efficiency and stability. Additionally, we add a Dropout layer to alleviate the overfitting problem caused by the small number of samples in the training set, ensuring that the model is more applicable to small sample classification problems. We compare the CBAM-CN with several common effective image recognition methods, and the experimental results demonstrate its effectiveness.

The proposed method in this paper can be analogously applied to other problems of small sample image classification and background texture interference. Further research can experimentally investigate such problems.

Author Contributions: C.W., Y.L. (Yaqiu Liu), P.W. and Y.L. (Yunlei Lv) conceived and designed the experiments, analyzed the data, and wrote the paper. All authors have read and agreed to the published version of the manuscript.

Funding: This research received no external funding.

Data Availability Statement: The data presented in this study are available on request from the corresponding author. The data are not publicly available due to privacy.

Conflicts of Interest: The authors declare no conflict of interest.

References

1. Lee, S.H.; Lum, W.C.; Boon, J.G.; Kristak, L.; Antov, P.; Pędzik, M.; Rogoziński, T.; Taghiyari, H.R.; Lubis, M.A.R.; Fatriasari, W.; et al. Particleboard from agricultural biomass and recycled wood waste: A review. *J. Mater. Res. Technol.* **2022**, *20*, 4630–4658. [CrossRef]
2. Yang, C.; Lai, W.; Su, J.; He, W.; Gao, Z. Applied Research on Prediction Methods of Properties of Particleboard Based on Data-Driven Methods. *J. Biobased Mater. Bioenergy* **2021**, *15*, 1–9. [CrossRef]
3. Honarvar, F.; Varvani-Farahani, A. A review of ultrasonic testing applications in additive manufacturing: Defect evaluation, material characterization, and process control. *Ultrasonics* **2020**, *108*, 106227. [CrossRef] [PubMed]
4. Plessis, A.; Yadroitsava, I.; Yadroitsev, I. Effects of defects on mechanical properties in metal additive manufacturing: A review focusing on X-ray tomography insights. *Mater. Des.* **2020**, *187*, 108385. [CrossRef]
5. Yang, J.; Li, S.; Wang, Z.; Dong, H.; Wang, J.; Tang, S. Using Deep Learning to Detect Defects in Manufacturing: A Comprehensive Survey and Current Challenges. *Materials* **2020**, *13*, 5755. [CrossRef]
6. He, T.; Liu, Y.; Yu, Y.; Zhao, Q.; Hu, Z. Application of deep convolutional neural network on feature extraction and detection of wood defects. *Measurement* **2020**, *152*, 107357. [CrossRef]
7. Zhang, H.; Wang, Y.; Yu, C. Research on Key Technology of Online Detection for Particleboard. In Proceedings of the 2021 IEEE International Conference on Electronic Technology, Communication and Information (ICETCI), Changchun, China, 27–29 August 2021; pp. 512–515.
8. Huang, Y.; Qiu, C.; Wang, X.; Wang, S.; Yuan, K. A Compact Convolutional Neural Network for Surface Defect Inspection. *Sensors* **2020**, *20*, 1974. [CrossRef]
9. Liu, S.; Jiang, W.; Wu, L.; Wen, H.; Liu, M.; Wang, Y. Real-Time Classification of Rubber Wood Boards Using an SSR-Based CNN. *IEEE Trans. Instrum. Meas.* **2020**, *69*, 8725–8734. [CrossRef]
10. Wu, X.; Qiu, L.; Gu, X.; Long, Z. Deep Learning-Based Generic Automatic Surface Defect Inspection (ASDI) with Pixelwise Segmentation. *IEEE Trans. Instrum. Meas.* **2021**, *70*, 1–10. [CrossRef]
11. Ding, F.; Zhuang, Z.; Liu, Y.; Jiang, D.; Yan, X.; Wang, Z. Detecting Defects on Solid Wood Panels Based on an Improved SSD Algorithm. *Sensors* **2020**, *20*, 5315. [CrossRef]
12. Mazzia, V.; Salvetti, F.; Chiaberge, M. Efficient-CapsNet: Capsule network with self-attention routing. *Sci. Rep.* **2021**, *11*, 14634. [CrossRef]
13. Sun, K.; Yuan, L.; Xu, H.; Wen, X. Deep Tensor Capsule Network. *IEEE Access* **2020**, *8*, 96920–96933. [CrossRef]
14. Pawan, S.J.; Rajan, J. Capsule networks for image classification: A review. *Neurocomputing* **2022**, *509*, 102–120. [CrossRef]
15. Shakhnoza, M.; Sabina, U.; Sevara, M.; Cho, Y.-I. Novel Video Surveillance-Based Fire and Smoke Classification Using Attentional Feature Map in Capsule Networks. *Sensors* **2022**, *22*, 98. [CrossRef]
16. Wang, W.; Yang, X.; Li, X.; Tang, J. Convolutional-capsule network for gastrointestinal endoscopy image classification. *Int. J. Intell. Syst.* **2022**, *37*, 5796–5815. [CrossRef]
17. Li, D.; Zhang, M.; Kang, T.; Li, B.; Xiang, H.; Wang, K.; Pei, Z.; Tang, X.; Wang, P. Fault diagnosis of rotating machinery based on dual convolutional-capsule network (DC-CN). *Measurement* **2022**, *187*, 110258. [CrossRef]
18. Sreekala, K.; Cyril, C.P.D.; Neelakandan, S.; Chandrasekaran, S.; Walia, R.; Martinson, E.O. Capsule Network-Based Deep Transfer Learning Model for Face Recognition. *Wirel. Commun. Mob. Comput.* **2022**, *2022*, 2086613. [CrossRef]
19. Khanal, J.; Tayara, H.; Zou, Q.; To Chong, K. DeepCap-Kcr: Accurate identification and investigation of protein lysine crotonylation sites based on capsule network. *Brief. Bioinform.* **2021**, *23*, bbab492. [CrossRef]
20. Lu, E.; Hu, X. Image super-resolution via channel attention and spatial attention. *Appl. Intell.* **2022**, *52*, 2260–2268. [CrossRef]
21. Ju, A.; Wang, Z. Convolutional block attention module based on visual mechanism for robot image edge detection. *EAI Endorsed Trans. Scalable Inf. Syst.* **2021**, *9*, e4. [CrossRef]
22. Li, G.; Fang, Q.; Zha, L.; Gao, X.; Zheng, N. HAM: Hybrid attention module in deep convolutional neural networks for image classification. *Pattern Recognit.* **2022**, *129*, 108785. [CrossRef]
23. Huang, Y. Research on dynamic routing algorithm based on gaussian mixture model. *MATEC Web Conf.* **2022**, *355*, 03017. [CrossRef]
24. Zhang, Q.; Wu, B.; Sun, Z.; Ye, Y. Gating augmented capsule network for sequential recommendation. *Knowl.-Based Syst.* **2022**, *247*, 108817. [CrossRef]

forests

MDPI

Article

Visualization Simulation of Branch Fractures Based on Internal Structure Reconstruction

Meng Yang [1,2,*], Yi Zhang [1] and Benye Xi [3]

1 School of Information Science and Technology, Beijing Forestry University (BFU), Qinghua East Road 35, Haidian District, Beijing 100083, China
2 Engineering Research Center for Forestry-Oriented Intelligent Information Processing of National Forestry and Grassland Administration, Beijing 100083, China
3 The College of Forestry, Beijing Forestry University (BFU), Qinghua East Road 35, Haidian District, Beijing 100083, China
* Correspondence: yangmeng@bjfu.edu.cn

Abstract: This paper presents a visualization algorithm for wood fracture simulation based on wood science and wood internal structure reconstruction. The algorithm can simulate a reasonable and realistic wood fracture effect. First, the 3D point-cloud data of the bark structure are obtained using a laser scanner, and the cross-section of the branch is obtained by voxelization of the surface mesh model. Then, the outer contour of the cross-section is shrunk inward to reconstruct the annual rings and wood fiber bundles, and reasonable internal structures of branch 3D models are generated. The internal structure consists of a hierarchical model composed of several ring-like annual rings, and each annual ring is divided into a series of continuous fan rings. On the basis of the reconstruction results, the wood fracture surface model generated by the parameter control can be mapped to the irregularly shaped 3D branch model. In this research, the internal structure of branches and the shape of annual rings on the fracture surface of branches are analyzed to provide a reliable fracture model for different branch fractures of trees. In addition, the realistic fractured tree branch model generated by this algorithm can be widely applied in fields such as animation film special effects, game scene simulation, virtual reality scene construction, and mechanical research on broken tree branches.

Keywords: branch fractures; internal structure reconstruction; 3D point cloud; visualization simulation

1. Introduction

Specific fields such as film special effects, real-time gaming, and geometric modeling often require highly detailed models to create realistic scenes. Woody plants are an essential element of nature, and their 3D simulation plays a significant role in constructing natural scenes. In real-life situations, the shape of a tree can vary, and almost every tree has broken branches. Different factors, such as the tree species and environmental conditions, can affect branch breaking in different ways. Therefore, it is crucial to reconstruct branch fractures realistically and effectively to better simulate real-life scenes.

Natural disasters such as wind, rain, hail, blizzards, and tornadoes can also occur in the simulation of natural scenes. These disasters can cause damage to trees, resulting in broken branches. For instance, strong winds can blow branches, heavy snow on branches can break them, and tornadoes can uproot entire trees. The cracks in the broken branches caused by different natural disasters have varying effects. To create a more realistic 3D natural scene, it is essential to simulate the natural phenomenon of these branch fractures.

Tree branch fracture is caused by mechanical factors and involves both wood science and physics. Therefore, simulating the generation of fractured tree branches can be an effective way to study wood's resistance to external forces. Additionally, visualizing the fracture of tree branches can provide an intuitive representation of the tree's structural failure, which can aid in the development of strategies for garden protection.

Wood fracture simulation is a crucial aspect of natural landscape simulation. Wood is a complex biological organism, and the internal structure of branches varies among different tree species, making it difficult to simulate the fracture of wood and branches using traditional methods. Fractured branch models collected through 3D scanning instruments often have many voids due to the unevenness of the wood spurs at the fracture site. Using random noise to generate the fracture surface of the wood leads to a lack of realism in the simulation, as it oversimplifies the complex properties of wood. Therefore, there is a pressing need for a 3D plant technique for geometric modeling that can simulate detailed and realistic branch fracture models, allowing for a better analysis of the internal morphological structure and function of woody plants.

This paper introduces a novel approach for reconstructing the internal structure of tree rings and wood fiber bundles using the external contours of branches. The primary objective of this research is to develop a method for generating precise and evenly distributed cross-sectional slices and fragment models of tree rings, which can significantly enhance the creation of detailed and lifelike tree models. Moreover, the resulting model of fractured branches is particularly valuable for simulating the response of trees to external forces and evaluating their mechanical characteristics.

2. Related Work

In current graphics research, there is a focus on the dynamic simulation of trees and the development of material fracture algorithms. However, there are few simulation algorithms that are effective in capturing the unique properties of wood and simulating its fracture phenomenon. Furthermore, there are limited direct simulations of tree branch fracture. Wood is a complex biological organism that is porous, anisotropic, and viscoelastic, possessing exceptional mechanical properties. Previous research on tree fracture has mainly focused on two areas: computer graphics and wood fracture mechanics. In computer graphics, the focus has been on simulating trees in a realistic manner from a visualization perspective, including the modeling of trees and their interactions with various environmental factors such as external forces, wind, and rain.

2.1. Research on Tree Modeling Algorithm

In the early days of tree modeling research, procedural modeling methods were commonly used. One example is the work by Fernández et al. [1], who studied the function–structure model of *Pinus radiata* and were able to reproduce the development and growth of the species using mechanical formulas. Ancelin et al. [2] proposed a numerical model designed to simulate the biomechanical behavior of growing trees. The model is based on the transfer matrix method and is adjusted incrementally to calculate the evolution of trunk biomechanics during growth. Weber et al. [3] proposed an algorithm to create a tree model. The algorithm can adjust the morphological characteristics such as the rotation angle interval between branches, the length of the sub-branch, and the number of branches through multiple parameters. Initially, by learning a framework, the 3D shape, camera, and texture of an object are recovered from a single image [4]. Then, Hu et al. [5] proposed a self-supervised mesh reconstruction (SMR) method to enhance the 3D mesh attribute learning process. By requiring only contour mask annotations, SMR can be trained in an end-to-end fashion and can generalize to reconstruct natural objects. With the development of 3D scanning technology, many 3D point cloud-based tree modeling methods have emerged. This method generates point clouds that allow for 3D models that are more consistent with real trees. For example, Livny et al. [6] proposed a method to automatically generate realistic tree models from scanned 3D point clouds of trees. This work is based on pre-generated classified tree information and is able to handle large area scans and generate models automatically. Liu et al. [7] proposed TreePartNet, a neural network aimed at reconstructing tree geometry from point clouds obtained by scanning real trees.

In addition to research on modeling algorithms for whole trees, there are many studies on tree growth animation. For example, Kratt et al. [8] studied the animation simulation of wood growth according to the principle of tree cambium growth, including an algorithm for generating bark folds. Xiao and Chen [9] developed a model of the plant leaf drying phenomenon. The model is driven by the differential contraction of leaf tissue, and it also simulates the leaf vein system on plant leaves.

2.2. Research on the Interactive Simulation Algorithm of Trees and Environment

There are several geometric representations of plant models. Different representations have their own advantages and disadvantages in simulating the interaction of trees with environmental factors. Quigley et al. [10] proposed an interactive real-time animation method. This method represents the tree model as multiple articulated rigid bodies. By setting the stiffness of different branch node rigid bodies to avoid the animation effect of excessive softness and bending when subjected to external forces. The tree animation is processed in real time on a tree scene with tens of thousands of branches or a wooded scene composed of many trees with complex branch structures, such as wind and collision. Diener et al. [11] proposed a method to simulate in real time a complex scene of tens of thousands of trees drifting with the wind under a user-controlled wind field, where the branch nodes calculate the acceleration, velocity, and displacement of the object on the basis of the applied external forces, mass matrix, damping matrix, and stiffness matrix.

Yang et al. [12] used an extended trigonal spring model to represent the tree structure. Using the theory of nonrigid tree leaf surfaces with hydrophilic properties to simulate branch and raindrop interactions. Xie et al. [13] used a leaf model based on a mass–spring model representation to simulate the effect of hail striking and tearing leaves.

Pirk et al. [14] proposed a biologically sound approach to simulate tree combustion. Their tree combustion model can establish a link between the description of fine mechanisms (pyrolysis of wood, mass loss of logs, insulation effect of charcoal, and temperature change due to water evaporation) and the description of macroscopic effects (forest fire spread phenomenon), which can realistically simulate the effect of tree combustion.

Jernej et al. [15] decomposed the model volume mesh into several subdomains using the finite element method. A simplified deformable model is constructed for each subdomain to achieve a real-time simulation of deformed objects. Li et al. [16] proposed a simulation method based on the theory of material mechanics that can set stable anisotropic material parameters and can accurately simulate the deformation effect of plant twisting and bending. Bohan et al. [17] proposed a method based on the power law relationship between branch length and its diameter and between length and natural vibration frequency to automate the hardness setting of tree models according to plant biomechanics, which can simulate the dynamic effects of trees with richer layers.

2.3. Fracture Modeling and Animation in Graphics

For the simulation of material fracture, materials are usually considered as ideal brittle or isotropic materials. James et al. [18] analyzed the stress tensor calculated using a finite element model. They modeled where the cracks should start on the model volume mesh and in which direction they should propagate. Pfaff et al. [19] proposed a method for adaptive crack propagation on a thin sheet model. This method dynamically reconstructs a high-quality triangular mesh and adaptively maintains the details of the simulation. Chen et al. [20] proposed a user-friendly method for designing and controlling the effect of fracture surfaces. The method refines the low-resolution fracture surface mesh into a high-resolution and detail-rich fracture surface according to the user-set material strength field. Desbenoit et al. [21] proposed an interactive approach to model cracks by mapping editable 2D curve patterns onto a 3D model. Hädrich et al. [22] proposed a new method to simulate wood as an anisotropic material. This method uses the shape-matching method as the basis for modeling the isotropic properties of wood. A fiber model based on the Cosserat rod theory is used to generate fracture.

2.4. Image Detection of Wood Annual Rings

The annual rings of wood can reflect important information such as the age of the tree and the growth environment of the tree. Tree rings not only are a dating tool, but also provide information on environmental factors during the generation of tree rings, serving as a proxy for environmental changes. The width of tree rings is considered a function of the total number of cells and the radial size of cells, while cell growth rate is a function of other climatic factors such as temperature, soil moisture, and light conditions. It is this functional relationship that transforms the statistical correlation between tree ring growth and external climate factors into a causal relationship [23,24]. Existing work has processed wood cross-section images through image enhancement [25], image processing [26], edge detection [27], and other methods, obtaining information such as the number of annual rings by detecting annual rings.

Previous methods approximated the branch fracture surface as a smooth elliptical surface, and they lacked certain rationality and realism. The effect of the branch fracture surface depends on the material difference of different species of wood and the material strength of the internal annual structure of the branch. In this paper, a better model of the branch fracture surface is obtained by reconstructing the internal structure of branches.

3. Materials and Methods

As shown in Figure 1, the result of parametric modeling [28] is a cylinder with a variable-height branch fracture surface. The cross-section is still a circle, uniformly divided into N_{ring} circles and N_{sec} sectors. However, the shapes of real branches and the 3D branch models used in the animation are different. According to actual observations and knowledge of wood science, the internal structure of tree branches has irregularly shaped annual rings with uneven width, and real wood fiber bundles are also irregular thin rectangles. Therefore, in order to generate a more realistic branch fracture surface that meets the animation requirements, a reasonable and effective method of mapping the fracture surface generated on the regular cylinder to the 3D branch model is needed. As shown in Figure 2, this paper obtains real fracture branches, then further generates bark point-cloud data and internal voxelized data, and finally generates a complete tree broken model through mapping.

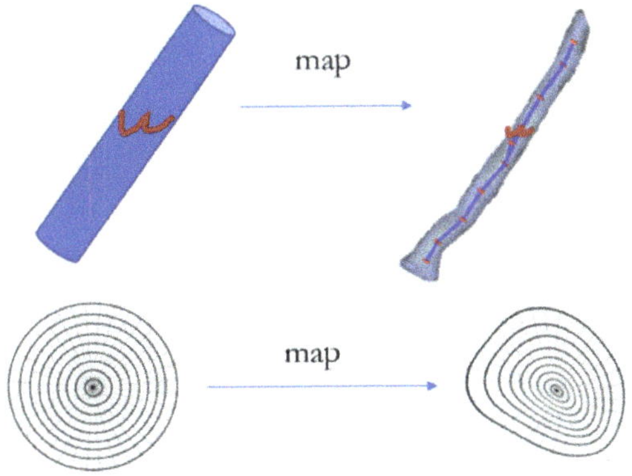

Figure 1. Schematic diagram of the method of mapping the fracture surface generated on a regular cylinder to a 3D branch model.

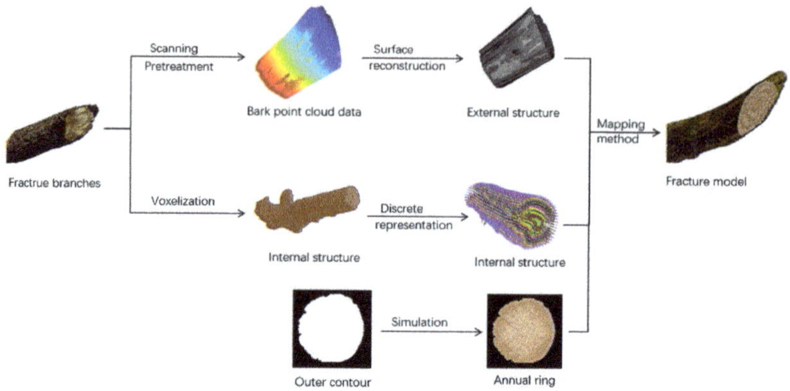

Figure 2. Flowchart of this method.

3.1. Branch Voxelization Model

According to related research in wood science and botany, the structure of the stem of a branch woody plant is composed of four parts: bark, cambium, essential part, and pith, from outside to inside [29–31]. The voxelization method converts the geometric form representation of an object (vertices and facets) into a voxel representation that is closest to the object at a given resolution N. The voxelization method is used to convert the branch surface mesh into volume data. The spatial voxels used to represent a 3D model are similar to the two-dimensional pixels used to represent an image. It can be understood as an extension from a two-dimensional pixel unit to a three-dimensional cube unit. The internal distribution and arrangement of trees are similar to the arrangement of plant cells. The voxelized branch model can obtain a branch shape composed of many voxels, which is similar to the arrangement of cells inside the branch. According to the arrangement and combination of voxels to form the various parts of the branch woody plant structure, the changes in the shape and position of different parts can be better realized. The outermost voxels can form the shape of annual rings on the branch fracture surface. The outermost voxels can be used to deform and control the shape of the fractured part, generate different fractured surfaces, and form the shape of annual rings on the fractured surface.

In order to obtain the contours of each layer of the 3D branch model. The branch model needs to be divided into multiple segments in the axial direction. The voxelization method [32,33] divides the model space into $N \times N \times N$ grids, and sets the grid to 0 or 1 depending on whether the grid is covered by the model. The voxelization method eventually generates volume data of size $N \times N \times N$ consisting of 0 s and 1 s, where 0 denotes that the voxel is not in the model, and 1 denotes that the voxel is in the model. Each voxel represents a particle.

In order to obtain a clearer contour map of the branch interface, the 3D model of the branch should be voxelized using a high-resolution N. When $N = 512$, volume data with 134,217,728 values are generated. However, since the axial length of the branch model is much larger than its cross-sectional width and height, if the whole branch is voxelized directly, it is not possible to generate high-quality volume data even with a larger resolution of 512. As shown in Figure 3, a part of the voxelized branch data is shown. It can be seen that the cross-section of the branch consists of only a few hundred voxels, and the width and height of each cross-section are a few tens of voxels wide. Therefore, in order to obtain a clear cross-section, the long branch model should be divided into N_{model} segments of the same axial length before voxelization, with each segment being slightly longer than the cross-sectional width of the branch model.

Figure 3. Voxelization result of a tree model.

3.2. Branch Outer Bark Structure Simulation

Bark is an important part of woody plants. The bark is located in the outermost part of woody plants and is in direct contact with the real environment. Influenced by different environmental factors, the structure and shape of the bark are very different from the internal structure of the branch. By using real tree branch fracture models to obtain point-cloud data, a detailed and complete outer skin structure of the fractured tree branch can be obtained, resulting in a more realistic model of the fractured tree branch. Using the PRINCE 775 two-color laser handheld 3D scanner of SCANTECH, the fractured tree branches were placed on the platform, and the handheld scanner scanned the fractured tree branches 360° to obtain the point-cloud data of the external contour of the branches. The point-cloud data were used to simulate the bark structure, which was more consistent with the real bark shape. The obtained external point-cloud data were preprocessed, noisy points and free points were removed, and point-cloud data were simplified to generate more accurate data, as shown in Figure 4a. The average distance density of this bark point-cloud data was calculated as 0.1367 points per cubic meter. The point-cloud data at the fracture surface were removed, only the bark data were retained, and the mesh structure was reconstructed using the point-cloud data, as shown in Figure 4b. The bark point-cloud data are denoted as p.

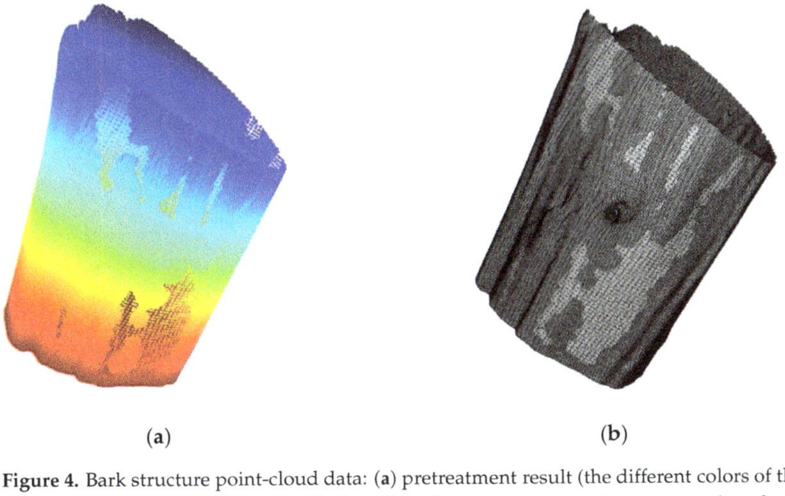

Figure 4. Bark structure point-cloud data: (**a**) pretreatment result (the different colors of the point-cloud data are used to facilitate the display effect; (**b**) reconstruction of the generated mesh structure.

3.3. Branch Cross-Sectional Annual Ring Simulation

Taking the resolution $N = 512$, after voxelizing the branch model, $512 \times N_{model}$ cross-sections of the branch model were obtained. The length and width values of each cross-sectional map were 512. Figure 5a shows the extracted outer contour drawing. The black color indicates the outer part of the branch, and the white color indicates the inner part of the branch.

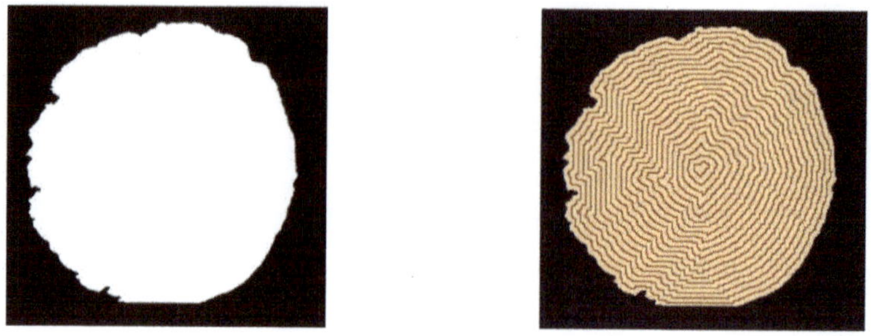

(a) (b)

Figure 5. Broken branch annual rings simulation results: (**a**) outer contour drawing; (**b**) simulation-generated annual ring cross-section.

To get the outline of a circle such as a growth ring, a method similar to the erosion algorithm in image processing was used [34,35]. Erosion is performed on the white part of the black-and-white section; each time the white part is corroded, the circle shrinks. By recording this circle of corrupted white pixels, the shape of a chronicle is obtained. Iterative etching is performed until all white pixels in the cross-section are set to black, and the annual wheel is obtained from the outside to the inside.

The erosion method uses a filter kernel of size K to traverse all pixels in the cross-sectional view, and the filter kernel can cover $K \times K$ pixels. When the pixel in the center of the filter kernel is white and there are M_K black pixels currently covered by the filter kernel, the white pixel value in the center is marked and stored. After traversing all pixels, the stored white pixels are modified to the specified color. For example, when the size of the filter core $K = 3$ and the user set parameter $M_K = 1$, for white pixels, if one of the eight adjacent pixels located at top, bottom, left, right, top left, bottom left, top right, and bottom right is a black pixel, it is stored and set to a uniform color in the new erosion result graph.

According to the knowledge of wood science, it is known that annual rings can be divided into earlywood and latewood, with earlywood having a light color and latewood having a dark color. The pixel color of the odd numbered circles was set to dark brown, and the pixel color of the even numbered circles was set to light brown; the dark circles were considered latewood, and the light circles on the inner side were considered earlywood. The dark outer circle and the adjacent light inner circle form an annual ring, and the cross-sectional distribution of annual rings based on the outer contour of the branch can be obtained as shown in Figure 5b. Firstly, a filtering kernel with $K = 1$ is used to corrode the external contour map of the tree cross-section, and the stored white pixel values are set to dark colors with RGB values of (30, 92, 186). Then, a filter kernel with $K = 3$ for corrosion traversal is used, the stored white pixel values are set to light colors with RGB values of (184, 215, 248), and filters of sizes 1 and 3 are alternatively used to check the external contour map of the tree cross-section for corrosion until there are no white pixel points in the external contour map of the tree cross-section.

The width of the true annual rings on wood reflects the environmental conditions of the year. When the environmental conditions such as temperature, moisture, and sunlight are suitable for the growth of trees in a year, the annual rings produced in that year will be thicker, and vice versa. Therefore, the width of the annual rings on the cross-section of wood can be wide or narrow, but it is not narrower from the inside to the outside, and there is no obvious pattern. As shown in Figure 6, a cross-sectional map of size 256 × 256 was processed to obtain a more realistic simulation of the annual rings in the cross-sectional area of the branch by reasonably designing the size of the filter kernel used in the corrosion process. First, we alternated the filter kernels of size 3 and 5 to generate a few thin annual rings in the outermost circle. Then, only the filter kernel of size 3 was used to generate a few annual rings with the narrowest width. Next, the kernels of size 3 and 5 were alternated to generate a few wider annual rings. Then, the kernels of size 3 and 7 were alternated to generate a few wider annual rings, resulting in a gradual change in the width of the annual rings. Lastly, the inner annual rings with the widest width at the center of the branch cross-section were generated by alternating the filter kernels of size 3 and 9 and the filter kernels of size 3 and 11. A total of 24 annual rings were generated.

Figure 6. Simulation results of annual rings in cross-section of branches with different annual ring widths.

In the internal reconstruction of some particularly irregular outer contours, in order to conform to the real situation of wood, it should be ensured that the innermost annual rings that are very close to a perfect circle can be obtained, Therefore, the outer contour and the innermost annual ring contour are extracted at the same time, and the annual rings are generated from the outside to the inside and from the inside to the outside respectively. As shown in Figure 7, since the tree growth process is initially less affected by external factors, early growth rings that are approximately perfect circles can be generated by using the outer contour of the innermost ring. When the tree grows to a certain year, the shape of the annual ring will change due to the influence of specific external factors. The outer contour can be used to generate the late growth ring that matches the later growth.

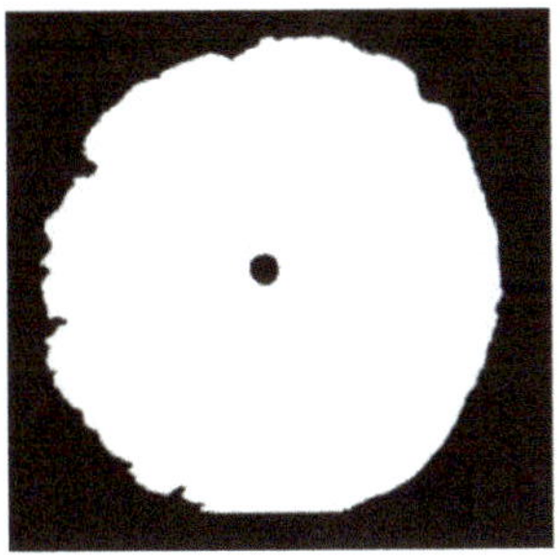

(**a**)

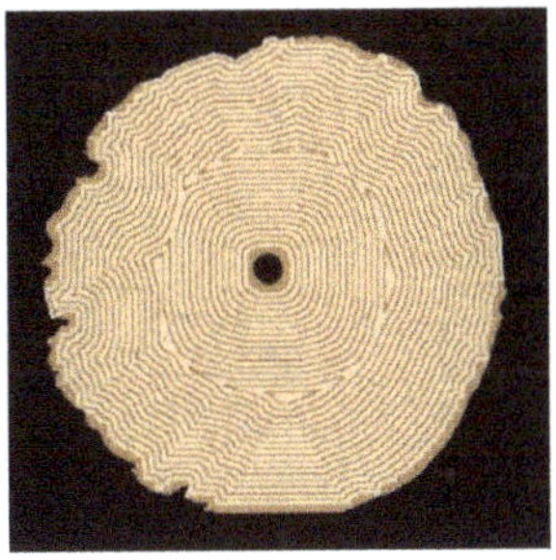

(**b**)

Figure 7. Simulation results of annual rings in branch cross-sections generated with different contours: (**a**) outer contour drawing; (**b**) simulation-generated annual ring cross-section.

During tree growth, the formation of annual rings in the same circle is easily affected by sunlight intensity and temperature. It is also related to the influence of the local environment and climate. When trees grow in different directions in different environments, it is easy to form eccentric growth rings, as shown in Figure 8. During the simulation process, the range covered by each corrosion ring is varied in the x- and y-directions, in order to achieve different widths of the same annual ring in different directions and generate eccentric annual ring images.

Figure 8. Simulation results of tree rings in branch cross-sections with different widths of the same annual ring.

The tree cross-section annual ring simulation was carried out on 512 sections of the volume data of a branch model, and the calculated results were stored in a volume data structure with a color. The visualization effect is shown in Figure 9. The longitudinal sections of the wood interior were visually very similar to real wood.

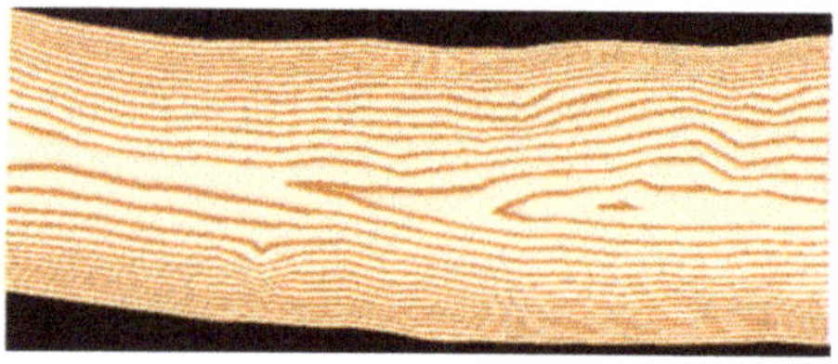

Figure 9. Longitudinal section of reconstruction of timber interior structure.

3.4. Discrete Representation of Branch Model

The tree branch model is discretized into particles with information about the index value of the annual rings in which they are located, the sector location, and the number of layers in the axial direction. In this way, the information about the internal structure of wood can be stored with a small amount of data. Therefore, a method is proposed to discretize the branch model into particles on the basis of the reconstructed internal structure of wood.

The internal structure of a branch reconstruction is stored in colored body data sections. Each section i has $N_{ring_{(i)}}$ annual rings, and each annual ring can be discretized into N_{sec} particles. Thus, the number of particles to be discretized from the wood internal reconstruction data with resolution N_{model} can be expressed as follows:

$$Sum_{discrete_{particles}} = \sum_{i}^{N_{model}} N_{sec} N_{ring_{(i)}}. \quad (1)$$

The method of discretizing an annual cycle into N_{sec} particles is as follows: before each profile erosion operation, the center position of the current branch cross-section is calculated. The center position is equal to the average position of all white dots in the current section. Then, every $(360/N_{sec})$ degrees, a straight line is emitted from the center outward, and the position of the first black point in the path of each straight line is found along the direction from the center to the outside, which is recorded as the position of a discrete particle. The result of discretizing a cross-section into particles on the basis of annual structure of the wood cross-section is shown in Figure 10, with the discrete particles in red. Figure 11 shows a section of a branch represented by discrete particles, with different colors of particles at different annual rings. Figure 12 shows the surface mesh of the branch reconstructed from the discrete particles. The surface mesh of the "bark" was reconstructed using the position information of the particles located in the outermost annual rings of all sections. The surface meshes of the two ends of the branch were reconstructed using the position information of all particles in the first and last sections.

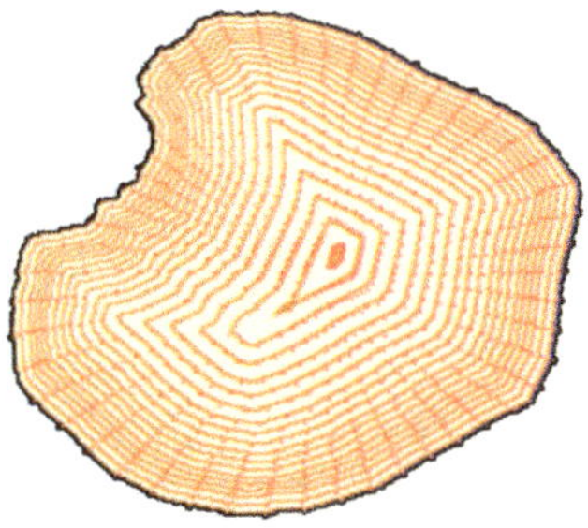

Figure 10. Discrete particles represent a cross-section of the wood.

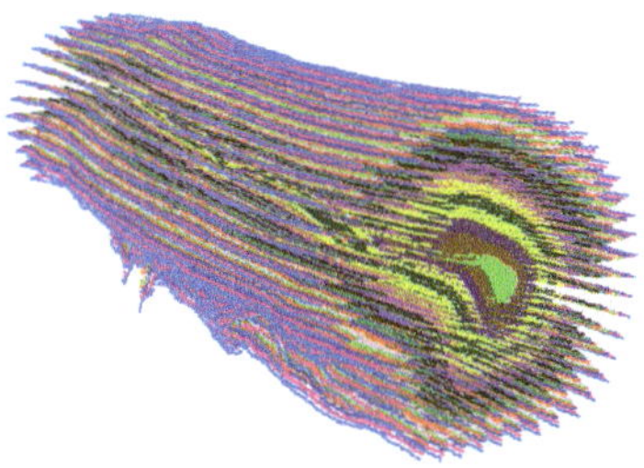

Figure 11. Discrete particles represent the wood; the color of the particle indicates the annual ring it is in.

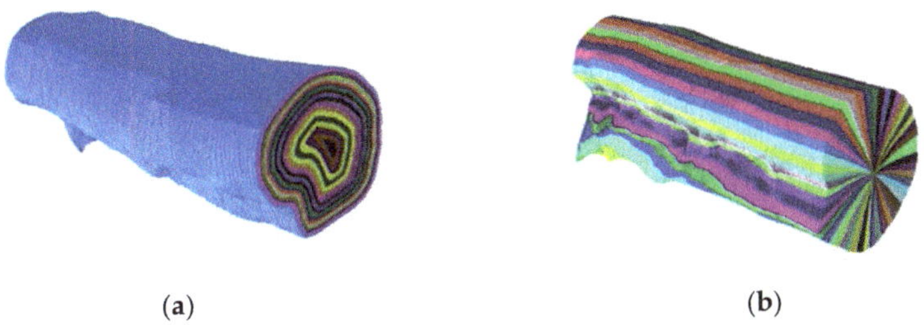

Figure 12. Branch model for discrete particle reconstruction, where the color of the particle indicates the annual ring it is in: (**a**) different annual rings; (**b**) different sectors.

4. Mapping Method

By mapping the vertex information of the fracture surface generated on the regular cylinder to the internal wood particles reconstructed according to the external contour of the branch, the fracture surface that fits the contour of the branch and conforms to the internal structural rules of the wood could be obtained. The method in [25] was used to generate a vertex V_{ij} at the i annual ring and the j sector on the regular cylinder, and its height was $H_{V_{ij}}$, where i and j are greater than 0 but less than 40, and $H_{V_{ij}}$ = 16.8 cm. α_{map} denotes the ratio of the axial length of the dendritic mesh model to the total number of sections N_{model} occupied by the dendritic voxel data in the axial direction, where N_{model} = 512. The algorithm in this paper was used to get a discrete particle P_{ijk} located on the k section, where k is greater than 0 but less than N_{model}. The vertex of the new fracture surface obtained by mapping is represented by $P_{ijk'}$. V'_{ij} represents the vertices at the i

annual ring and the j sector of the new fracture surface obtained through mapping. V'_{ij} is calculated using Equations (2) and (3):

$$k' = \left\lfloor \alpha_{map} H_{V_{ij}} \right\rfloor, \tag{2}$$

$$V'_{ij} = P_{ijk'}. \tag{3}$$

The mesh model of the new fracture surface k' was constructed according to all V'_{ij} obtained by calculation. The bark point-cloud data sections were divided according to the total number of sections. The vertex V'_{ij} on the outermost ring of the fracture plane was connected with the bark point-cloud data point p on the same section layer to form a complete branch fracture model.

5. Results

5.1. Experimental Environment

The algorithm in this paper was run in the following hardware and software experimental environment: 16 GB memory, Windows 10 64 bit operating system, CPU AMD Ryzen 7 5800H, GPU NVIDIA GeForce RTX 3060, Visual Studio 2017, Python 3.6, and OpenGL 3.3.

5.2. Experimental Results

Cross-sectional tree ring images of tree branches were generated using methods similar to corrosion algorithms. Different filtering kernels could be used to generate rings with different widths. Adding colors with the same RGB values as the real image to each ring could generate a more realistic cross-sectional ring image of fractured branches. Figure 13 shows a comparison of the tree ring contour extraction method based on the image processing method [25] and the tree ring reconstruction results in this paper. The first line is the tree ring reconstruction of *China fir* with a thoracic diameter of 36 cm, and the second line is the tree ring reconstruction of *Cryptomeria fortune* with a thoracic diameter of 24 cm.

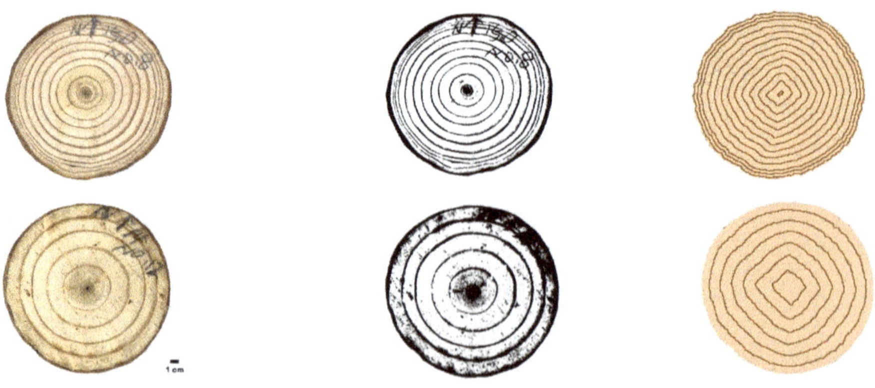

(a) (b) (c)

Figure 13. The comparison of the method of extracting annual rings based on image processing method and the reconstruction results in this paper: (**a**) real tree rings; (**b**) image processing method [25]; (**c**) the method of this paper.

The internal contours were constructed using 3 × 3 filter kernels for different sizes of cross-sectional maps, as shown in Table 1. A 128 × 128 cross-sectional map can generate 41 circles of contours, a 256 × 256 cross-sectional map can generate 80 circles of contours, and a 512 × 512 cross-sectional map can generate 156 circles of contours. A larger resolution takes a longer time to build the internal contour.

Table 1. The comparison results of annual rings generated from different size cross-sectional maps.

Resolution	Contour (#)	Time (s)
32 × 32	11	0.1628
64 × 64	21	0.5469
128 × 128	41	3.9118
256 × 256	80	29.1836
512 × 512	159	240.4738

The outer contour was used to simulate the annual ring. After each erosion algorithm, the pixel value of the generated contour and the value of the remaining white pixels were calculated, and then the next erosion was performed. At the same time, according to the number of corrosions, a specific color was given to the corresponding contour to form the final effect diagram of the annual ring section. To process contour images of different sizes, the pixel values that needed to be calculated were different. A larger image resolution required larger pixel values to be calculated, thus taking longer.

The length and width of the collected tree branch were 16.8 cm and 4.5 cm, respectively. The scanned point-cloud data contained 101,422 points with a point density of 0.1367 points per cubic meter. Scanning to obtain point-cloud data of the bark structure of broken branches could preserve complete detail information, generate realistic bark surface structure, voxel the fractured branches, and add ring information to the fracture surface to better simulate the fractured branches; however, the details on the fracture surface were missing. Figure 14b shows the rendered branch fracture model, and Figure 14a shows the real branch fracture picture. It can be seen that the overall simulation effect was good, but there was still a certain gap between the simulation and the real picture in the details. Figure 14c shows the fractured branch model generated by parametric modeling [26], depicting a branch fracture surface with uneven height on a cylinder, but its cross-section was still a circle. Compared with the fractured branch model generated in this paper, the outer surface of the branch was uneven. The bark fracture was similar to the real fracture branch after adding texture, appearing realistic. The real annual rings could be seen at the fracture surface, conforming to the real branch shape. The model generated by parametric modeling was still a cylinder. However, at the fracture surface, the protrusions simulated in this paper were not obvious enough. Using parametric modeling and the fracture surface generated by filtering, it can be seen that the protrusions were more obvious, but the generated positions were still relatively random.

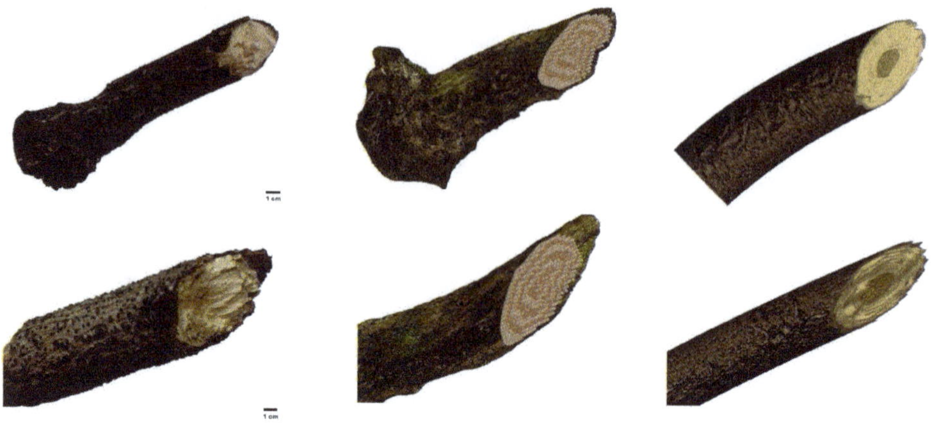

(a) (b) (c)

Figure 14. Rendered production of broken tree branch model compared with the real picture: (**a**) real branch fracture picture; (**b**) rendered branch fracture model; (**c**) other work [28].

6. Discussion

In this paper, we proposed a 3D plant technique to reconstruct the internal annual rings and wood fiber bundle structure on the basis of the external profile of tree trunk or branch, The reconstructed internal structure can be used to map the fracture surface and to represent the inhomogeneous material strength inside the branch. This method is an early computer graphics method to simulate wood fracture according to wood structure and wood science theory. The method incorporates basic features of wood fracture, such as the internal hierarchical structure of wood, including the annual rings and wood fiber bundles, into the parametric modeling approach. The reconstructed model plays an important role in the study and analysis of the internal morphological structure of woody plants. At the same time, the simulation of branch fracture also plays a certain role in the corresponding research of forestry. The branch fracture model is the embodiment of the details of the tree model, and the function of woody plants is accurately calculated and analyzed using the branch fracture structure model. Therefore, the simulation algorithm can be used not only in the fields of entertainment and animation games, but also to promote the research and application of plant morphology and structure in smart agriculture.

7. Conclusions

Although the algorithm in this paper achieved the expected results of wood fracture simulation, there were some limitations in its design and experimental process. The design of the method for extracting internal points when performing discretization is still inadequate. The circumference of the annual rings increases from inside to outside according to its position in the cross-section. The circumference of the outermost annual wheel is the longest, and the circumference of the innermost annual wheel is the shortest. The circumference of the outer annual rings is often tens of times longer than that of the inner annual rings. In the discretization result, the points of the outer annual rings are sparsely distributed, and the points of the inner annual rings are densely distributed. According to the viewpoint proposed in this paper, each annual ring is regarded as composed of N_{sec} wood fiber bundles. The cross-sectional area of wood fiber bundles in the outer annual rings is much larger than that of wood fiber bundles in the inner annual rings. However, according to actual observations of wood and theories in wood science and botany, the cross-sectional area of wood fiber bundles within the same branch cross-section should be essentially the same. For example, the number of segments of each annual rings should be set according to the circumference of the rings, with a longer circumference necessitating the divisions of more segments.

Future work can build on the algorithm in this paper to obtain more realistic and accurate wood fracture simulation results, and future research can be conducted in the following directions:

1. Bark fracture can be simulated. The fracture effect of the bark and the wood fracture effect of the xylem are very different. Using the bark gullies and grain, the bark fracture effect can be further perfected. When the bark breaks, cracks tend to spread along these gullies and form other fracture structures.
2. The effect of fracture brought by structures such as knots on wood and forks on branches can be considered. For example, knots in branches and trunks can add significant complexity to reconstructing the internal structure of wood. It is necessary to design an algorithm that can detect the position of nodes or branches and reprocess the internal structure of the wood there.
3. The real internal structure of the wood obtained from the CT scan can be compared and evaluated with the results of this paper to improve the algorithm.
4. The obtained internal structure can be used to set the strength of the wood internal material. The wood fracture animation can be generated in combination with the material point method to handle anisotropic materials.

Author Contributions: Conceptualization, Y.Z., M.Y. and B.X.; methodology, Y.Z., M.Y. and B.X.; validation, Y.Z., M.Y. and B.X.; formal analysis, Y.Z., M.Y. and B.X.; investigation, Y.Z., M.Y. and B.X.; resources, Y.Z., M.Y. and B.X.; writing—original draft preparation, Y.Z.; writing—review and editing, Y.Z., M.Y. and B.X.; visualization, Y.Z.; supervision, Y.Z., M.Y. and B.X.; project administration, M.Y. and B.X.; funding acquisition, M.Y. and B.X. All authors have read and agreed to the published version of the manuscript.

Funding: This work was supported by the National Key Research and Development Program of China (No. 2021YFD2201203).

Data Availability Statement: Not applicable.

Acknowledgments: The authors thank the peer reviewers for their valuable contributions to this research. Thank you to Yang (Liuming Yang) for providing us with assistance in the topic. Thank you to Liu's team (Yuwan Liu) for providing us with point cloud data. Thank you to Wang (Pei Wang) for providing us with professional knowledge and answers.

Conflicts of Interest: The authors declare no conflict of interest.

References

1. Fernández, M.; Norero, A.; Vera, J.; Perez, E.C. A functional–structural model for radiata pine (*Pinus radiata*) focusing on tree architecture and wood quality. *Ann. Bot.* **2011**, *108*, 1155–1178. [CrossRef]
2. Ancelin, P.; Fourcaud, T.; Lac, P. Modelling the biomechanical behaviour of growing trees at the forest stand scale. Part I: Development of an incremental transfer matrix method and application to simplified tree structures. *Ann. For. Sci.* **2004**, *61*, 263–275. [CrossRef]
3. Weber, J.; Penn, J. Creation and rendering of realistic trees. In Proceedings of the 22nd Annual Conference on Computer Graphics and Interactive Techniques, New York, NY, USA, 15 September 1995; pp. 119–128.
4. Kanazawa, A.; Tulsiani, S.; Efros, A.; Malik, J. Learning category-specific mesh reconstruction from image collections. In Proceedings of the European Conference on Computer Vision (ECCV), Munich, Germany, 8–14 September 2018; pp. 371–386.
5. Hu, T.; Wang, L.; Xu, X.; Liu, S.; Jia, J. Self-supervised 3D mesh reconstruction from single images. In Proceedings of the IEEE/CVF Conference on Computer Vision and Pattern Recognition, Nashville, TN, USA, 20–25 June 2021; pp. 6002–6011.
6. Livny, Y.; Pirk, S.; Cheng, Z.; Yan, F.; Deussen, O.; Cohen-Or, D.; Chen, B. Texture-lobes for tree modelling. *ACM Trans. Graph. (TOG)* **2011**, *30*, 1–10. [CrossRef]
7. Liu, Y.; Guo, J.; Benes, B.; Deussen, O.; Zhang, X.; Huang, H. TreePartNet: Neural decomposition of point clouds for 3D tree reconstruction. *ACM Trans. Graph.* **2021**, *40*, 232:1–232:16. [CrossRef]
8. Kratt, J.; Spicker, M.; Guayaquil, A.; Fiser, M.; Pirk, S.; Deussen, O.; Hart, J.C.; Benes, B. Woodification: User-Controlled Cambial Growth Modeling. *Comput. Graph. Forum* **2015**, *34*, 361–372. [CrossRef]
9. Xiao, H.; Chen, X. Modeling and simulation of curled dry leaves. *Soft Matter* **2011**, *7*, 10794–10802. [CrossRef]

10. Quigley, E.; Yu, Y.; Huang, J.; Lin, W.; Fedkiw, R. Real-time interactive tree animation. *IEEE Trans. Vis. Comput. Graph.* **2017**, *24*, 1717–1727. [CrossRef]
11. Diener, J.; Rodriguez, M.; Baboud, L.; Revéret, L. Wind projection basis for real-time animation of trees. *Comput. Graph. Forum* **2009**, *28*, 533–540. [CrossRef]
12. Yang, M.; Wu, E.H. Approach for physically-based animation of tree branches impacting by raindrops. *J. Softw.* **2011**, *22*, 1934–1947. [CrossRef]
13. Xie, J.; Yang, M.; Liang, W.; Cui, Y.; Qi, X.; Yang, G. Simulation of hail impact on leaves based on mass-spring model. *J. Graph.* **2019**, *40*, 32.
14. Pirk, S.; Jarząbek, M.; Hädrich, T.; Michels, D.L.; Palubicki, W. Interactive wood combustion for botanical tree models. *ACM Trans. Graph. (TOG)* **2017**, *36*, 1–12. [CrossRef]
15. Barbič, J.; Zhao, Y. Real-time large-deformation substructuring. *ACM Trans. Graph. (TOG)* **2011**, *30*, 1–8. [CrossRef]
16. Li, Y.; Barbič, J. Stable anisotropic materials. *IEEE Trans. Vis. Comput. Graph.* **2015**, *21*, 1129–1137. [CrossRef]
17. Wang, B.; Zhao, Y.; Barbič, J. Botanical materials based on biomechanics. *ACM Trans. Graph. (TOG)* **2017**, *36*, 1–13. [CrossRef]
18. O'brien, J.F.; Bargteil, A.W.; Hodgins, J.K. Graphical modeling and animation of ductile fracture. In Proceedings of the 29th Annual Conference on Computer Graphics and Interactive Techniques, San Antonio, TX, USA, 23–26 July 2002; pp. 291–294.
19. Pfaff, T.; Narain, R.; De Joya, J.M.; O'Brien, J.F. Adaptive tearing and cracking of thin sheets. *ACM Trans. Graph. (TOG)* **2014**, *33*, 1–9. [CrossRef]
20. Chen, Z.; Yao, M.; Feng, R.; Wang, H. Physics-inspired adaptive fracture refinement. *ACM Trans. Graph. (TOG)* **2014**, *33*, 1–7. [CrossRef]
21. Desbenoit, B.; Galin, E.; Akkouche, S. Modeling cracks and fractures. *Vis. Comput.* **2005**, *21*, 717–726. [CrossRef]
22. Hädrich, T.; Scheffczyk, J.; Palubicki, W.; Pirk, S.; Michels, D.L. Interactive Wood Fracture. *SCA (Posters)* **2020**, 5–7.
23. Fritts, H.C. *Tree Rings and Climate*; Academic Press: London, UK, 1976; pp. 1–567.
24. Shao, X. Several Advances in Tree Ring Chronology. *Quat. Res.* **1997**, *3*, 265–271.
25. Wang, Y.F.; Feng, H.L.; Fang, Y.M. Improving the tree-age measurement by enhancing the tree-ring image with bilateral filter. *J. For. Eng.* **2017**, *2*, 109–114.
26. Xue, Q.; Yang, Y. The Application Advances of Image Processing and Pattern Recognition Technology in Dendrochronology. *Chin. Agric. Sci. Bull.* **2015**, *31*, 24–29.
27. Han, Q.Y.; He, J.D. Segmentation Algorithm of Tree Ring Image Based on Region Growing. *J. Agric. Mech. Res.* **2006**, *4*, 204–206.
28. Yang, L.; Yang, M.; Yang, G. Modeling fractures and cracks on tree branches. *Comput. Graph.* **2019**, *80*, 63–72. [CrossRef]
29. Sapala, A.; Runions, A.; Routier-Kierzkowska, A.L.; das Gupta, M.; Hong, L.; Hofhuis, H.; Verger, S.; Mosca, G.; Li, C.; Hay, A.; et al. Why plants make puzzle cells, and how their shape emerges. *Elife* **2018**, *7*, e32794. [CrossRef]
30. Jackson, M.; Xu, H.; Duran-Nebreda, S.; Stamm, P.; Bassel, G.W. Topological analysis of multicellular complexity in the plant hypocotyl. *Elife* **2017**, *6*, e26023. [CrossRef]
31. Montenegro-Johnson, T.D.; Stamm, P.; Strauss, S.; Topham, A.T.; Tsagris, M.; Wood, A.T.; Smith, R.S.; Bassel, G.W. Digital single-cell analysis of plant organ development using 3DCellAtlas. *Plant Cell* **2015**, *27*, 1018–1033. [CrossRef] [PubMed]
32. Min, P. binvox[CP/OL]. 2004. Available online: https://www.patrickmin.com/binvox/ (accessed on 11 May 2020).
33. Nooruddin, F.S.; Turk, G. Simplification and repair of polygonal models using volumetric techniques. *IEEE Trans. Vis. Comput. Graph.* **2003**, *9*, 191–205. [CrossRef]
34. Dougherty, E. *Mathematical Morphology in Image Processing*; CRC Press: Boca Raton, FL, USA, 2018.
35. Shih, F.Y. *Image Processing and Mathematical Morphology: Fundamentals and Applications*; CRC Press: Boca Raton, FL, USA, 2009.

forests

MDPI

Article

Constructing Coupling Model of Generalized B-Spline Curve and Crown (CMGBCC) to Explore the 3D Modeling of Chinese Fir Polymorphism

Zeyu Cui [1,2,3], Huaiqing Zhang [1,2,3,*], Yang Liu [1,2,3], Jing Zhang [1,2,3], Tingdong Yang [1,2,3], Yuanqing Zuo [1,2,3] and Kexin Lei [1,2,3]

1 Institute of Forest Resource Information Techniques, Chinese Academy of Forestry, Beijing 100091, China
2 Key Laboratory of Forest Management and Growth Modelling, NFGA, Beijing 100091, China
3 National Long Term Scientific Research Base of Huangfengqiao Forest Monitoring and Simulation in Hunan Province, Beijing 100091, China
* Correspondence: zhang@ifrit.ac.cn

Abstract: Crown simulation based on basis spline (b-spline) interpolation is a compatible method to simulate tree polymorphism at present. However, there are two problems when it simulates the crown: the first problem is that the derivative value at the top point needs to be given manually, and the second is that the type of value point needs to be collected equidistantly. To solve the above problems and realize convenient and accurate tree polymorphism simulation, this study took Chinese fir as the study object, set the crown morphological feature as the model value point, and constructed a coupling model of generalized B-spline curve and crown (CMGBCC) as the constraint condition of the crown shape to simulate the polymorphism in the process of a tree three-dimensional (3D) model. The position and size of the distribution on the 3D model of the branches were constrained by the curve, and the 3D modeling of a Chinese fir polymorphism was constructed. According to the collection of Chinese fir-type value points in the sample plot, the study realized the detailed types of value points' precise simulation for three polymorphisms of the Chinese fir crown, including natural pruning, crown displacement, and crown shape difference. At the same time, the different withered existence states of the branches were considered preliminarily. Compared to the 3D model with the field survey data, indicating that constructed models could simulate the difference in tree crown morphology precisely, the branch models were separated by convenience to simulate the process of Chinese fir growth. In the process of construction, CMGBCC did not need to add the derivative value in a manual way and could collect the type of value points according to the characteristics of the crown morphological changes completely. Compared to the results of the crown curve constructed, which were based on generalized B-spline (GB-spline) interpolation and b-spline interpolation, it showed that the number of crown value points collected by the GB-spline interpolation method decreased by 18% on average. The precision of the crown shape constraint was improved by 7.63% compared to b-spline interpolation. The 3D modeling of tree polymorphism was combined with the relationship between tree morphology and environment. At the same time, it was convenient to simulate the behavior of forest management measures, such as pruning.

Keywords: polymorphism; 3D model of Chinese fir; generalized B-spline interpolation; withered state of branch; natural pruning; crown displacement; crown shape difference

1. Introduction

The three-dimensional (3D) modeling of trees is the 3D construction process for tree morphological structures. In general, the morphological structure of trees with the same species follows a common feature. However, for each individual, the influences from the surrounding environment are different: the morphology of the individual has different manifestations, such as natural pruning [1,2], crown displacement [3,4], and crown shape difference: these are important manifestations of a tree 3D model polymorphism. Meanwhile, in the process of 3D modeling, the expression of the interaction between the morphological structure and environment is limited by tree polymorphism.

Relevant studies have shown that spatial factors such as stand density and space competition have a direct impact on tree crown morphology [1,5]. Air pollution has an inhibitory effect on growth, thereby affecting the morphological structure [6]. Climatic factors such as light, drought, and temperature also influence the morphological structure by interfering with the growth rate further [7–9]. In addition, other factors such as topography and altitude can influence the growth and change the tree morphology through differences in soil water content [10]. In summary, the tree's morphological structure is influenced by various environmental stress factors. The mechanisms of these various factors overlap and intersect with each other. The spatial structure can comprehensively reflect the effects of various environmental factors, such as spatial competition, light, and terrain, on the crown's morphological structure [11–14]. Therefore, this study combined the spatial structure to explore the performance of tree polymorphism.

At present, for the 3D modeling research of tree polymorphism, Ma et al. used to age and spatial structure parameters as variables to construct an under branch height (UBH) model, simulating the occurrence process of natural pruning through the tree growth visual simulation and constrained crown contour based on basis spline (b-spline) interpolation [15]. Subsequently, researchers constructed a UBH model based on spatial structure parameters and tree heights to simulate natural pruning with the influence of different spatial structures [16,17]. For the whole crown shape, current research has mainly utilized the crown curve for constraint simulation [18,19], including the crown curve power function (POW) [20] and spline interpolation [15]. The shape value points were substituted into those two functions to fit the crown's outer contour shape. In addition, the effect of space resource competition on the tree branch growth, branch distribution, and crown shape was considered, and environmental factors were applied to express tree polymorphism [21]. Additionally, Palubicki et al. combined the lighting and plant growth mechanism to construct a 3D tree model [22]. These studies explored tree polymorphism and polymorphism modeling from different perspectives.

Indeed, evidence exists that Chinese fir is a uniaxially branched conifer species, and its crown morphology exhibits different characteristics at different ages. The number of branches, length, angle, and distribution structure of branches all follow certain rules [23]. This morphological structure is affected by the environment, showing an irregular diversity, which is suitable for polymorphism 3D modeling research. With the introduction of a generalized B-spline (GB-spline) curve, the accuracy of Chinese fir's crown shape simulation was improved; b-spline interpolation is one of the most popular methods, which must input an equal distance of type value points.

The main objective of this study was twofold: (1) To construct a 3D Chinese fir polymorphism model (3D-CFPM) to predict tree morphological structures using detailed shape point data (tree height, crown width, UBH, height at the maximum crown width (HMCW) and other crown shape mutation points) of the sample plot survey; (2) to explore a GB-spline curve when constructing a coupling model of a generalized B-spline curve and crown (CMGBCC), and simulating the different Chinese fir crown shape polymorphisms under natural pruning, crown displacement, and crown shape differences. These results were validated by the existing b-spline curve and instrument field measurement to verify the effectiveness of CMGBCC when constraining the Chinese fir crown shape.

2. Materials and Methods

2.1. Data

All fieldwork was conducted at the Chinese fir plantation area in the Shanxia Forest Farm of the Subtropical Forestry Experiment Center, Jiangxi Province, China (114°37′~114°41′ E, 27°43′~27°46′ N, 85~300 m altitude). It belongs to the terrain of the low hill, with an average annual temperature of about 17.5 °C and an annual rainfall of 1100–1700 mm, belonging to a subtropical monsoon humid climate. Considering the influence of spatial structure and age on the trees' morphological structure, a total of 239 Chinese firs in 8 plots of 20 m × 20 m were selected for data collection. The selected range had an age span of 12–30 years and different stand densities. The relative position of each Chinese fir was measured in the sample plot to analyze its spatial structure; the basic type of value points was measured for each tree, including tree height, diameter at breast height (DBH), and crown width in different directions (include four directions: east, west, south, north), UBH, and HMCW. Additionally, based on the measurement of the crown morphological mutation points of each tree, detailed type value points were obtained. The distribution of eight plots and the distribution of trees in each plot were as shown in Figure 1.

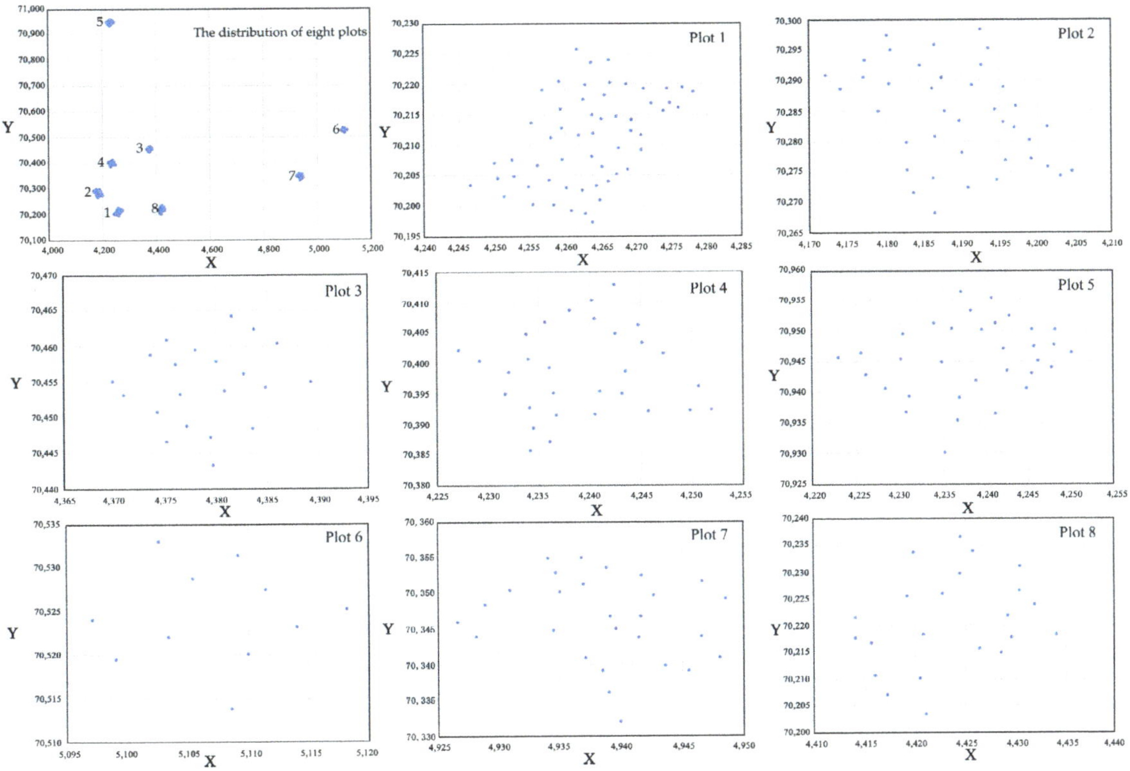

Figure 1. The distribution of eight plots and the distribution of trees in each plot.

The data statistics of the basic type of value points for trees in the sample plot are shown in Table 1. Number 1–8 are eight plots' reference which are measured.

Table 1. Statistics of the basic type of value points of trees in the sample plot.

Type	Height/m	DBH/cm	UBH/m	HMCW/m	Crown Width/m	Age/a
Max	30.4	36.5	14.6	18.1	4.0	30
Min	8.8	8.3	4.5	7.2	0.4	12
Average	17.8	23.6	8.0	13.9	1.7	24

2.2. Constructing CMGBCC

B-spline interpolation is a commonly used interpolation function for fitting the shape and outline of objects [24–27], which has good local controllability. The crown curve is a function curve that describes the outer contour of the tree crown. Ma built the crown curve of the Chinese fir based on b-spline interpolation in the B-spline interpolation. Compared with the interpolation methods such as POW, it was suitable for expressing the irregularity of the crown [15]. However, the b-spline interpolation to construct the crown curve has certain defects; these include the fact that the derivative value at the top point of the tree is unknown, and the crown value point collection needs to meet the equidistant collection during fitting. Regarding these two defects, the former causes the need to increase the artificial input parameter values in the process of constructing the crown curve, which cannot effectively confirm the rationality of the parameter values; while the equidistant collection type value points greatly limit the record, the position of the crown shape changes, which may cause the simulation results to under-represent the crown's morphology. Therefore, based on the b-spline interpolation to construct the crown curve, this study introduced the periodic basis spline (pb-spline) and GB-spline interpolation into the study of crown curve construction to solve the current shortcomings gradually.

Owing to the fact that the envelope range of the crown curve was the outer contour of the longitudinal section of the tree crown, the endpoints at the top of the tree were continuous and closed. In the actual measurement, the slope at the top point of the tree was difficult to obtain accurately, and b-spline interpolation fitting could only be based on derivative values given empirically. The pb-spline interpolation did not need to provide this value; therefore, this study introduced pb-spline interpolation on the basis of b-spline.

Since the top point of the crown curve was closed and continuous, according to the characteristics of the continuous curve function, we could define the value of point a where the left curve approached the vertex as f(a), and the value of point b where the right curve approached the vertex as f(b). The curve function f(x) was assumed as follows:

$$\begin{cases} f(a) = f(b) \\ f'(a) = f'(b) \\ f''(a) = f''(b) \end{cases}, \tag{1}$$

The expression of the cubic polynomial: $g_k(t)$ of the k-th interpolation only related to the base point t and was expressed as:

$$g_k(t) = \varphi\left(\frac{t - t_k}{h}\right), \tag{2}$$

where h is the max value of the interpolation of t_{i+1} and t_i which are two adjacent points and $\varphi(t)$ is the cubic b-spline system, the for which expression was:

$$\varphi(t) = \frac{1}{6}\begin{cases} (t+2)^3 \quad t \in [-2,-1] \\ 1+3(t+1)+3(t+1)^2-3(t+1)^3 \quad t \in [-1,0] \\ 1+3(1-t)+3(1-t)^2-3(1-t)^3 \quad t \in [0,1], \\ 2(2-t)^3 \quad t \in [1,2] \\ 0 \mid t \mid \geq 2 \end{cases} \tag{3}$$

By combining the values of $g_i(t)$, $g'_i(t)$, and $g''_i(t)$ at each base point in Equation (1), the following conclusions could be drawn:

$$\begin{cases} a_0 + 4a_1 + a_2 = a_n + 4a_{n+1} + a_{n+2} \\ a_0 - a_2 = a_n - a_{n+2} \\ a_0 - 2a_1 + a_2 = a_n - 2a_{n+1} + a_{n+2} \end{cases}, \tag{4}$$

where a_0, a_1, and a_2 presented the three points on the left curve that approached the vertex. Additionally, a_n, a_{n+1}, and a_{n+2} were three points on the right curve that approached the vertex.

Simultaneously, the three equations of Equation (4) could be solved:

$$\begin{cases} a_0 = a_n \\ a_1 = a_{n+1}, \\ a_2 = a_{n+2} \end{cases} \tag{5}$$

At this time, there were n unknown parameters of the b-spline and n linear equations:

$$\begin{bmatrix} 4 & 1 & & & & 1 \\ 1 & 4 & 1 & & & \\ & 1 & 4 & 1 & & \\ & & \cdots & & & \\ & & & 1 & 4 & 1 \\ 1 & & & & 1 & 4 \end{bmatrix} \begin{bmatrix} a_1 \\ a_2 \\ a_3 \\ \&\ldots \\ a_{n-1} \\ a_n \end{bmatrix} = 6 \begin{bmatrix} f_1 \\ f_2 \\ f_3 \\ \&\ldots \\ f_{n-1} \\ f_n \end{bmatrix}, \tag{6}$$

Since the coefficient matrix of the equation system was a strongly dominant diagonal matrix, according to the relevant theorem of numerical calculation, the linear equation system had a unique solution. However, the coefficient matrix of this equation system did not satisfy the tridiagonal; therefore, the improved chasing method was used to solve it [28].

The introduction of pb-spline interpolation solved the problem of manually inputting the derivative value at the top point but did not solve the critical issue of the equidistant collection of type value points. Therefore, on this basis, the interpolation function was further improved to GB-spline interpolation, constructing CMGBCC.

For GB-spline interpolation, there was the Cox-de Boor recurrence formula [29,30]:

$$\begin{gathered} B_{i,0}(t) = \begin{cases} 1 & t \in [t_i, t_{i+1}] \\ 0 & \text{otherwise} \end{cases} \\ B_{i,k}(t) = \frac{t-t_i}{t_{i+k}-t_i} B_{i,k-1}(t) + \frac{t_{i+k+1}-t}{t_i+k_+-t_{i+1}} B_{i+1,k-1}(t) \end{gathered}, \tag{7}$$

When k = 3, the cubic GB-spline curve could be expressed as:

$$B_{i,3}(t) = \begin{cases} \dfrac{(t-t_{i-2})^3}{(t_{i+1}-t_{i-2})(t_i-t_{i-2})(t_{i-1}-t_{i-2})} & t \in [t_{i-2}, t_{i-1}] \\ \dfrac{t-t_{i-2}{}^2(t_i-t)}{(t_{i+1}-t_{i-2})(t_i-t_{i-1})(t_i-t_{i-2})} + \dfrac{(t-t_{i-2})(t-t_{i-1})(t_{i+1}-t)}{(t_{i+1}-t_{i-2})(t_{i+1}-t_{i-1})(t_i-t_{i-1})} & \\ + \dfrac{(t_{i+2}-t)t-t_{i-1}{}^2}{(t_{i+2}-t_{i-1})(t_{i+1}-t_{i-1})(t_i-t_{i-1})} & t \in [t_{i-1}, t_i] \\ \dfrac{t_{i+2}-t^2(t-t_i)}{(t_{i+2}-t_{i-1})(t_{i+2}-t_i)(t_{i+1}-t_i)} + \dfrac{(t-t_{i+2})(t-t_{i+1})(t-t_{i-1})}{(t_{i+2}-t_{i-1})(t_{i+1}-t_i)(t_{i+1}-t_{i-1})} & \\ + \dfrac{(t-t_{i-2})t_{i+1}-t^2}{(t_{i+1}-t_i)(t_{i+1}-t_{i-1})(t_{i+1}-t_{i-2})} & t \in [t_i, t_{i+1}] \\ \dfrac{(t_{i+2}-t)^3}{(t_{i+2}-t_{i+1})(t_{i+2}-t_i)(t_{i+2}-t_{i-1})} & t \in [t_{i+1}, t_{i+2}] \\ 0 \text{ otherwise} & \end{cases}, \tag{8}$$

$B_{i,3}(t)$ was calculated and the value at each point was:

$$B_{i,3}(t_j) = \begin{cases} \dfrac{(t_{i-1}-t_{i-2})^2}{(t_{i+1}-t_{i-2})(t_i-t_{i-2})} & j = i-1 \\ \dfrac{(t_i-t_{i-2})(t_{i+1}-t_i)}{(t_{i+1}-t_{i-1})(t_{i+1}-t_{i-2})} + \dfrac{(t_i-t_{i-1})(t_{i+2}-t_i)}{(t_{i+1}-t_{i-1})(t_{i+2}-t_{i-1})} & j = i, \\ \dfrac{(t_{i+2}-t_{i+1})^2}{(t_{i+2}-t_i)(t_{i+2}-t_{i-1})} & j = i+1 \\ 0 \text{ otherwise} & \end{cases} \tag{9}$$

The shape value points obtained by the field measurement were brought into the calculation, and the crown curve of the tree could, therefore, be obtained. When the mutation position of the crown shape was recorded in more detail. The detailed shape points is shown in Figure 2a, and the simulation of CMGBCC is shown in Figure 2b.

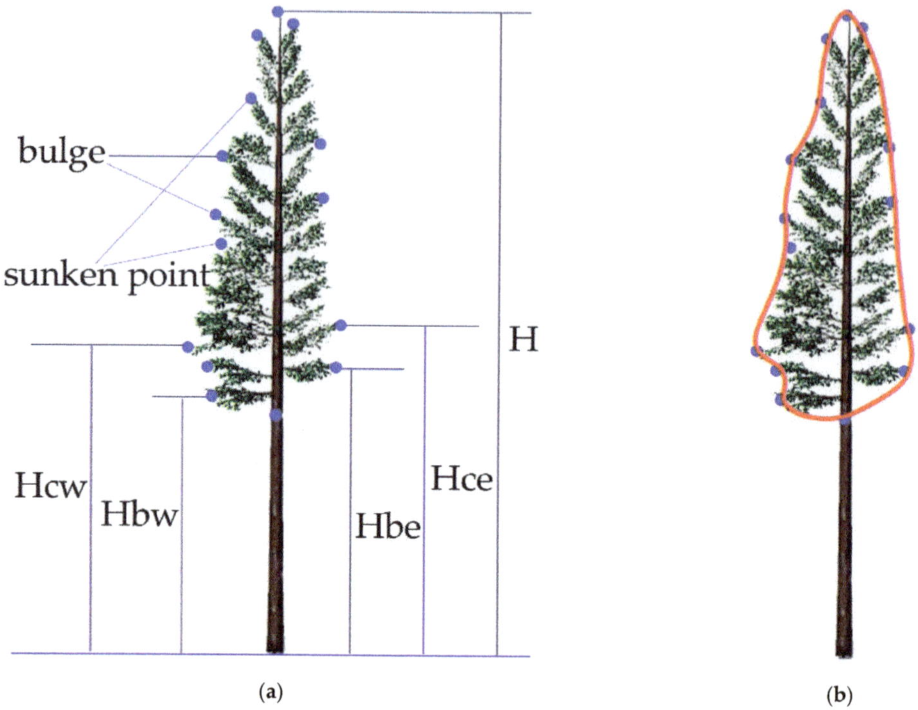

Figure 2. The simulation of CMGBCC.

In Figure 2a, we have shown some characteristic points of tree crown morphology. H is the height of the tree, Hcw is the HMCW on the west side, Hbw is the UBH on the west side, Hce is the HMCW on the east side, and Hbe is the UBH on the east side. These five parameters are the basic value points of the tree crown in the viewing angle of one side. The bulge is the outwardly bulging part of the crown, and the sunken point is the inwardly concave part of the crown. In Figure 2b, we showed the result of CMGBCC. The red lines represent the crown curve simulated by CMGBCC.

2.3. Constructing Coupling Model of 3D Chinese Fir Trunk, Branches Model and Withered State

The composition structure of Chinese fir could be decomposed into the trunk, the first-level branches, second-level branches, leaves, and so on. The crown morphology would directly influence the tree morphology; the morphological structure of the first-level branches and the position distribution on the trunk were the direct factors that influenced the crown morphology. At the same time, the amount of calculation on the subsequent loading model was considered, and this study dissected the Chinese fir model as the trunk model and first-level branch model.

SpeedTree is currently a more professional and complete tree 3D modeling software [31]. The trunk of the Chinese fir is straight; therefore, the Chinese fir shaving equation could be taken for the 3D model of the trunk as the outer contour constraint [32–34]. Then, the skeleton of the 3D trunk model was constructed in the speed tree, and the Chinese fir bark texture map was added.

The theoretical basis for the construction of a first-level branch model was the morphological and structural characteristics of the first-level branch. Because the overall morphological structure of the Chinese fir conformed to the fractal theory and demonstrated self-similarity, the morphological structure of the first-level branch was similar to the overall morphological structure of the tree. The distribution characteristics of the second-level branch were the basis of the first-level branch. The elevation angle of the second-level branch on the Chinese fir basically conformed to normal distribution; this was mainly concentrated in the range of 50°~100°, and the azimuth angle was mainly concentrated at an interval of 135°~225° and 315°~45°, which was distributed in the horizontal direction. The leaves were arranged approximately symmetrically on the branches [35].

Because the first-level branch was the most direct unit to determine the shape of the tree crown, considering the amount of model loading and rendering, models in the model library were composed of first-level branches.

At the same time, the tree grew all the time. with branches and leaves growing new and dying with age [36]. Therefore, the construction of the branch model considered the influence of growth and designed branches with different withered states.

The rules considered in the modeling process are shown in Table 2.

Table 2. The rules of constructing branch models.

The Level of Branch	Index	Description
first-level branch	withered state	amount of needle cover on branches
first-level branch	elevation angle	the angle between the branch and the vertical direction of the trunk
first-level branch	branch length	length of branch
first-level branch	base diameter	branch base diameter
second-level branch	elevation angle	the angle between the branch and the vertical direction of the primary branch
second-level branch	azimuth	the angle between the branches in the horizontal direction of the primary branch
second-level branch	undershoot	length between the branch point and base

Then, the different types of first-level branch models by these rules were constructed. The withered states were controlled by the number of leaves on the branches, which included the branch that had no leaves, the branch that had limited leaves, the branch that had 50% leaves, and the branch that had whole leaves. The first-level branch's elevation angle was controlled by the elevation angle of the branch on the trunk. The branch length was controlled by the branch matching point and branch point. The base diameter was controlled by the age of the branch.

The bark and leaf texture maps of the Chinese fir are shown in Figure 3a,b.

(a) (b)

Figure 3. The bark and leaf texture maps of Chinese fir.

Figure 3a is the bark texture map of Chinese fir, Figure 3b is the leaf texture map of Chinese fir

2.4. Simulations of Three Types of the Tree Polymorphism

Three types of tree polymorphisms, including natural pruning, crown displacement, and crown shape difference, were simulated and expressed in the 3D model by the distribution of branches and the constraints of the crown curves.

Natural pruning was the result of the dead branches on trees causing the UBH to move upward; therefore, natural pruning could be simulated by the combination of the UBH model [16,17,37] and the loading of the 3D branch models. Due to the different strengths of the crown affected by space competition in different directions, the corresponding UBHs were also different. Therefore, the UBHs in various directions were calculated by the UBH models of the Chinese fir in different directions [17], and the obtained parameters were transferred to the branch models for loading. The loading positions of the branches in different directions were adjusted in this module, and the simulation of natural pruning was realized.

$$\frac{\mathrm{Hb} = \mathrm{H}}{\left(1 + e^{0.047+0.003\mathrm{DBH}-0.044\mathrm{PH}\ -\ 0.206\mathrm{PV}}\right)}, \tag{10}$$

In the formula, Hb represented the UBH in a single direction, PH was the horizontal spatial structure parameter in a single direction, and PV was the vertical spatial structure parameter in a single direction.

Natural pruning represents the distribution of the lowest living branches in different directions, while the dead branches below them were the result of historical natural pruning, which also has an important impact on tree morphology. Therefore, the loading position of dead branch models in the corresponding direction was taken from the position of the lowest living branches as the reference. According to the research on the dead branches and leaves of Chinese fir, the dead branch models were loaded down randomly for about three rounds to simulate the survival state of the branches to the current natural pruning result state.

The simulation process of natural pruning is shown in Figure 4.

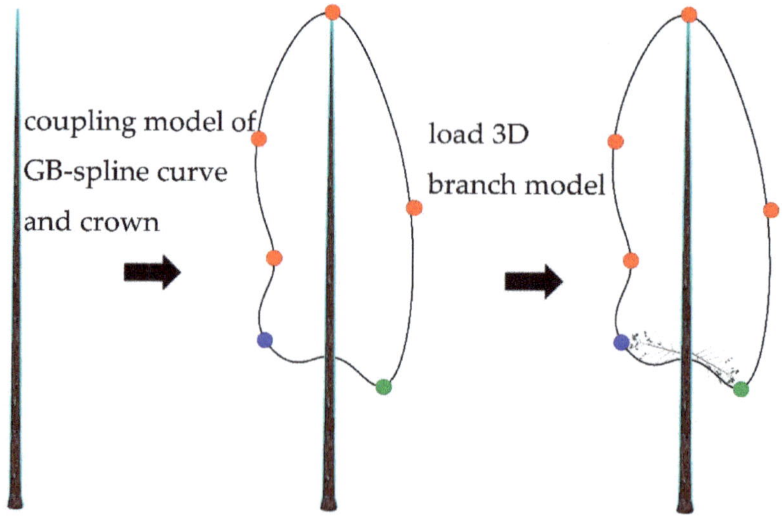

Figure 4. The simulation process of natural pruning.

In Figure 4, the blue point and green point are the lowest positions of the living branches in different directions, while the red points are other living branches' positions.

Crown displacement is the phenomenon that the crown shape is asymmetric, concave, or convex to one side when the tree crown in different directions is affected by different environments [3,4,38,39]. It is a common polymorphic form in the crown of the stand. The crown displacement phenomenon was simulated based on the crown curve's constraints on the outer contour shape of the tree crown.

Since the result of the crown displacement phenomenon was that the crown shifted to one side as a whole, it could be simulated by the crown curve and fitted by the basic value points. The basic value point data parameters obtained by measurements and calculations were transferred to the crown curve building module; the tree crown shape was constrained by the outer contour, and the 3D branch models were loaded. When the widths of the tree crowns in the corresponding east–west and north–south directions were quite different, the model could simulate the crown displacement phenomenon. The simulation process is shown in Figure 5.

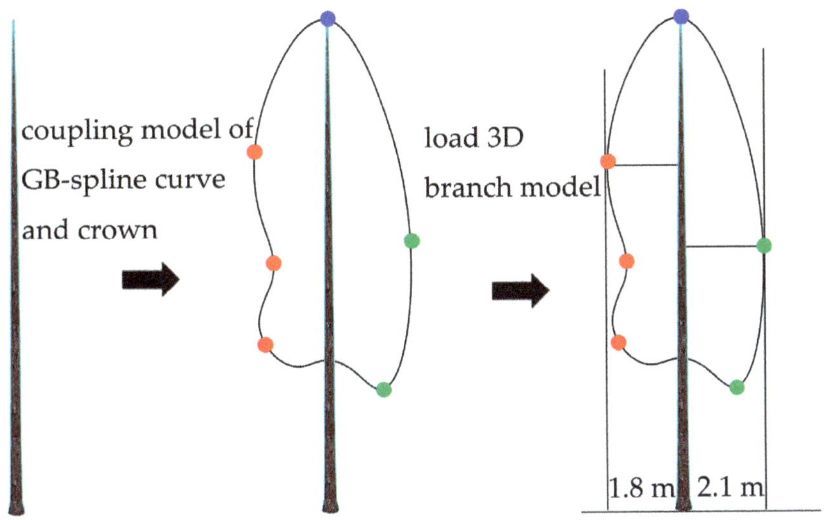

Figure 5. The simulation process of crown displacement.

In Figure 5, the red points were left-type value points, the green points were right-type value points, and the blue point was the top point in the tree. The widest left crown was 1.8 m, and the widest right crown was 2.2 m.

The crown shape difference refers to the interaction of the tree crown with the environment. Based on the basic crown shape law of these tree species, it presented irregular bumps [39]. Compared with the crown displacement, it showed the details of the crown shape at length. Therefore, the simulation of the crown shape difference was based on the simulation of crown displacement, and this simulation was realized by fitting the crown curves to the detailed shape value points. This process is shown in Figure 6.

In Figure 6, the red points are type value points, the blue circles are the sunken positions on the crown shape difference, and the green circles are the bulge positions on the crown shape difference.

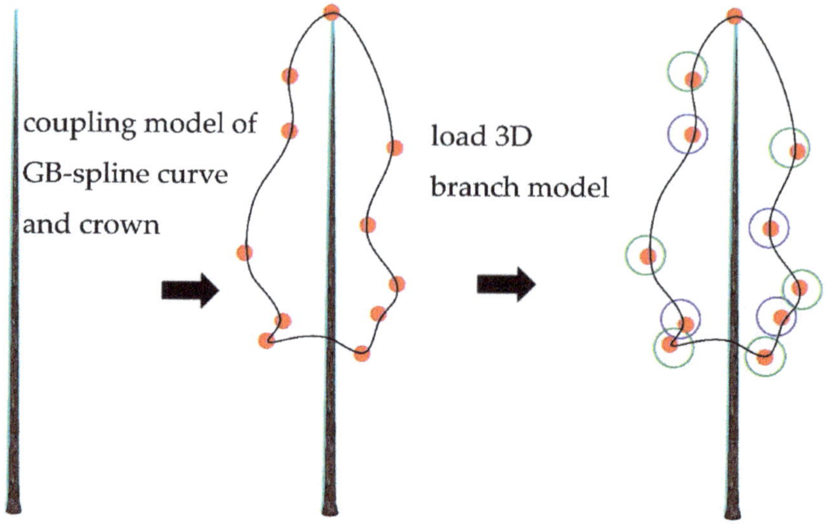

Figure 6. The simulation process of crown shape difference.

3. Results

3.1. 3D Models of Chinese Fir Trunk and Branch

3.1.1. 3D Models of Chinese Fir Trunk

The trunk of the Chinese fir is straight, and the shape of the trunk varies a little between different ages. Therefore, based on the experimental data, the tree height's distribution range in the experimental area was between 8.8 and 30.4 m, with an interval of 5 m, and the closest interval median was taken. At the tree height value, the initial trunk model was constructed at each interval. The model properties are shown in Table 3.

Table 3. The trunk model properties.

Tree Height Range	Model Height/m	DBH/cm
8.0~9.0	8.8	13.5
14.0~15.0	14.5	20
20.0~21.0	20.5	27.3
26.0~27.0	26.4	29.7
30.0~31.0	30.4	31.2

Taking the tree height of 20.5 m as an example, the constructed 3D model of the trunk is shown in Figure 7.

3.1.2. 3D Models of Chinese Fir Branch

Since the first-level branch directly influenced the overall shape of the Chinese fir, the branch model in the model library only included the first-level branch model. From the research on the Chinese fir branch morphology and structure, it could be seen that the Chinese fir crown could be divided into three layers according to the elevation angle of the branches and arranged from low to high positions, while the number of leaves on the constructed branch model ranged from less to more. In this study, considering that dead branches had a certain influence on tree polymorphism, the modeling was carried out according to the index parameters in Table 2.

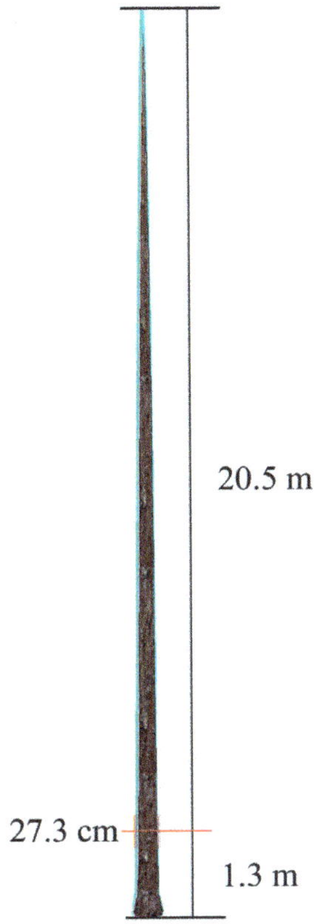

Figure 7. The 3D model of the Chinese fir trunk.

According to these indexes, the models were constructed at an interval of 10°, and the elevation angle of the primary branch was between 40° and 100°. Among them, the first-level branch accounted for 80% to 85% of all branches between 60° and 90°. Since the dead branches generally followed the bottom-up law, the elevation angle distribution law of the Chinese fir branches generally decreased from the bottom to the top. The new branches at the top of the tree were mainly distributed between 40° and 60°, which were constructed to cover all the leaves; between 90° and 100°, mainly dead first-level branches with very few leaves survived.

The branch models and model properties for different withered states are shown below. Figure 8 shows the dead first-level branch model, Figure 9 shows the branch model with a very small number of leaves surviving, Figure 10 shows the branch model with 50% dead leaves, and Figure 11 shows the branch model without dead and dropped leaves.

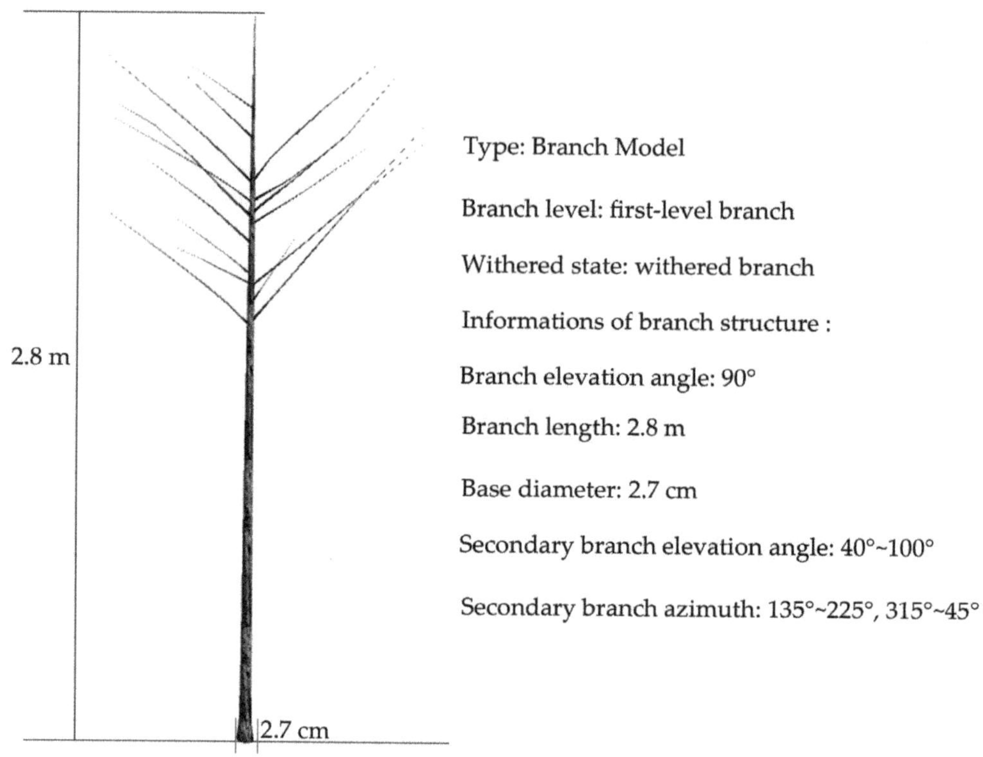

Figure 8. The dead first-level branch model.

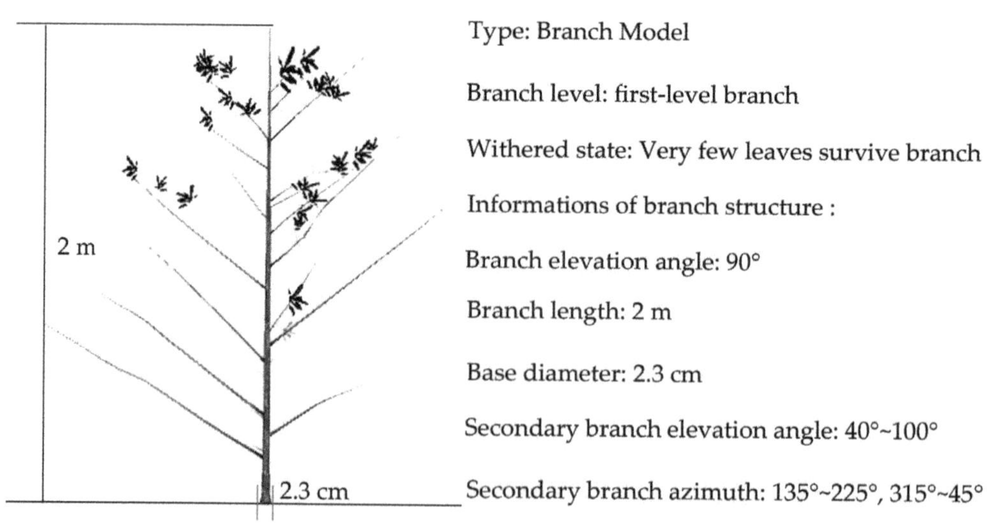

Figure 9. The branch model with a very small number of surviving leaves.

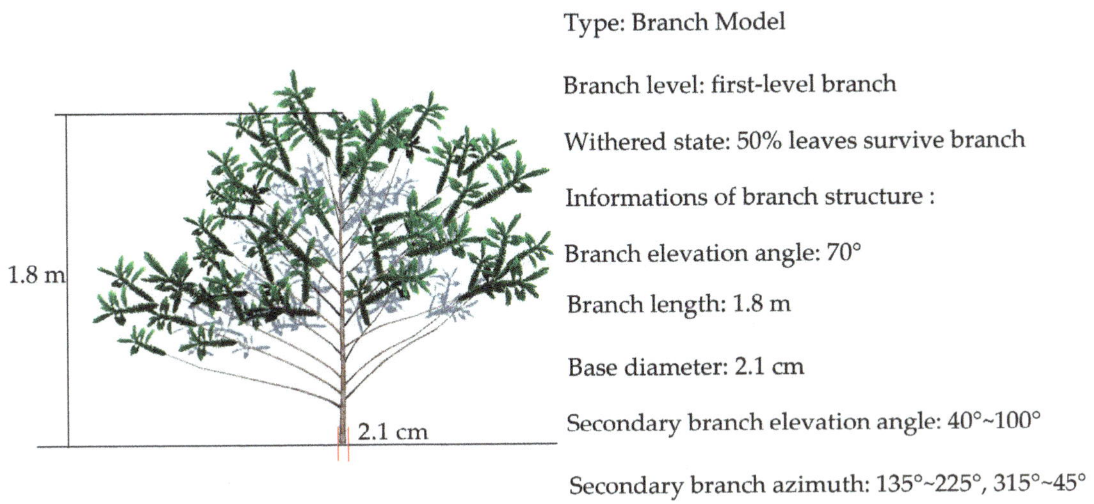

Figure 10. The branch model with 50% dead leaves.

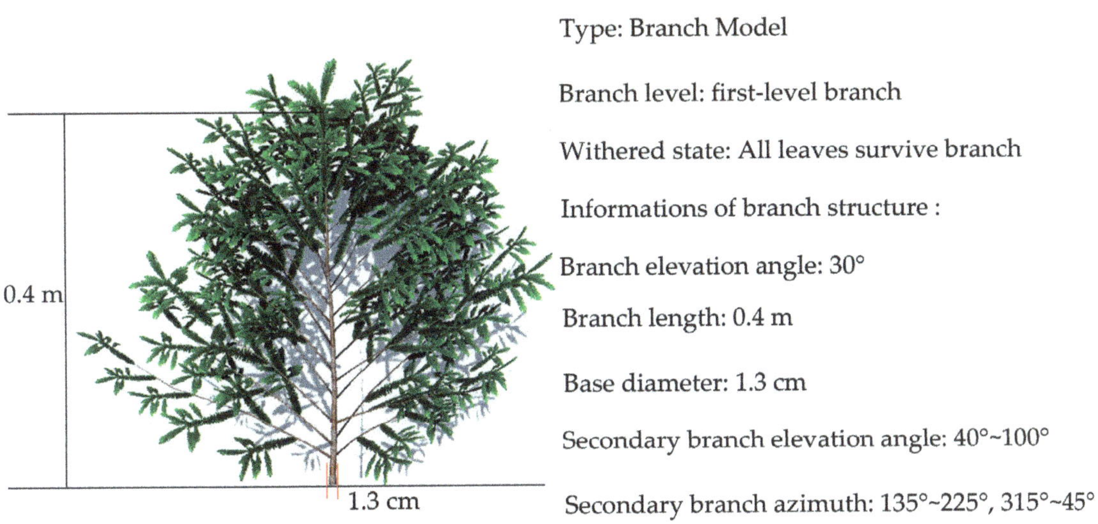

Figure 11. The branch model without dead and dropped leaves.

3.2. The Result of CMGBCC

The 3D-CFPM was obtained by CMGBCC according to the distribution structure of Chinese fir branches and the constraints of the fitted crown curve. Taking a spatial structure unit as an example, the 3D-CFPM in the spatial structure unit is shown in Figure 12.

Number 1 to 7 are seven trees in a spatial structure unit. The red circle No. 1 is the center tree, the blue circle No. 2 is the surrounding tree on the east side, the blue circles No. 3 and No. 4 are the surrounding tree on the north side, the blue circle No. 5 is the surrounding tree on the west side, and the blue circles No. 6 and No. 7 represent the surrounding tree on the south side.

Figure 12. 3D-CFPM in spatial structure unit.

The 3D model of the central Chinese fir was constructed in the spatial structure unit, the value of the crown shape is shown in Figure 13a, and the result of CMGBCC is shown in Figure 13b.

(a) (b)

Figure 13. The result of CMGBCC.

In Figure 13a, the west is on the left, and the east is on the right. The bottom blue point is the position under the high branch, the red point is the lowest dead branch position, and the red line is the constructed crown curve. It can be seen from the figure that the space competition on the left side was less than right; therefore, the natural pruning on the left was less than that on the right side, which then caused the UBH to be lower on the left; at the same time, due to the less competition on the left side, the branch growth conditions were better than the right side, and the tree crown presented a phenomenon of crown displacement to the left. On the right side, due to competition reasons, there was a concave in the upper part of the UBH, which resulted in a special crown shape, showing the difference in crown shape. The natural pruning, crown displacement, and crown shape differences were simulated on Chinese fir in the figure. In Figure 13b, we visualized the shape of the tree crown.

The method with b-spline interpolation could constrain the crown shape too. In the this it needed to set a derivative value at the top point and isometric acquisition type value points. However, in this method, the derivative value at the top point did not need to be set, the influence of human error on the model was reduced, and the efficiency of the model fitting was improved. In addition, the type of value points collection was more flexible, which could be collected according to the actual situation of the crown shape. Only a small amount of collection was required in the case of no mutation, and more collection was required in the case of severe morphological changes. However, the tree morphology usually followed a rule, and morphological changes that did not conform to the rule only took place at the location of environmental interaction; therefore, most of the time, there was no mutation. This not only reduced the collection amount of type value points in general but also made CMGBCC more consistent with the actual situation. The comparison of the two methods is shown in Table 4.

Table 4. The comparison of the two methods.

Index	b-Spline Interpolation	CMGBCC
Derivative value at the top point	Add manually and uncontrollable factors of the model	It does not need to manually input the derivative value at the top point, simplifying the model construction process and reducing the error caused by humans
Collection rules of type value points	Equal distance acquisition	It can be collected according to the crown shape change rules
Collection interval of type value points	Equal distance acquisition	Acquisition at any distance
Model complexity	The model form is relatively simple	The model form is relatively complex

3.3. *3D Modeling and Verification of Chinese Fir Polymorphism*

The three polymorphisms of the Chinese fir were simulated by CMGBCC constraints and branch model loading, and the 3D-CFPM was also constructed.

Taking plot 1 as an example, there were 56 Chinese firs in plot 1, and this basic type of value data for the Chinese fir are shown in Table 5.

Table 5. Statistics of forest trees in sample plot 1.

Type	Height/m	DBH/cm	UBH/m	HMCW/m	Crown Width/m
Max	28.3	34.2	14.2	17.0	3.8
Min	9.6	9.3	4.9	8.6	0.4
Average	17.0	20.4	8.3	13.1	1.9

According to the detailed type of value point data that were obtained by combining the basic type of value point data and the crown shape mutation point data, the Chinese firs in plot 1 were simulated, and the positions of the Chinese fir were distributed according to the relative positions obtained by measurements. The real photo was taken at observation point 1, as shown in Figure 14a, and the simulation results and the visualization of the crown curve are shown in Figure 14b.

(a) (b)

Figure 14. Comparison of real photos and simulation results of observation point 1 in sample plot 1.

The crown curve of the front row of the Chinese firs was visualized and expressed. There were seven Chinese firs in the front row. It could be seen that the UBH of each tree varied with the influence of the surrounding competition. These different results were simulated and expressed for the natural pruning situation. The visualization results of the crown curve directly showed the crown displacement phenomenon and crown shape difference phenomenon in the Chinese fir crown, both of which were related to the spatial structure and expressed the crown deviation in the direction of greater competition and less competition in the same tree. This more clearly simulated the crown displacement phenomenon. At the same time, the mutation points of the crown shape were expressed, and the crown shape difference was simulated.

Figure 14a is photos of Chinese fir morphology collected during field measurements, Figure 14b is the simulation results of Chinese fir morphology at this angle. Number 1 to 7 are seven Chinese firs in the front row, they are numbered from left to right. It's same for Figures 15 and 16.

The simulation results of observation positions 2 and 3 in sample plot 1 were compared with the real situation, as shown in Figures 15 and 16.

From Figures 15 and 16, it can also be seen that the six trees in the front row of observation point 2 and the nine trees in the front row of observation point 3 had polymorphisms in their crown morphology due to different environmental influences in different directions. This study achieved a good simulation of the polymorphism of the Chinese fir.

The number of crown-type value points collected by CMGBCC was compared with the method based on b-spline interpolation. In this study, 3713 type value points were collected on 239 Chinese firs, while 4528 points were collected in the same trees by the previous method, which was collected every 1 m, reducing the number by 18%. The crown curves constructed by these two methods were compared. The crown curve regions constructed by the two methods were subjected to the operation of an intersection inversion. The obtained region was part of the crown curve whose range was wrongly divided and where there was less missing when divided based on the b-spline interpolation. Compared with the previous methods, the average result of the verified result showed that the error area that was missing or more than the true crown curve was reduced by 7.63%.

(a) (b)

Figure 15. Comparison of real photos and simulation results of observation point 2 in sample plot 1.

(a) (b)

Figure 16. Comparison of real photos and simulation results of observation point 3 in sample plot 1.

4. Discussion

The most obvious innovation in this study was to construct the CMGBCC model. In mathematics, at the progress of simulating GB-spline interpolation, the CMGBCC model has two advantages: (1) it has simplified steps for assigning the derivative value at the top point, and (2) there is no requirement for the equidistant collection of type value points. First, the GB-spline interpolation was suitable for the crown shape fitting as it could provide precise simulation results for the crown shape features. Compared with the b-spline interpolation, the number of type value points required in GB-spline interpolation decreased by 18%, and the accuracy of the crown shape constraint increased by 7.63%. Second, after fitting the crown shape curve, polymorphisms such as natural pruning, crown displacement, and crown shape difference could be better expressed. Third, adding different states of the withered branch to coincide with the real tree branch growth helped to improve the performance of the tree polymorphism. Therefore, the CMGBCC model improved the b-spline interpolation to construct the crown curve from a basic principle, which raised the crown curve fitting degree with true shape and optimized the crown curve construction process simultaneously. From the early 3D tree modeling to the current 3D tree polymorphism modeling [40,41], related research has considered the influence of environmental interactions and has added environmental factors to the 3D tree modeling process gradually. In the forest stand, the development of the tree's polymorphism was influenced by the environment widely. It is an important reason that makes it difficult to restore a real tree's morphology, and most 3D tree models cannot express the forest growth effectively. Based on the above points, CMGBCC provided a good solution. This study could be applied to other tree species, and the crown shape could be simulated by CMGBCC based on simply collecting corresponding type value points.

The 3D tree polymorphism model is the basis for the visual simulation of forest growth and management [42,43], and it is also the direct object of related research. For the simulation of forest growth and management visualization, it is necessary to represent the relationship between tree growth, environmental factors, and the tree growth allometric results caused by environmental factors to simulate reality. At the same time, the process of tree growth is accompanied by the new branches' occurrence and the old branches' death. Currently, related 3D tree modeling research has not considered withered branches and has not been represented in the simulation of tree growth visualization. In the 3D-CFPM, we considered the deficiencies of the 3D tree polymorphism model in simulating the branch withered status. In addition, for the forest management visual simulation, calculating the management indicators needs to use morphological and structural parameters. The 3D tree polymorphic model can provide relatively reliable basic parameters with morphological changes, which could make forest management visual simulation results closer to the real management situation.

In the future, we aim to combine the model with a constructed underlying algorithm, generate a large number of models in real-time to fill the model library and continue to explore the branch withered process to quantify the branches' withered state at different heights to further improve the tree polymorphism simulation performance.

5. Conclusions

In this study, we proposed a 3D modeling method of the Chinese fir polymorphism, which, based on the analysis of the shortcomings of the 3D tree polymorphism simulation, constructed CMGBCC and considered the withered branch state effects of the tree morphology in the modeling process. This model improved the simulation of tree morphology, which is conducive to optimizing the basic model of tree growth and forest management visual simulation in forestry practice; therefore, simulation can be more in line with forestry laws.

The following are the highlights of this study:

- Constructing CMGBCC improved the fitting function for the crown curve.
- We considered the influence of different withered state branches on the whole shape in 3D tree modeling.
- We simulated three types of polymorphisms, including natural pruning, crown displacement, and crown shape difference.

Author Contributions: Conceptualization, Z.C. and H.Z.; methodology, Z.C.; software, Z.C. and Y.L.; validation, Z.C., Y.L. and J.Z.; formal analysis, Z.C., T.Y. and K.L.; investigation, Z.C., Y.Z. and K.L.; resources, H.Z.; data curation, Z.C. and Y.L.; writing—original draft preparation, Z.C.; writing—review and editing, Z.C., Y.L., H.Z., J.Z. and T.Y.; visualization, Z.C. and Y.Z.; supervision, H.Z.; project administration, H.Z.; funding acquisition, H.Z. All authors have read and agreed to the published version of the manuscript.

Funding: This research was funded by the National Natural Science Foundation of China, grant number 32071681 and the Foundation Research Funds of IFRIT, grant number CAFYBB2021ZE005-2.

Data Availability Statement: Not applicable.

Acknowledgments: We are grateful to Huacong Zhang at the Experimental Center of Subtropical Forestry and Linlong Wang at the Research Institute of Forestry Policy and Information Techniques Chinese Academy of Forestry for help with data collection.

Conflicts of Interest: The authors declare no conflict of interest.

References

1. Ogawa, K. Size Dependence of Leaf Area and the Mass of Component Organs during a Course of Self-Thinning in a Hinoki (*Chamaecyparis obtusa*) Seedling Population. *Ecol. Res.* **2003**, *18*, 611–618. [CrossRef]
2. Wang, C.S.; Tang, C.; Hein, S.; Guo, J.J.; Zhao, Z.G.; Zeng, J. Branch Development of Five-Year-Old Betula Alnoides Plantations in Response to Planting Density. *Forests* **2018**, *9*, 42. [CrossRef]
3. Busing, R.T. Disturbance and the Population Dynamics of Liriodendron Tulipifera: Simulations with a Spatial Model of Forest Succession. *J. Ecol.* **1995**, *83*, 45. [CrossRef]
4. Fleck, S.; Mölder, I.; Jacob, M.; Gebauer, T.; Jungkunst, H.F.; Leuschner, C. Comparison of Conventional Eight-Point Crown Projections with LIDAR-Based Virtual Crown Projections in a Temperate Old-Growth Forest. *Ann. For. Sci.* **2011**, *68*, 1173–1185. [CrossRef]
5. Mizunaga, H.; Umeki, T. The Relationships between Tree Size and Spatial Distribution of Foliage-Mass within Crowns of Japanese Cedars (*Cryptomeria japonica*) in Young Plantations. *For. Ecol. Manag.* **2001**, *149*, 129–140. [CrossRef]
6. Locosselli, G.M.; de Camargo, E.P.; Moreira, T.C.L.; Todesco, E.; Andrade, M.d.F.; de André, C.D.S.; de André, P.A.; Singer, J.M.; Ferreira, L.S.; Saldiva, P.H.N.; et al. The Role of Air Pollution and Climate on the Growth of Urban Trees. *Sci. Total Environ.* **2019**, *666*, 652–661. [CrossRef] [PubMed]
7. Petritan, A.M.; von Lüpke, B.; Petritan, I.C. Influence of Light Availability on Growth, Leaf Morphology and Plant Architecture of Beech (*Fagus sylvatica* L.), Maple (*Acer pseudoplatanus* L.) and Ash (*Fraxinus excelsior* L.) Saplings. *Eur. J. For. Res.* **2009**, *128*, 61–74. [CrossRef]
8. Fang, K.; Chen, D.; Gou, X.; D'Arrigo, R.; Davi, N. Influence of Non-Climatic Factors on the Relationships between Tree Growth and Climate over the Chinese Loess Plateau. *Glob. Planet. Change* **2015**, *132*, 54–63. [CrossRef]
9. Buechling, A.; Martin, P.; Canham, C. Climate and competition effects on tree growth in Rocky Mountain forests. *J. Ecol.* **2017**, *105*, 1636–1647. [CrossRef]
10. Coomes, D.A.; Allen, R.B. Effects of Size, Competition and Altitude on Tree Growth. *J. Ecol.* **2007**, *95*, 1084–1097. [CrossRef]
11. Tucker, G.F.; Lassoie, J.P.; Fahey, T.J. Crown Architecture of Stand-Grown Sugar Maple (*Acer saccharum* Marsh.) in the Adirondack Mountains. *Tree Physiol.* **1993**, *13*, 297–310. [CrossRef]
12. Turkington, R.; Goldberg, D.E.; Olsvig-Whittaker, L.; Dyer, A.R. Effects of Density on Timing of Emergence and Its Consequences for Survival and Growth in Two Communities of Annual Plants. *J. Arid Environ.* **2005**, *61*, 377–396. [CrossRef]
13. Davies, O.; Pommerening, A. The Contribution of Structural Indices to the Modelling of Sitka Spruce (*Picea sitchensis*) and Birch (*Betula* Spp.) Crowns. *For. Ecol. Manag.* **2008**, *256*, 68–77. [CrossRef]
14. Marozau, A.; Kotszan, U.; Nowakowska, J.; Berezovska, D.; Moser, W.K.; Hsiang, T.; Oszako, T. The Influence of the Provenance and Spatial Structure on the Growth of European Silver Fir (*Abies alba* Mill.) of Autochthonous Origin in a Forest Plantation in the Białowieża Forest. *Forests* **2022**, *13*, 831. [CrossRef]
15. Ma, Z.; Zhang, H.; Li, Y.; Yang, T.; Chen, Z.; Li, S. Visual Simulation of Chinese Fir Crown Growth Based on Spatial Structure. *For. Res.* **2018**, *13*, 150–157. (In Chinese) [CrossRef]
16. Zhu, N.; Zhang, H.; Cui, Z.; Yang, T.; Li, Y.; Liu, H. Visual Simulation of Chinese Fir under Branch Height in Consideration of Spatial Structure. *J. Nanjing For. Univ. Nat. Sci. Ed.* **2022**, *46*, 51–57. (In Chinese) [CrossRef]
17. Cui, Z.; Zhang, H.; Zuo, Y.; Yang, T.; Liu, Y.; Zhang, J.; Wang, L. The Distribution of under Branch Heights in Various Directions of the Three-Dimensional Chinese Fir Model. *J. Nanjing For. Univ. Natural Sci. Ed.* **2022**, *46*, 81–87. (In Chinese) [CrossRef]
18. Crecente-Campo, F.; Marshall, P.; LeMay, V.; Diéguez-Aranda, U. A Crown Profile Model for Pinus Radiata D. Don in Northwestern Spain. *For. Ecol. Manag.* **2009**, *257*, 2370–2379. [CrossRef]
19. Chen, Y.; Dong, C.; Wu, B. Crown Profile Modeling and Prediction Based on Ensemble Learning. *Forests* **2022**, *13*, 410. [CrossRef]
20. Liu, M.; Zhang, H.Q.; Lu, K.N. Research on Three-Dimensional Simulation of Tree's Morphology Based on Tree-Crown Growth Model. In Proceedings of the 2010 International Conference on Computational Intelligence and Software Engineering, Wuhan, China, 10–12 December 2010; IEEE: New York, NY, USA, 2010. [CrossRef]
21. Runions, A.; Lane, B.; Prusinkiewicz, P. Modeling Trees with a Space Colonization Algorithm. *Nat. Phenom.* **2007**, *6*, 63–70.
22. Palubicki, W.; Horel, K.; Longay, S.; Runions, A.; Lane, B.; Měch, R.; Prusinkiewicz, P. Self-Organizing Tree Models for Image Synthesis. *ACM Trans. Graph.* **2009**, *28*, 1–10. [CrossRef]
23. Suzuki, T.; Ohsaki, K.; Sato, H. A representation method for todo-fir shapes using computer graphics. *J. Jpn. For. Soc.* **1992**, *74*, 504–508. (In Japanese)
24. Eilers, P.H.C.; Marx, B.D. Flexible Smoothing with B-Splines and Penalties. *Stat. Sci.* **1996**, *11*, 89–102. [CrossRef]
25. Wu, Z.; Zhou, M.; Wang, X. Interactive Modeling of 3D Tree with Ball B-Spline Curves. *Int. J. Virtual Real.* **2009**, *8*, 101–107. [CrossRef]

26. Liu, X.; Wang, X.; Wu, Z.; Zhang, D.; Liu, X. Extending Ball B-Spline by B-Spline. *Comput. Aided Geom. Des.* **2020**, *82*, 101926 [CrossRef]
27. Harmening, C.; Hobmaier, C.; Neuner, H. Laser Scanner–Based Deformation Analysis Using Approximating B-Spline Surfaces *Remote Sens.* **2021**, *18*, 3551. [CrossRef]
28. Sun, Y. Existence of Triple Positive Solutions for a Third-Order Three-Point Boundary Value Problem. *J. Comput. Appl. Math.* **2008**, *221*, 194–201. [CrossRef]
29. Böhm, W. Generating the Bézier Points of B-Spline Curves and Surfaces. *Comput. Des.* **1981**, *13*, 365–366. [CrossRef]
30. Beccari, C.V.; Casciola, G. A Cox-de Boor-Type Recurrence Relation for C1 Multi-Degree Splines. *Comput. Aided Geom. Des.* **2019**, *75*, 101784. [CrossRef]
31. Fritsch, D.; Kada, M. Visualisation Using Game Engines. *Geo-Informations-Systeme* **2004**, *2004*, 32–36.
32. Tang, X.; Pérez-Cruzado, C.; Fehrmann, L.; Álvarez-González, J.G.; Lu, Y.; Kleinn, C. Development of a Compatible Taper Function and Stand-Level Merchantable Volume Model for Chinese Fir Plantations. *PLoS ONE* **2016**, *11*, e0147610. [CrossRef]
33. Zhang, S.; Sun, J.; Duan, A.; Zhang, J. Variable-Exponent Taper Equation Based on Multilevel Nonlinear Mixed Effect for Chinese Fir in China. *Forests* **2021**, *12*, 126. [CrossRef]
34. Duan, A.; Zhang, S.; Zhang, X.; Zhang, J. Development of a Stem Taper Equation and Modelling the Effect of Stand Density on Taper for Chinese Fir Plantations in Southern China. *PeerJ* **2016**, *2016*, e1929. [CrossRef] [PubMed]
35. Lu, K.; Zhang, H.; Liu, M. Study on Plant Architecture of Cunningham ia lanceolata Based on Measured Data. *For. Res.* **2011**, *24*, 132–136. (In Chinese) [CrossRef]
36. Sun, H.; Zhang, J.; Duan, A.; He, C. Estimation of the Self-Thinning Boundary Line within Even-Aged Chinese Fir (*Cunninghamia lanceolata* (Lamb.) Hook.) Stands: Onset of Self-Thinning. *For. Ecol. Manag.* **2011**, *261*, 1010–1015. [CrossRef]
37. Zhu, W.; Liu, Z.; Jia, W.; Li, D. Modelling the Tree Height, Crown Base Height, and Effective Crown Height of Pinus Koraiensis Plantations Based on Knot Analysis. *Forests* **2021**, *12*, 1778. [CrossRef]
38. Vovides, A.G.; Berger, U.; Grueters, U.; Guevara, R.; Pommerening, A.; Lara-Domínguez, A.L.; López-Portillo, J. Change in Drivers of Mangrove Crown Displacement along a Salinity Stress Gradient. *Funct. Ecol.* **2018**, *32*, 2753–2765. [CrossRef]
39. Longuetaud, F.; Piboule, A.; Wernsdörfer, H.; Collet, C. Crown Plasticity Reduces Inter-Tree Competition in a Mixed Broadleaved Forest. *Eur. J. For. Res.* **2013**, *132*, 621–634. [CrossRef]
40. Tanveer, M.; Thomas, A.; Wu, X.; Zhu, H. Simulate forest trees by integrating l-system and 3d cad files. In Proceedings of the 2020 3rd International Conference on Information and Computer Technologies (ICICT), San Jose, CA, USA, 9–12 March 2020. [CrossRef]
41. Wang, Z. Cloud Theory and Fractal Application in Virtual Plants. *Int. J. Intell. Syst. Appl.* **2011**, *3*, 17–23. [CrossRef]
42. Tohru, N.; Tohru, N.; Mitsuo, M. Visualization of Optimized Solution Space Using a Simulation System for the Development of Local Forest Management Planning. *Ann. For. Res.* **2016**, *59*, 117–128. [CrossRef]
43. Peng, F.; Zheng, H.; Lu, S.; Shi, Z.; Liu, X.; Li, L. Growth Model and Visualization of a Virtual Jujube Tree. *Comput. Electron. Agric.* **2019**, *157*, 146–153. [CrossRef]

remote sensing

MDPI

Article

A Novel Framework for Stratified-Coupled BLS Tree Trunk Detection and DBH Estimation in Forests (BSTDF) Using Deep Learning and Optimization Adaptive Algorithm

Huacong Zhang [1,2,3], Huaiqing Zhang [1,3,*], Keqin Xu [2], Yueqiao Li [2], Linlong Wang [1,3], Ren Liu [2], Hanqing Qiu [1,3] and Longhua Yu [2]

1 Institute of Forest Resource Information Techniques, Chinese Academy of Forestry, Beijing 100091, China; zhang718736170@163.com (H.Z.)
2 Experimental Center of Subtropical Forestry, Chinese Academy of Forestry, Fenyi 336600, China
3 Key Laboratory of Forestry Remote Sensing and Information System, NFGA, Beijing 100091, China
* Correspondence: zhang@ifrit.ac.cn

Abstract: Diameter at breast height (DBH) is a critical metric for quantifying forest resources, and obtaining accurate, efficient measurements of DBH is crucial for effective forest management and inventory. A backpack LiDAR system (BLS) can provide high-resolution representations of forest trunk structures, making it a promising tool for DBH measurement. However, in practical applications, deep learning-based tree trunk detection and DBH estimation using BLS still faces numerous challenges, such as complex forest BLS data, low proportions of target point clouds leading to imbalanced class segmentation accuracy in deep learning models, and low fitting accuracy and robustness of trunk point cloud DBH methods. To address these issues, this study proposed a novel framework for BLS stratified-coupled tree trunk detection and DBH estimation in forests (BSTDF). This framework employed a stratified coupling approach to create a tree trunk detection deep learning dataset, introduced a weighted cross-entropy focal-loss function module (WCF) and a cosine annealing cyclic learning strategy (CACL) to enhance the WCF-CACL-RandLA-Net model for extracting trunk point clouds, and applied a (least squares adaptive random sample consensus) LSA-RANSAC cylindrical fitting method for DBH estimation. The findings reveal that the dataset based on the stratified-coupled approach effectively reduces the amount of data for deep learning tree trunk detection. To compare the accuracy of BSTDF, synchronous control experiments were conducted using the RandLA-Net model and the RANSAC algorithm. To benchmark the accuracy of BSTDF, we conducted synchronized control experiments utilizing a variety of mainstream tree trunk detection models and DBH fitting methodologies. Especially when juxtaposed with the RandLA-Net model, the WCF-CACL-RandLA-Net model employed by BSTDF demonstrated a 6% increase in trunk segmentation accuracy and a 3% improvement in the F_1 score with the same training sample volume. This effectively mitigated class imbalance issues encountered during the segmentation process. Simultaneously, when compared to RANSAC, the LSA-RANCAC method adopted by BSTDF reduced the *RMSE* by 1.08 cm and boosted R^2 by 14%, effectively tackling the inadequacies of RANSAC's filling. The optimal acquisition distance for BLS data is 20 m, at which BSTDF's overall tree trunk detection rate (*ER*) reaches 90.03%, with DBH estimation precision indicating an *RMSE* of 4.41 cm and R^2 of 0.87. This study demonstrated the effectiveness of BSTDF in forest DBH estimation, offering a more efficient solution for forest resource monitoring and quantification, and possessing immense potential to replace field forest measurements.

Keywords: BSTDF; WCF-CACL-RandLA-Net; LAS-RANSAC; BLS; DBH; tree trunk detection; forestry inventory

1. Introduction

Diameter at breast height (DBH, 1.30 m high) is one of the key indices for forest inventory, reflecting forest growth conditions and carbon storage distribution, and constituting an essential component of foundational forestry data [1]. Light detection and ranging (LiDAR), an active remote sensing technology, can provide high-resolution descriptions of forest trunk structures and possesses significant advantages in DBH estimation [1–6]. Currently, LiDAR systems mainly include terrestrial laser scanners (TLS), airborne laser scanners (ALS), and mobile laser scanners (MLS). High-density point cloud data acquired by TLS enables the measurement of millimeter-precision 3D tree structures, such as tree height, DBH, and crown diameter [7]. However, in practical applications, single-site TLS scanning exhibits a relatively high trunk non-detection rate (17.4–32%) [8,9], while multi-site scanning necessitates higher processing costs and is unsuitable for large-scale forest parameter acquisition. ALS can rapidly obtain extensive forest point clouds, but in dense stands, the light beams are often obstructed by tree crowns [4], which hinders tree trunk detection and DBH estimation. MLS systems offer powerful tools for addressing tree occlusion and immobility issues, effectively reducing time and labor costs [10]. However, the accuracy of MLS data is typically lower than that of multi-site TLS data [8], and the application conditions of MLS are susceptible to the GNSS signal, terrain, and other objective environmental factors [9]. With the rapid reduction in size and weight of laser scanners and the development of GNSS technology, the emergence of Backpack LiDAR systems (BLS) has compensated for the limitations of other LiDAR technologies. As a backpack-style form of MLS, BLS collects data by walking and exhibits higher scanning efficiency [5], making it highly suitable for large-scale forest tree trunk detection and DBH information extraction.

At present, there are numerous methods for estimating DBH based on LiDAR, including least squares (geometric) circle fitting (LS) [11,12], cylindrical fitting (CF) [13], Hough transform (HT) [14], convex hull algorithm (CHA) [11], and random sample consensus (RANSAC) [3]. These methods are prone to interference from factors such as noise and foliage occlusion during the fitting process, which can lead to reduced fitting accuracy. Consequently, optimizing the fitting methods to enhance accuracy and robustness is crucial. In the DBH fitting process, typical steps include "ground point identification—trunk localization—point cloud slicing at DBH—DBH estimation" [2,8,10,15–17]. In this process, separating trunk point cloud data from non-trunk point cloud data is one of the key steps that aim to eliminate noise interference and prevent the misestimation of DBH. However, achieving accurate and efficient tree trunk detection in forest BLS data still presents challenges. Visual interpretation is the most accurate method for eliminating non-trunk point clouds, yet this approach requires skilled professionals and is more labor-intensive. Utilizing a computer to extract the linear spatial features of tree trunk point clouds for cylindrical detection has proven efficient [18,19], yet its precision diminishes in the face of complex forest spatial structures. As such, the development of a tree trunk detection framework or method adaptable to forest scenarios has become pressing. After all, tree trunk detection lays the groundwork for calculating forest resources, such as DBH [20], volume estimation, and individual tree segmentation [21].

Computers based on deep learning technology possess strong memory and learning capabilities [16,20–23], effectively addressing the limitations of traditional DBH estimation methods in extracting forest DBH information, such as their difficulty in point cloud noise recognition, insufficient robustness, and low extraction efficiency. By learning trunk features, repeatable and scalable trunk recognition and localization tasks with large amounts of data can be achieved. Currently, mainstream deep learning methods such as PointNet [23,24], KPConv [25], DCNN [26], and PointCNN [27] have been applied to point cloud segmentation. Most of these methods use the farthest point sampling (FPS) to retain point cloud spatial features as much as possible, but the processing speed is slow for larger datasets (1.0×10^5 and above). In summary, for tree trunk detection and DBH estimation in complex forests, researchers can improve processing capabilities and extraction accuracy by focusing on dataset creation, deep learning models, and DBH fitting algorithms.

In a bid to overcome the challenges mentioned earlier, we put forth a novel framework for deep learning-based tree trunk detection and diameter at breast height (DBH) estimation, leveraging BLS hierarchical coupling (BSTDF). This innovation facilitated effective tree trunk identification and DBH estimation from MLS point cloud data. Furthermore, we delved into the impact of varying MLS scanning distance levels on the precision of tree trunk detection and DBH estimation. The main contributions of the proposed framework are as follows:

(1) In light of the substantial volume of point cloud data in forest scenarios, we introduced a novel approach to constructing a deep learning tree trunk detection dataset, which is predicated on hierarchical coupling. This approach effectively curtails the data volume necessary during the tree trunk detection process;
(2) To rectify the issue of uneven class accuracy in point cloud segmentation within the RandLA-Net model for tree trunk detection scenarios, we introduced an enhanced RandLA-Net semantic segmentation model with dual modules (WCF and CACL). By increasing the model's attention towards the target classes and employing more flexible learning strategies, we improved the segmentation accuracy for the tree trunk class, ultimately leading to an elevated detection rate;
(3) We enhanced the RANCAC algorithm by implementing an adaptive approach and a least squares optimization algorithm. These meticulously devised strategies served to optimize the fitting model parameters and boost the convergence of model error, thereby significantly elevating the accuracy of the DBH fitting within forest scenarios;
(4) Lastly, having conducted a comparative analysis of the precision of tree trunk detection and DBH fitting at different MLS scanning distance levels, we proposed a recommended distance for MLS scanning, which offers a balance between data quality and scanning efficiency.

2. Materials and Methods

2.1. Study Area

The study area for this paper encompasses a series of forest stands on both sides of the main road in the Tree Garden of the Subtropical Forestry Experiment Center, part of the Chinese Academy of Forestry. The study area has an average slope of 5°, with higher elevation in the south and lower elevation in the north. Covering a total area of 65,000 m^2 (650 m $\times$ 100 m) and extending in a north–south direction, it is located in the northwestern part of Fenyi County, Jiangxi Province (Figure 1A,B). With abundant forest resources, Fenyi County has a forest coverage rate of 65.76%. The main tree species in the area include *Chinese fir*, *Masson pine*, *Schima superba*, *camphor*, and *Machilus thunbergii*, which are typical tree species in southern China. The regional vegetation is characterized by mid-subtropical evergreen broad-leaved forests, with a frost-free period of 240–307 days. The annual average rainfall ranges from 1500 to 1700 mm, while the annual average temperature lies between 11.6 and 19.3 °C. The lowest temperature in January is 3.2 to 4.7 °C, and the highest temperature in July is 29.8 to 38.3 °C.

2.2. Data

2.2.1. Field Survey Data

Field surveys were undertaken in April and May of 2022. A comprehensive inventory was completed on a tree-by-tree basis within the study area, using the Real-Time Kinematic Global Navigation Satellite System (RTK-GNSS) to determine the location of each sample tree. Our inventory criteria were set for trees with a diameter at breast height (DBH) greater than 5 cm, and we measured and recorded tree height, DBH, crown width, tree species, and location information. Both the DBH and crown width were gauged using a tape measure, while tree height was measured with a laser rangefinder and various environmental factors of the plot were recorded. Figure 1C illustrates the spatial distribution characteristics of trees within the study area. Table 1 presents the statistical features of trees in the study area. Figure 2A displays the scatter distribution of broadleaf tree DBH and trunk fitting within the study area, while Figure 2B exhibits the scatter distribution of conifer tree DBH and trunk fitting within the study area. The data revealed that there were 2303 trees in the study area, comprising 47 broad-leaved species and 15 coniferous species. The principal broad-leaved species included Michelia, camphor, birch, Eucalyptus, and osmanthus, while the main coniferous species consisted of cypress, Cryptomeria, Taxodium, and Thuja. The average DBH values of various tree species ranged from 9.08 cm to 32.77 cm. In this paper, field measurement results will serve as validation data for the experimental results.

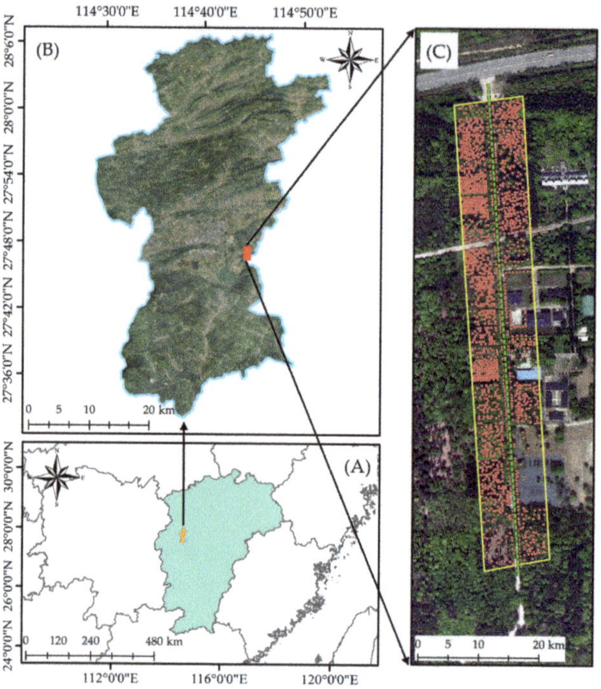

Figure 1. Study area: (**A**) the location of the study area in Jiangxi province; (**B**) the location of the study area in Fenyi County; (**C**) the red represents the distribution of trees, and the green dotted line represents the one-way route of LiDAR data acquisition.

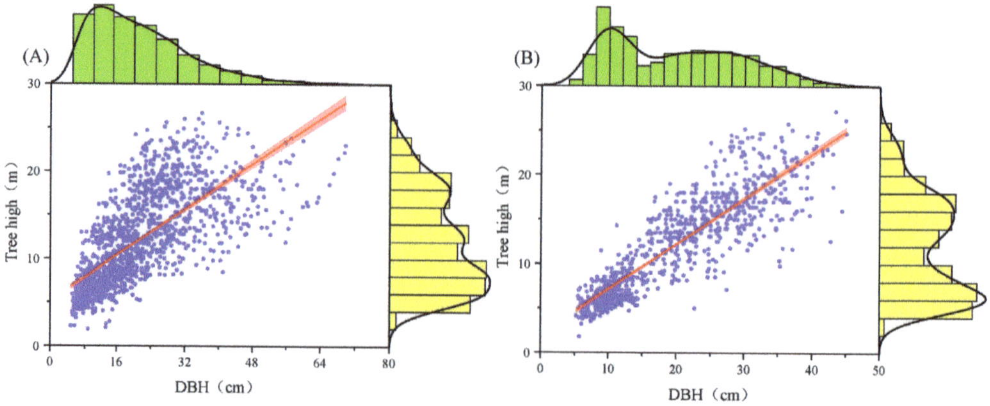

Figure 2. Distribution statistics of tree height and DBH of conifers and broad-leaved trees in the study area: (**A**) broad-leaved trees; (**B**) coniferous trees.

Dividing the dataset is a foundational step in training deep learning models. In our research, we categorized the trees surveyed within the entire research area into training, validation, and testing zones based on their functional regions.

Within the training zone, there were a total of 453 trees, with an average DBH of 21.34 cm and an average standard deviation of DBH at 9.83 cm. The validation zone comprised 220 trees, with an average DBH of 19.24 cm and an average standard deviation of DBH of 9.52 cm. The testing zone included 1851 trees, with an average DBH of 20.78 cm and an average standard deviation of DBH of 11.20 cm. The types of trees and the diameter distribution across all three functional zones were relatively balanced, providing representative samples. Table 1 displays the data features of the trees in each functional zone within the research area.

Table 1. Descriptive statistics of field measurements (species, number, DBH, and TH) in this study.

Dataset Function Partition Type	Area (m^2)	Species Name of Tree	Number of Trees	DBH (cm)		TH (m)	
				Mean	Std	Mean	Std
Training sample area	19,600	*Yulania denudata (Desrousseaux) D. L. Fu; Magnolia officinalis Rehd.et Wils., Michelia L.; Cupressus funebris Endl.; Pinus massoniana Lamb.; Cryptomeria japonica var. sinensis Miquel.*	453	21.34	9.83	11.28	3.68
Validation sample area	5000	*Betula luminifera H.Winkl.; Osmanthus sp., Michelia L.; Pinus massoniana Lamb.; Taxus wallichiana var. chinensis (Pilg.) Florin.*	220	19.24	9.52	12.33	6.17
Experimental sample area	45,400	*Michelia L., Osmanthus sp.; Cinnamomum camphora (L.) presl; Nageia nagi (Thunberg) Kuntze; Taxodium distichum (L.) Rich.; Taxodium distichum var. imbricatum (Nuttall) Croom.*	1851	20.78	11.20	12.39	6.13
Total	65,000	*Michelia L., Osmanthus sp.; Cinnamomum camphora (L.) presl; Cupressus funebris Endl.; Taxus wallichiana var.chinensis (Pilg.) Florin; Taxodium distichum (L.) Rich.*	2303	21.22	6.58	12.76	3.06

2.2.2. BLS Data

The BLS used in this study was the RIEGL miniVUX 1UAV (Figure 3C), equipped with the RIEGL miniVUX 3D laser scanner and Trimble APX-15 GNSS Inertial Navigation OEM (Figure 3B). The measurement trajectory of the BLS is shown in Figure 1C. The scanner system was initialized within the plot, and the plot was covered by a single pass of a rectangular transect. The BLS system's positioning relies on SLAM systems, GNSS positioning, and an inertial measurement unit. During the data collection process, the operator walked at a speed of 5 km/h, generating a high-density BLS data point cloud of approximately 600 points/m^2.

2.3. Methods

In this study, we created a stratified coupled tree trunk detection deep learning dataset based on BLS data and improved the WCF-CACL-RandLA-Net model and LSA-RANSAC algorithm to achieve tree trunk detection (Objective 1) and DBH estimation (Objective 2) within the study area. The flowchart of this study is shown in Figure 4.

Figure 3. BLS data collection equipment: (**A**) operation method under the forest; (**B**) GNNS receiver for enhanced positioning; (**C**) LiDAR scanner RIEGL miniVUX-1UAV; (**D**) point cloud in the research area.

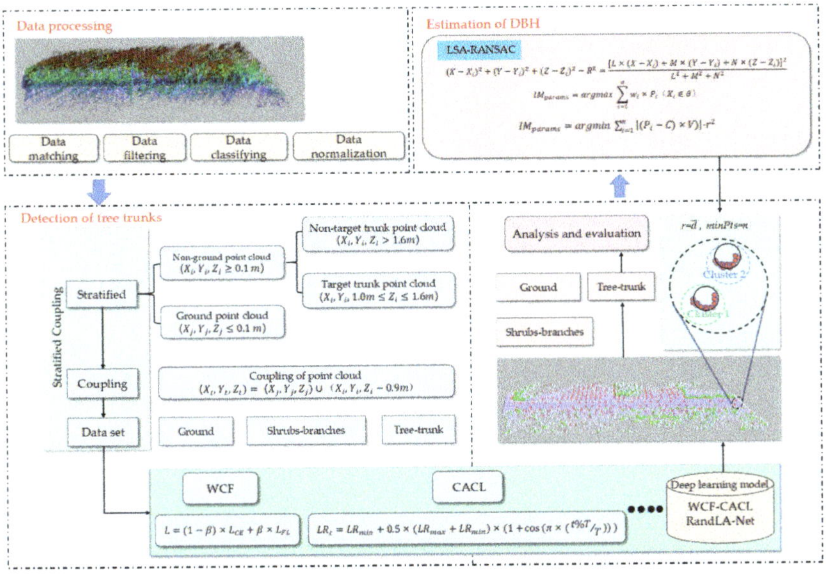

Figure 4. The flowchart of this study.

2.3.1. Creation of the Tree Trunk Detection Deep Learning Dataset Based on Stratified Coupling

Using BLS as the base data and referencing the basic requirements of the RandLA-Net model for the dataset, we applied a novel stratified coupling method to create the tree trunk detection deep learning dataset to reduce data volume and enhance model learning efficiency. Through this method, we reduced the point cloud count from the original forest point cloud of 1.43×10^8 to the stratified coupling dataset of 6.14×10^6.

Step 1: BLS data preprocessing. We set the initial data input as (X_i, Y_i, Z_i). However, we noted that the point cloud contained many noise points, which were filtered out using a density-based criterion. The filtered point cloud data were classified into a ground point cloud and a vegetation point cloud. Finally, we normalized the vegetation point cloud using a high-precision DEM model.

Step 2: Stratification of the normalized point cloud. We stratified the normalized point cloud into three categories, including a ground point cloud layer (X_j, Y_j, Z_j), a target point cloud layer (X_t, Y_t, Z_t), and a non-target point cloud layer (X_n, Y_n, Z_n).

$$(X_j, Y_j, Z_j) = (X_i, Y_i, Z_i < 0.1\ m) \tag{1}$$

$$(X_t, Y_t, z_t) = (X_i, Y_i, 1.0\ m \leq Z_i \leq 1.6\ m) \tag{2}$$

$$(X_n, Y_n, Z_n) = (X_i, Y_i, Z_i < 1.0\ m \cup Z_i > 1.6\ m) \tag{3}$$

Step 3: Coupling of the stratified point clouds. To ensure the spatial structural continuity of the stratified point clouds and improve the deep learning model's training efficiency, we coupled the stratified point clouds. The coupled point cloud output was set as (X, Y, Z).

$$(X_t, Y_t, Z_t) = (X_j, Y_j, Z_j) \cup (X_t, Y_t, Z_t - 0.9\ m) \tag{4}$$

Step 4: Dataset partitioning. The total number of BLS point clouds after stratified coupling was 6.14×10^6, divided into the training dataset (TDS), validation dataset (VDS), and partition dataset (PDS), with quantities of 2.12×10^6, 1.02×10^4, and 4.01×10^6, respectively. In the training and validation data, we manually annotated three types using

CloudCompare 2.11.3, including (1) tree trunk point clouds (Tree-trunk), (2) shrub and tree branch point clouds (Shrubs-branches), and (3) ground and understory vegetation point clouds (Ground). Statistics show that in TDS, the number of Tree-trunk, Shrubs-branches, and Ground were 6.08×10^4, 1.53×10^5, and 1.90×10^6, respectively. In the VDS, the numbers were 0.95×10^4, 1.74×10^4, and 7.23×10^4, respectively. Figure 5 displays the dataset creation process and Figure 5G shows the spatial distribution of TDS, VDS, and PDS dataset partitioning.

Step 5: Data augmentation. Point cloud data augmentation refers to the transformation and processing of the original point cloud data to increase the training samples, enhancing the model's generalization capability and accuracy. Data augmentation helps to address issues such as data scarcity, uneven sample distribution, and overfitting. In this study, we enhanced training samples using rotation and scaling, where rotation was performed clockwise around the X, Y, and Z axes by 180°, and the scaling factor was set to 0.8 and 1.2. After data augmentation, the training samples were expanded by 6 times to 1.27×10^7.

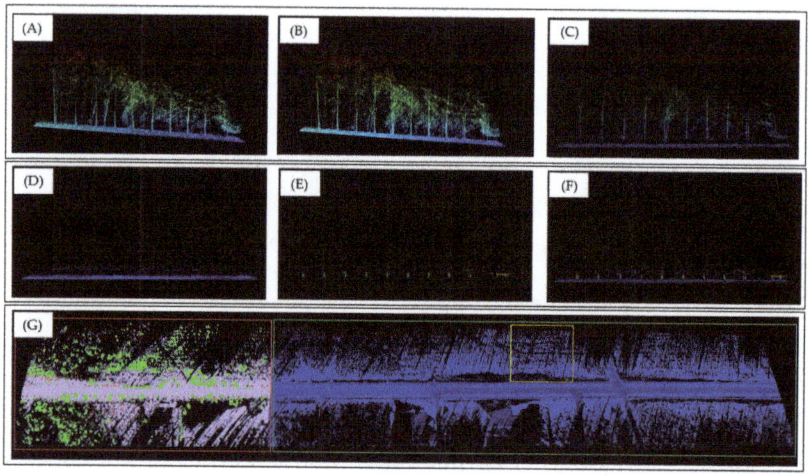

Figure 5. Data processing of this study: (**A**,**B**,**C**) preprocessing of data; (**D**,**E**,**F**) stratified and coupled of data; (**G**) classification of the data set (the red box represents TDS; the green box represents PDS; the yellow box represents VDS).

2.3.2. Construction of the WCF-CACL-RandLA-Net Model

In this study, the method used for tree trunk detection is the improved WCF-CACL-RandLA-Net model based on the RandLA-Net model. RandLA-Net [28] is an advanced deep learning (DL) model for large-scale point cloud semantic segmentation, which employs efficient random sampling (RS) and local feature aggregation (LFA). Its network structure is similar to the encoder–decoder network structure (first downsampling, then upsampling). To enhance the performance of the RandLA-Net model for tree trunk detection, we improved the loss function and learning strategy in RandLA-Net by introducing the weighted cross-entropy focal-loss module (WCF) and the cosine annealing cyclic learning strategy (CACL). WCF, through the integration of the weighted cross-entropy loss function (WCL) and the focal loss function (FL), strengthens the model's generalization ability while increasing the model's attention weight on target classes during the training process, which is beneficial for addressing class imbalance issues. CACL enables the dynamic adjustment of the model's learning rate throughout the training process. The cyclically varying learning decay strategy significantly enhances the exploration of model parameters, helps the model escape local optima, and avoids overfitting during training. Figure 6 displays the network framework of the WCF-CACL-RandLA-Net model.

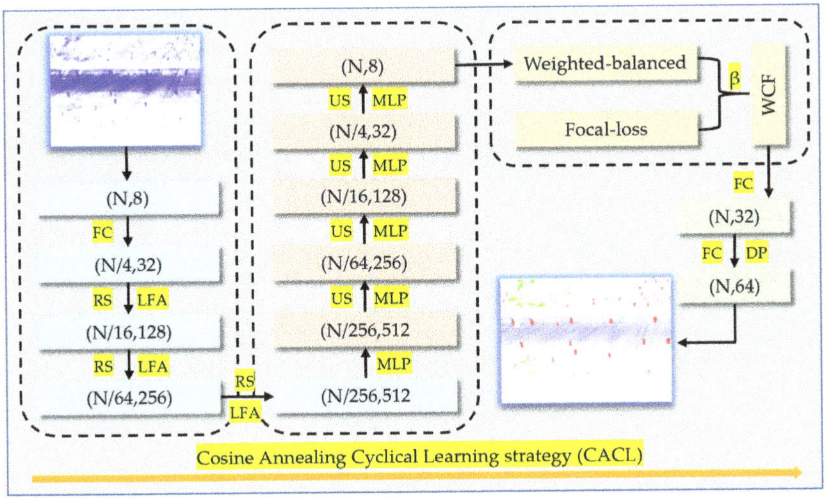

Figure 6. The network structure of WCF-CACL-RandLA-Net. (N, D) represents the number and characteristic dimension of points, respectively. FC represents the fully connected layer; LFA represents the local feature aggregation; RS represents the random sampling; MLP represents the shared multi-layer perceptron; US represents the upsampling; DP represents the dropout.

(1) Weighted Cross-Entropy Focal-Loss (WCF)

The dataset used for tree trunk detection exhibits features such as class imbalance and a smaller proportion of target classes. The weighted cross-entropy loss function employed by RandLA-Net is insufficient to focus on less represented classes. In the training samples of this study, the target point cloud accounts for 3% of the total number of point clouds trained. To address this issue, we proposed a loss function that combines weighted cross-entropy loss and focal loss to enhance the model's focus on underrepresented classes.

Weighted cross-entropy loss is a variation of the standard cross-entropy loss function, adjusting the importance of each class in loss computation by assigning different weights to different classes. Focal loss is a loss function designed to address the class imbalance by introducing a modulating factor that increases the loss of difficult-to-classify samples, thereby making the model more attentive to these challenging samples. We combine weighted cross-entropy loss and focal loss to form a new loss function (WCF).

$$L_{CE} = -\sum_{i=1}^{N}\sum_{c=1}^{C} W_c \times Y_{ic} \times log(P_{ic}) \tag{5}$$

$$L_{FL} = -\sum_{i=1}^{N}\sum_{c=1}^{C} Y_{ic} \times (1 - P_{ic})^{\gamma} \times log(P_{ic}) \tag{6}$$

$$L = (1 - \beta) \times L_{CE} + \beta \times L_{FL} \tag{7}$$

In the formula, "N" represents the number of point clouds; "C"represents the number of classes; "W_c" represents the weight of class "C"; "Y_{ic}" represents the actual label count of point cloud "i" belonging to class "C"; "P_{ic}" represents the predicted probability of point cloud "i" belonging to class "C"; "γ" represents the modulating factor, used to control the weight of easy and difficult samples; "L_{CE}" represents weighted cross-entropy loss; "L_{FL}" represents focal loss; and "β" represents the weight coefficient. By adjusting "β", we can find the optimal balance point between the two loss functions.

WCF can mitigate class imbalance issues and enhance the performance of the model when dealing with datasets exhibiting different class distribution characteristics. It takes full advantage of weighted cross-entropy loss and focal loss. Finally, by adjusting the weight coefficient "β" and the modulating factor "γ", the performance of the RandLA-Net can be further optimized.

(2) Cosine Annealing Cyclic Learning Strategy (CACL)

RandLA-Net employed a fixed linear decay learning strategy. Although this approach helped the model converge quickly in the early stages of training, it sometimes led to the model becoming trapped in local optima during the later stages, thereby affecting its generalization ability. To address this issue, this study introduced the cosine annealing cyclic learning strategy (CACL). CACL utilized a dynamic learning rate adjustment method, periodically adjusting the learning rate during training, enabling the model to explore the parameter space at different learning rates. Based on the cosine annealing scheduler, the learning rate fluctuated following a cosine function. CACL facilitated the RandLA-Net model's escape from local optima, accelerated convergence speed, and reduced sensitivity to the initial learning rate and learning rate decay, rendering the model more suitable for large-scale training datasets and complex training tasks.

$$LR_t = LR_{min} + 0.5 \times (LR_{max} + LR_{min}) \times \left(1 + cos\left(\pi \times \left(\frac{t\%T}{T}\right)\right)\right) \tag{8}$$

In the formula, "LR_t" represents the learning rate at training step "t", while "LR_{min}" and "LR_{max}" denote the minimum and maximum learning rates, respectively. "T" indicates the period of learning rate adjustment, with % symbolizing modulo operation.

The deep learning environment for this study included an I7-1900 processor with 32 GB of memory and an NVIDIA RTX A4000 16 G graphics card, utilizing CUDA 11.0 for GPU computing acceleration. The deep learning framework was TensorFlow-GPU 1.13.0, based on Python 3.5.6. The input TDS point number amounted to 1.27×10^7. The nearest point "K" value was set to 16, the initial learning rate to 0.01, and "T" to 4. The model featured 40 epochs, with 500 steps per epoch. The values of "c" and "d" were set to 0.0001 and 0.01, respectively. Random sampling rates for each layer were (4, 4, 4, 2), and the feature dimensions were (16, 64, 128, 256, 512).

2.3.3. DBH Estimation Method Based on LSA-RANSAC

The WCF-CACL-RandLA-Net model generates classified point clouds with three types of labels. However, further processing is required for the clustering of individual tree trunk point clouds and DBH calculations. In this study, we designed and implemented the least squares adaptive RANSAC DBH estimation method (LSA-RANSAC). LSA-RANSAC is a point cloud DBH calculation method that integrates the adaptive RANSAC algorithm and least squares optimization. LSA-RANSAC can find the optimal parameters and the corresponding values within the desired range and define the error function by calculating the distance between the cylindrical model and the point cloud data to obtain the best-fitting cylindrical model. LSA-RANSAC takes full advantage of the robustness of the RANSAC algorithm [3] in handling noise and outliers, as well as the parameter search capability and model error convergence of adaptive methods and least squares optimization algorithms, making the entire fitting process more stable and efficient. The implementation steps of DBH calculation based on LSA-RANSAC are as follows:

Step 1: Implement the density-based spatial clustering of applications with noise (DBSCAN) [29] without specified cluster numbers by setting the minimum number of cluster points (*Min-Pts*) and neighborhood radius (*Eps*) parameters.

Step 2: Fit each tree trunk class using the LSA- RANSAC algorithm. First, determine the use of cylinders for DBH fitting. The equation for the fitted cylindrical model is given in Formula 9. Figure 7 illustrates the spatial schematic of the fitted cylinder.

$$(X - X_i)^2 + (Y - Y_i)^2 + (Z - Z_i)^2 - r^2 = \frac{[L \times (X - X_i) + M \times (Y - Y_i) + N \times (Z - Z_i)]^2}{L^2 + M^2 + N^2} \tag{9}$$

In the formula, (X, Y, Z) represented a point on the fitted cylindrical axis; (L, M, N) was the vector in the L direction of the cylindrical axis; and "r" was the radius of the circle. The cylindrical equation could be determined by these seven parameters.

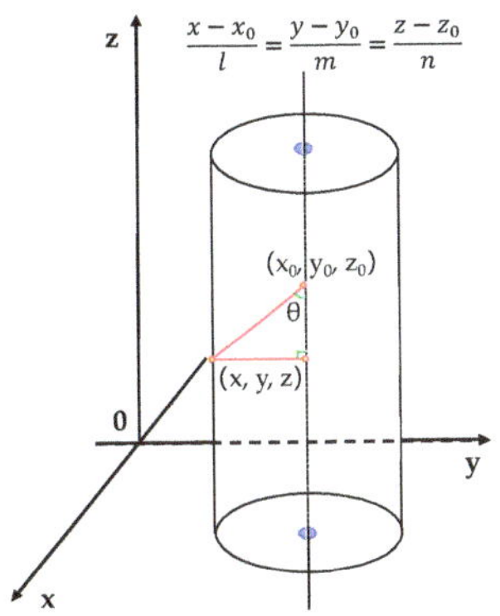

Figure 7. Schematic diagram of fitting cylindrical space of LSA-RANSAC.

The iterative fitting process of the LSA-RANSAC for the cylindrical model was as follows:

(1) Initialize parameters: the maximum distance from the point cloud to the model, the desired inlier probability, maximum iterations, and minimum inlier count were set. The best initial inlier count and the best initial model were initialized;
(2) Calculate the sample size based on the desired inlier probability;
(3) Determine the best initial model parameters iteratively: randomly select a sample point cloud from the point cloud according to the current sample size. The cylindrical model was fit using the selected sample point cloud, returning the model parameters and inlier indices. The inlier count was calculated and compared with the current best initial inlier count. If the inlier count was greater than the current best initial inlier count, the best initial inlier count, the best initial model, sample size, and iteration count were updated. The adjustment process adaptively updated the sample size and iteration count based on the current inlier count;
(4) Extract the best initial model parameters, IM_{params}: obtain the center point, axis direction, radius, and height of the cylinder from the best initial model acquired by the adaptive RANSAC algorithm;
(5) Define the fitting error function by calculating the distance between the best initial model parameter IM_{params} and the point cloud data;
(6) Perform nonlinear least-squares optimization on IM_{params}, using the Levenberg–Marquardt optimization method to solve for the minimum value of the fitting error function and obtain the best-fit cylindrical model parameters.

$$IM_{params} = argmax \sum_{i=1}^{n} w_i \times p_i(X_i \in \theta) \tag{10}$$

$$BM_{params} = argmin \sum_{i=1}^{n} (|X_i - C) \times V| - r^2 \tag{11}$$

In the formula, "IM_{params}" represented the best cylindrical model parameters, "n" represented the number of sample points, "X_i" represented the ith point in the point cloud, "w_i" represented the weight of the sample point, "$p_i(X_i \in \theta)$" represented the probability of the sample point "X_i" being fitted by the cylindrical model "θ". "C" represented the center point of the cylinder; "V" represented the axis direction of the cylinder, and "r" represented the radius of the cylinder. The final result was to find the best cylindrical parameter BM_{params} that minimized the sum of distances from all points to the cylinder surface.

Throughout the experiment, the segmented trunk point cloud results were used as input data. "*Min-Pts*" was set to 50, and "*Eps*" was set to 0.2. The maximum distance was set to 0.01; the desired probability was set to 0.50. The iteration count was capped at a maximum of 100, and concurrently, the floor for the inlier count was established at 15.

2.4. Comparison Method

The BSTDF framework primarily consists of two parts: tree trunk detection and DBH fitting. In this study, the KPConv [25], RandLA-Net [28], PointNet++ [30], and (the tree trunk extraction method based on vector features) VF [18] were chosen as the comparison method for the WCF-CACL-RandLA-Net model tree trunk detection deep learning model, utilizing the training samples from Section 2.3.1 for synchronous detection. Simultaneously, we chose locally optimal RANSAC [3], LS [11,12], and CHA [11] as the comparative method for LSA-RANSAC to estimate DBH. RANSAC achieves local optimality by iteratively selecting the cylindrical fitting model with the highest number of inliers. Consistent data is maintained throughout the entire fitting experiment process. The accuracy results are the average of three repeated experiments.

3. Results

3.1. Segmentation Results of Point Cloud Based on WCF-CACL-RandLA-Net Model

3.1.1. Evaluation Metrics

To assess the performance of RandLA-Net in trunk segmentation, we employed four overall benchmark metrics, including overall accuracy (OA), intersection over union (IoU), F_1 Score (F_1), and tree trunk detection rate (ER) to evaluate segmentation accuracy. "TP", "TN", "FP", and "FN", denoted true positives, true negatives, false positives, and false negatives, respectively. "n" represented the number of detected tree trunks, and "N" represented the actual number of surveyed trees in the sample plot.

$$OA = \frac{TN + TP}{TP + FN + TN + FP} \tag{12}$$

$$IoU = \frac{TP}{TP + FP} \tag{13}$$

$$R = \frac{TP}{TP + FN} \tag{14}$$

$$F_1 = \frac{2 \times (IoU \times R)}{IoU + R} \tag{15}$$

$$ER = \frac{n}{N} \times 100\% \tag{16}$$

3.1.2. Training Loss and Elapsed Time

During the training process of deep learning algorithms, we used training accuracy and loss functions to assess the convergence of the model training process, which might also reflect the error between the final segmentation results of the point cloud segmentation model and the true values. Figure 8A illustrates the increase in training accuracy over time. After 40 training iterations, both WCF-CACL-RandLA-Net and RandLA-Net training accuracies stabilized between 0.99 and 1.00. Within the same training duration, WCF-CACL-RandLA-Net demonstrated higher training accuracy and better convergence than RandLA-Net. Figure 8B depicts the change in loss functions for both models during the training period. As the models were trained, the loss functions gradually converged, and after 40 training iterations, the training losses for both models stabilized between 0 and 0.05. The WCF-CACL-RandLA-Net training process required 3 h and 2 min, while the RandLA-Net model training process took 3 h, indicating that WCF and CACL did not increase the training burden.

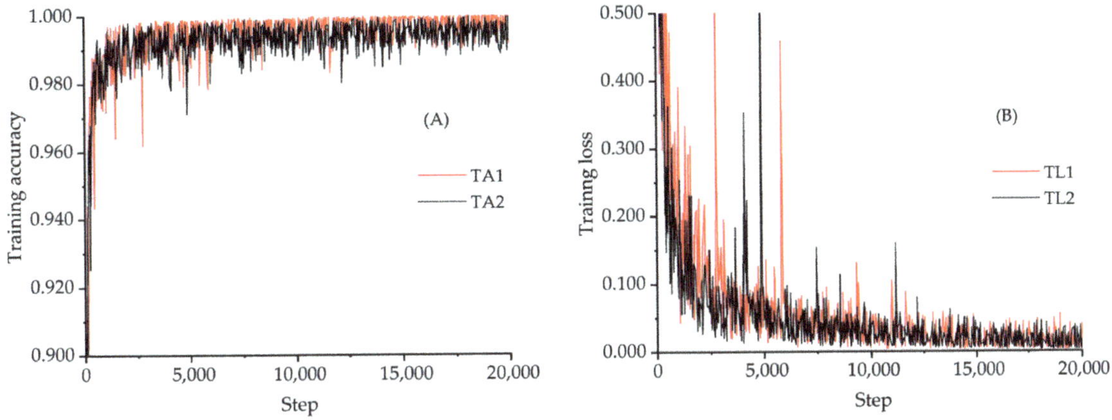

Figure 8. The training process of two deep learning models: (**A**) the change process of training accuracy; (**B**) the training loss change process. Black represents RandLA-Net. The red color in the figure represents WCF-CACL-RandLA-Net.

3.1.3. Quantitative and Qualitative Evaluation

As previously mentioned, Table 2 displays the quantitative and visual inspection evaluation results of the tree trunk detection process for both models. The *OA*, *IoU*, and F_1 for the WCF-CACL-RandLA-Net model were 0.98, 0.88, and 0.92, respectively. Compared to RandLA-Net, the OA of the WCF-CACL-RandLA-Net model did not change significantly due to the point cloud count of the ground category accounting for over 80% of the total category point cloud count. However, the segmentation accuracy metrics *IoU* increased from 0.85 to 0.88, and F_1 increased from 0.90 to 0.92, achieving higher segmentation accuracy; particularly for Tree-trunk and Shrubs-branches, the segmentation accuracy improved by 6% and 5%, respectively, while F_1 increased by 3% for both. The aforementioned results indicate that the WCF effectively improved the segmentation accuracy of the RandLA-Net model when processing imbalanced the point cloud datasets. During the training process, the WCF module emphasized the model's attention to trunk class point clouds, helping the model to perform well in identification and localization tasks even in sparse and occluded situations, thus, making the model's segmentation accuracy more balanced and enhancing its generalization ability. Simultaneously, CACL allowed the model to have more flexible learning rate variation and parameter exploration capabilities; through end-to-end training, it improved the segmentation performance of the RandLA-Net model for objects easily confused between trunks and leaves.

Table 2. Comparison of segmentation accuracy between two deep learning models.

Classification of the Point Cloud	*OA*		*IoU*		F_1	
	WCF-CACL RandLA-Net	**RandLA-Net**	**WCF-CACL RandLA-Net**	**RandLA-Net**	**WCF-CACL RandLA-Net**	**RandLA-Net**
Tree trunk	0.99	0.99	0.84	0.78	0.88	0.85
Shrub and branch	0.98	0.98	0.86	0.81	0.90	0.87
Ground	0.96	0.96	0.97	0.97	0.97	0.98
Total/Mean	0.98	0.98	0.88	0.85	0.92	0.90

3.1.4. Comparative Studies

In order to demonstrate the efficacy of the proposed method qualitatively for segmenting tree trunks within complex forest MLS point cloud scenarios, we designed a number of experiments and compared it with selected popular methods, including one traditional method (VF) [18] and three deep learning approaches (KPConv [25], RandLA Net [28], and PointNet++ [30]). We tabulated the precision results of various methods in Table 3, with selected visual result examples showcased in Figure 9. Notably, our proposed method currently registered the highest *mIoU* score of 0.89, followed by RandLA-Net, trailing by roughly 4%. KPConv sat slightly below RandLA-Net by around 2%, with the lowest trunk segmentation precision attributed to PointNet++. It is worth emphasizing that all four aforementioned methods substantially outperformed traditional feature vector-based trunk detection methods.All deep-learning methods delivered satisfactory results for tree trunk point clouds with pronounced trunk shapes. However, for complex location distributions, such as scenarios where multiple trees are distributed in queues with severe spatial overlap, both PointNet++ and KPConv tended to generate omissions in trunk detection, often misclassifying surrounding branches and leaves as trunks. While RandLA-Net does show some improvement in reducing trunk omission errors, the misclassification of branches and leaves as trunks still remains prevalent. The training and testing duration for PointNet++ and KPConv outlasts that of RandLA-Net and WCF-CACL-RandLA-Net. The traditional method, although quick, is plagued by severe omissions or debugging errors, primarily due to its strong dependence on the integrity of the tree trunk annular point cloud and the spatial features of adjacent woods. The network structure and parameters of deep learning methods are able to implicitly articulate the spatial interactions between trunks, branches, leaves, and ground point clouds, facilitating feature representation, such as segmentation. Despite the added computational complexity introduced by deep learning methods, such shifts should be tolerated when performing trunk extraction and DBH estimation in large-scale MLS point clouds [20].

Table 3. Quantitative evaluation of multiple trunk segmentation methods (*mOA*, *mIoU*, and *IoU* for each class; all Values in %). Best results in bold.

References	Time of Train	Time of Test	*IOU* of Tree Trunk	*IOU* of Shrub-Branch	*IOU* of Ground	*mIOU*	*mOA*
KPconv	**2.9**	0.80	0.71	0.79	**0.98**	0.83	0.97
PointNet++	3.50	1.10	0.60	0.71	**0.98**	0.76	0.95
Randlanet	**2.90**	**0.10**	0.78	0.81	0.97	0.85	**0.98**
VF	--	--	0.36	--	--	0.36	
Ours	2.93	**0.10**	**0.84**	**0.86**	0.97	**0.89**	**0.98**

3.1.5. Ablation Experiments

In the forthcoming experiments, we delved into the influence of the weighted cross entropy focal loss function module (WCF) and the cosine annealing cycle learning strategy (CACL) on the segmentation precision of the RandLA-Net model during the process of tree trunk identification, leveraging the technique of ablation studies. Empirical findings highlight the positive contributions of both WCF and CACL to the segmentation capabilities of the RandLA-Net model. As illustrated in Table 4, during the segmentation of the hierarchically-coupled tree trunk identification dataset, WCF effectively mitigated discrepancies in segmentation precision induced by imbalanced point cloud class quantities. With the utilization of the WC-RandLA-Net, the recognition precision for tree trunks and shrubs within the point cloud data eclipsed the standard RandLA-Net model by 5% and 3%, respectively. Moreover, when the mean intersection over union (*IoU*) exceeded 80%, the step count amounted to 2000 and 3500 for the CACL-RandLA-Net and the RandLA-Net, respectively. This evidence corroborates the notion that CACL accelerates the convergence of deep learning models and exerts a propitious influence on segmentation precision.

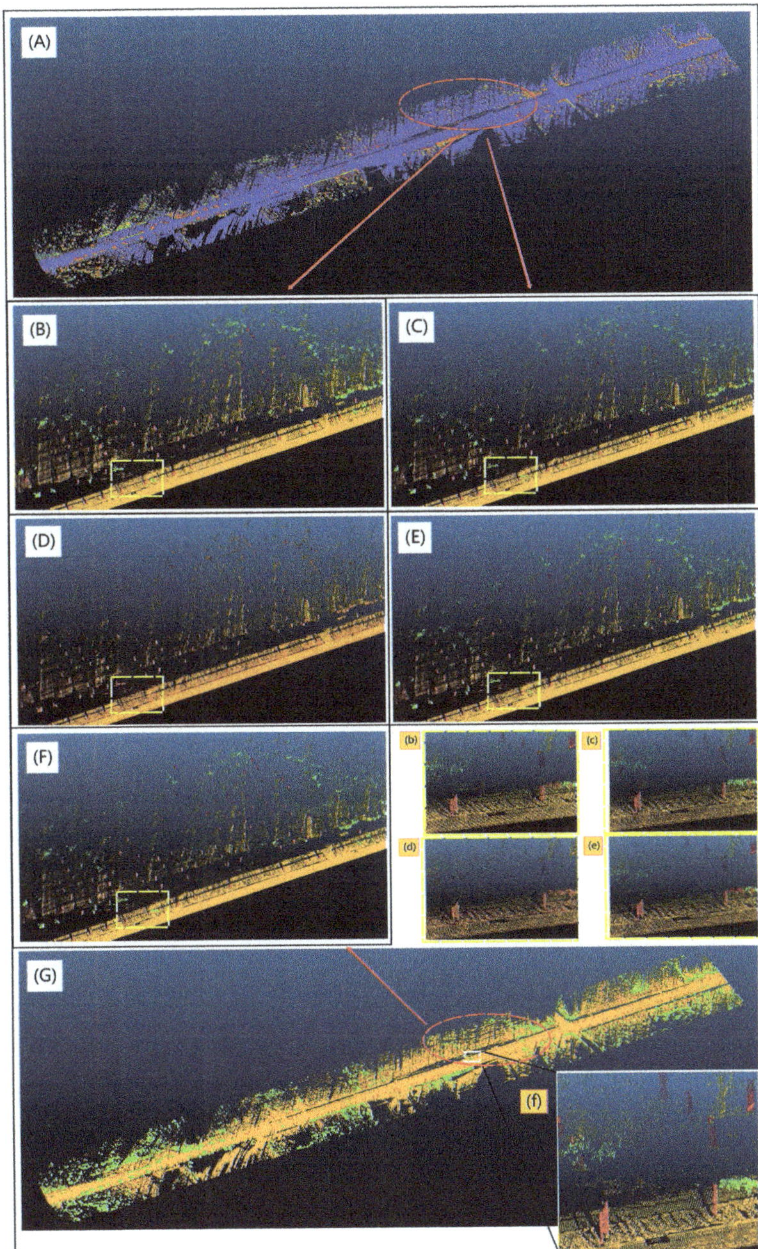

Figure 9. Segmentation results using different methods. Orange represents the ground, red represents the Tree-trunk category, and green represents the Shrub-branch category. (**A**) Point cloud to be segmented after Stratified-Coupled processing; (**B**,**b**) Segmentation results based on KPConv [25]; (**C**,**c**) segmentation results based on PointNet++ [30]; (**D**,**d**) segmentation results based on VF [18]; (**E**,**e**) segmentation results based on RandLA-Net [28]; (**F**,**f**) segmentation results based on WCF-CACL-RandLA-Net, (**G**) Research area segmentation results based on WCF-CACL-RandLA-Net.

Table 4. Comparison of ablation experimental results.

Model	*IOU* of Tree Trunk	*IOU* of Shrub-Branch	*IOU* of Ground	Mean *IOU*	Step of Mean *IOU* ≥ 0.8
RandLA-Net	0.78	0.81	0.97	0.83	3500
WCF-RandLA-Net	0.82	0.84	0.97	0.87	3500
CACL-RandLA-Net	0.79	0.82	0.97	0.86	2000
WCF-CACL-RandLA-Net	0.84	0.86	0.97	0.88	2000

3.1.6. Impact of Laser Scanning Distance Level on Tree Trunk Detection Rate (*ER*)

Similar to other tree trunk detection segmentation methods [8,9,15,16], irregular trunk shapes and point cloud scarcity caused by obstacles and laser scanning distance in the forest remain the primary factors affecting accurate measurements [23,31,32]. To determine the impact of scanning distance on *ER*, we divided trees into four distance levels (0–10 m, 10–20 m, 20–30 m, and 30–45 m) based on the horizontal distance between the trees and the scanning path (Figure 10) and compared the *ER* values within different distance levels. The results, as shown in Table 5, indicated that *ER* decreased as the distance level increased. The closer the samples were to the scanning path, the higher the *ER* value. The *ER* values for the 10 m, 20 m, 30 m, and 40 m levels were 97.70%, 83.36%, 68.09%, and 32.47%, respectively, with an average *ER* of 70.41%. The distance level with an *ER* above 80% was 20 m, and the average *ER* value at this distance level was 90.53%. Increasing the density of the LiDAR scanner's path in the forest was an effective way to improve *ER* values. Considering accuracy and efficiency, we recommend that the BLS data collection distance level for tree trunk detection and recognition should not exceed 20 m, and the optimal bandwidth (the parallel distance between two adjacent scanning paths) of multiple parallel paths should not exceed 40 m.

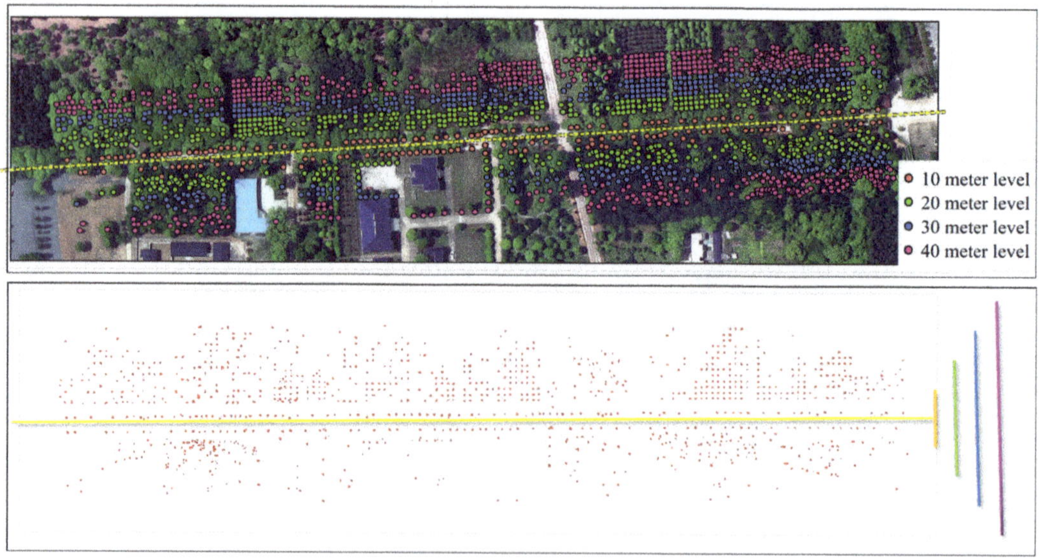

Figure 10. Distance level division and tree distribution map for tree trunk detection rate statistics.

Table 5. *ER* at different distance levels based on WCF-CACL-RandLA-Net segmentation results.

Evaluating Indicator	*ER* at Different Distance Levels			
	0–10 m	10–20 m	20–30 m	30–45 m
ER	97.70%	83.36%	68.09%	32.47%
Measured number of trees	174	535	468	627

3.2. DBH Estimation Results Based on LSA-RANSAC

3.2.1. Evaluation Metrics

The evaluation metrics for DBH fitting accuracy are the coefficient of determination (R^2), root mean square error (*RMSE*), and residuals between measured and estimated values (*Residual*).

$$R^2 = 1 - \frac{\sum_{i=1}^{n}(R)^2}{\sum_{i=1}^{n}(y_i - \overline{y})^2} \tag{17}$$

$$RMSE = \sqrt{\frac{\sum_{1}^{n}(y_i - \hat{y}_i)^2}{n}} \tag{18}$$

$$Residual = y_i - \hat{y}_i \tag{19}$$

In the formula, "y_i" represents the ith measured value, "$\hat{y}_i$" represents the ith predicted value, "$\overline{y}$" represents the average measured DBH, and "n" represents the number of detected trunks.

3.2.2. Fitting Accuracy Comparison

To verify the accuracy and effectiveness of the DBH estimation results, three subplots were set up within the plot area (Figure 11), including a coniferous forest sampling plot (CF: 50 m × 60 m), a broadleaf forest sampling plot (BF: 50 m × 60 m), and a coniferous-broad mixed forests sampling plot (CBF: 40 m × 60 m). The closure degrees of the three subplots were 0.90, 0.85, and 0.95, respectively, with a stand density of 666.7 trees/ha, 426.6 trees/ha, and 504 trees/ha. The DBH fitting results of all trees within the three subplots (Figure 11) were statistically analyzed (Table 6). The statistics showed that the coefficient of determination (R^2) for the LSA-RANSAC based DBH estimation was 0.77, 0.74, and 0.73, with RMSE values of 6.37 cm, 6.20 cm, and 6.92 cm. The residual values of all sample plots were uniformly distributed around the X-axis (Figure 12). In well-distributed coniferous and broadleaf forests, DBH estimation accuracy was higher than in randomly distributed mixed forests. The DBH estimation accuracy in coniferous forests was slightly higher than in broadleaf forests, which might be due to some trunks in broadleaf forests having bifurcations, causing cylinder overfitting and affecting DBH estimation accuracy. Compared to RANSAC, the average R^2 of LSA-RANSAC increased by 14%, and the average *RMSE* decreased by 1.08 cm. The lower DBH estimation accuracy of the RANSAC method was due to widespread cylinder underfitting. The LSA-RANSAC method effectively reduced error values during fitting by iteratively updating the selection of points and inliers and constraining the objective function of cylinder parameters, thus improving the DBH fitting accuracy of the RANSAC method. Although LSA-RANSAC showed effectiveness in DBH estimation, we still compared it with the least squares (LS) [11,12] and convex hull (CHA) [11] fitting methods. It is worth noting that the cylindrical fitting method based on RANSAC was generally significantly superior to the CHA and LS methods for estimating chest diameter. This is because the CHA and LS methods were more sensitive to missing cylindrical point clouds.

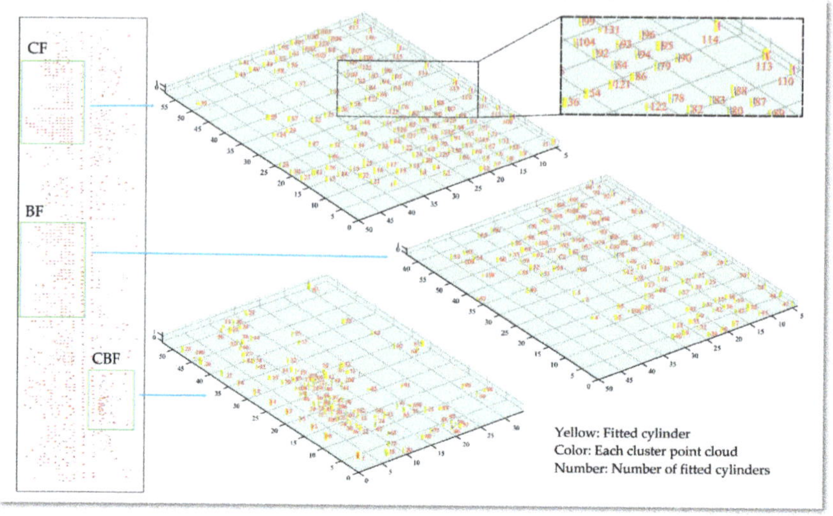

Figure 11. Tree distribution fitting DBH based on LSA-RANCAC.

Table 6. Precision statistics of evaluation indicators for DBH estimation models of three forest types.

Species Type of Foresty	Species Name of Tree	Number of Trees	LSA-RANSAC		RANSAC		CHA		LS	
			R^2	*RMSE*	R^2	*RMSE*	R^2	*RMSE*	R^2	*RMSE*
Coniferous forest	*Taxodium distichum (L.) Rich. andTaxodium distichum var. imbricatum (Nuttall) Croom*	129	0.77	6.37	0.60	7.10	0.39	10.58	0.53	10.23
Broadleaf forest	*Betula luminifera H.Winkl. and Liriodendron chinense (Hemsl.) Sargent.*	108	0.74	6.20	0.61	7.12	0.33	9.89	0.42	8.91
Coniferous–broad forests	*Cinnamomum camphora (Linn) Presl, Michelia maudiae Dunn, Cedrus deodara (Roxb.) G. Don and Abies fabri (Mast.) Craib*	106	0.73	6.92	0.59	8.53	0.20	11.19	0.23	13.70
Total/Mean		343	0.75	6.50	0.61	7.58	0.31	10.55	0.39	10.95

3.2.3. Analysis of Tree Trunk Detection Accuracy Based on LiDAR Scanning Distance Level

To evaluate the impact of scanning distance on DBH estimation, the estimated DBH values of the three subplots using LSA-RANSAC were statistically analyzed according to the distance level (Table 7), and the estimation accuracy of the DBH at different distance levels was compared. In the three forest types, the distance levels with DBH estimation accuracy above 80% were 30 m for the coniferous forest, 20 m for the broadleaf forest, and 10 m for the coniferous–broadleaf mixed forest. The DBH estimation accuracy in the mixed forest subplot was lower than that in the coniferous and broadleaf forest subplots due to the uniform distribution of trees in the coniferous and broadleaf forest subplots, while the trees in the broadleaf forest subplot were densely and centrally distributed. Overall, the R^2 of the DBH values estimated by the LSA-RANSAC method was negatively correlated with the LiDAR scanning distance level, while the *RMSE* was positively correlated. The average DBH estimation accuracy for the three forest types at distance levels within 20 m was above 80%. Within the 0–20 m range, the R^2 was 0.87, and the *RMSE* was 4.41 cm. Taking accuracy and efficiency into consideration, we also recommend that the range of BLS data collection for DBH estimation should not exceed 20 m, and the optimal bandwidth for multiple parallel paths should not exceed 40 m.

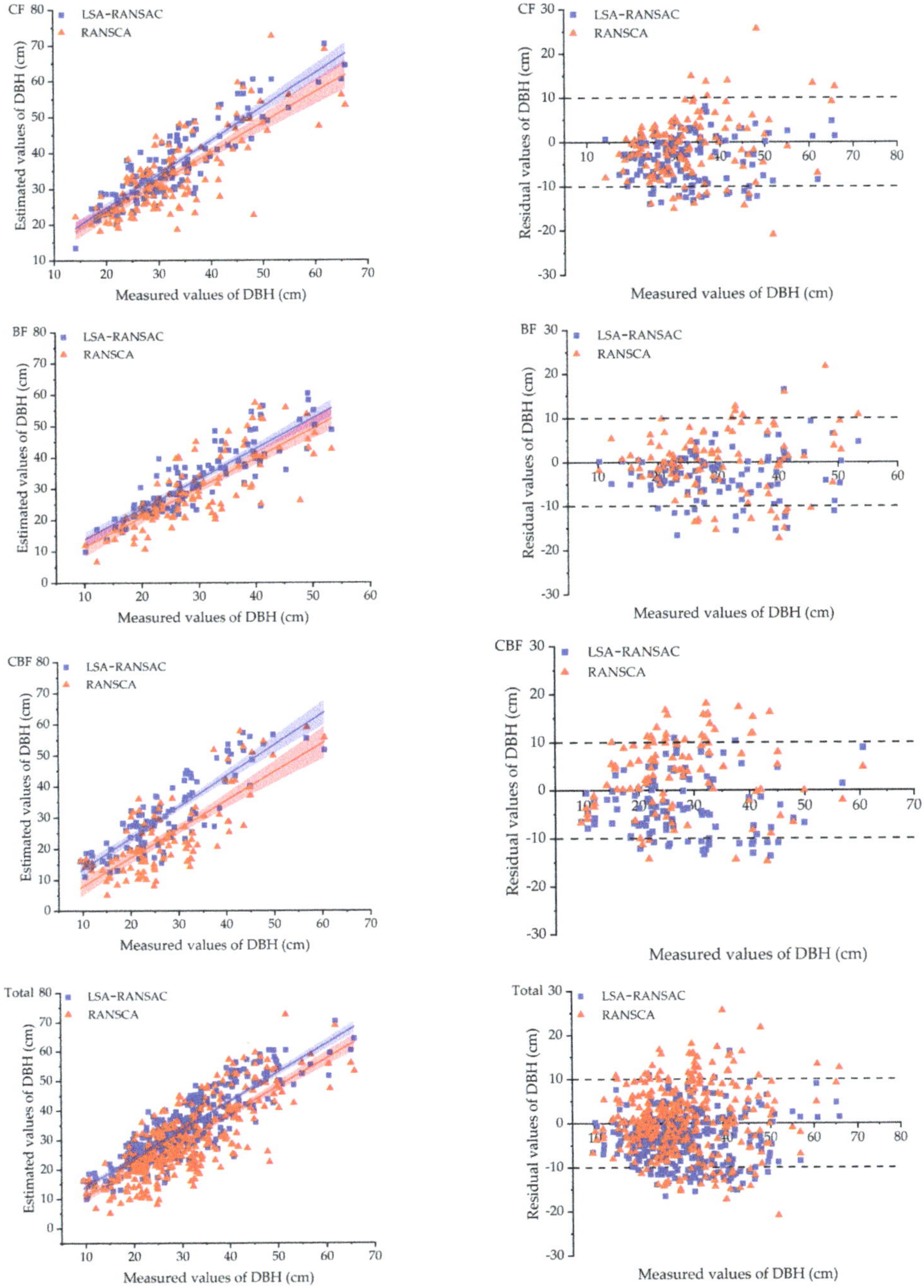

Figure 12. Comparison of the accuracy of DBH estimation results for three forest types based on LSA-RANCAC.

Table 7. DBH estimation accuracy of different distance levels in three forest types.

Species Type of Foresty	Evaluating Indicator	DBH Estimation Accuracy at Different Distance Levels			
		0–10 m	10–20 m	20–30 m	30–45 m
Coniferous forest	R^2	0.97	0.83	0.81	0.67
	RMSE	2.35	3.67	7.01	9.75
	Nmber of tree	10	51	43	25
Broad-leaved forest	R^2	0.98	0.82	0.61	0.76
	RMSE	2.53	5.04	8.56	6.99
	Nmber of tree	14	54	30	10
Coniferous–broad-leaved mixed forests	R^2	0.83	0.75	0.58	0.47
	RMSE	5.94	6.91	8.94	9.11
	Nmber of tree	42	48	9	7
Total/Mean	R^2	0.93	0.80	0.67	0.63
	RMSE	3.61	5.21	8.17	8.62
	Nmber of tree	66	153	82	42

4. Discussion

In this study, we focused on the importance of accurate assessment and scientific management of forest resources in the context of exacerbated global climate change and ecological environmental issues. We combined remote sensing technology and computer algorithms to propose a novel framework for forest tree trunk detection and DBH estimation based on BLS, called the BLS-based tree-trunk detection framework (BSTDF). Firstly, we created a reusable tree trunk detection deep learning dataset using a stratified coupling method. Secondly, we improved the learning strategy and loss function of the RandLA-Net model to adapt to tree trunk detection tasks in large-scale forests and address the issue of class imbalance in segmentation. Lastly, we introduced the LSA-RANSAC method to enhance the accuracy of DBH estimation. The effectiveness of the BSTDF was validated in a 65,000 m^2 forest at the Subtropical Forestry Experiment Center of the Chinese Academy of Forestry Sciences.

To fully harness the potential of mobile laser scanning (MLS), one of the primary focuses of this research lies in the construction of a tree trunk detection dataset for deep learning. In previous studies [20,23,25,33], the datasets used for forest point cloud segmentation predominantly comprised entire forest trees. While not all researchers regard direct segmentation of complete forest scenes as unacceptable, reducing the quantity of non-target point clouds is vital to enhancing segmentation efficiency and accuracy. Accordingly, we have adopted a novel hierarchically-coupled method to establish a dataset suitable for tree trunk detection based on deep learning. By dividing the full forest point cloud into four strata: ground point cloud layer (0.0 m to 0.1 m), non-target point cloud layer 1 (0.1 m to 1.0 m), target point cloud layer (1.0 m to 1.6 m), and non-target point cloud layer 2 (1.6 m and above), we couple the ground layer and target trunk layer based on coordinate calculations, thus, forming the foundational dataset for trunk detection. This data processing method dramatically trims the data volume within the detection point cloud, enhancing the learning efficiency of deep learning algorithms and providing substantial reference value in dataset creation.

In this study, we used the improved WCF-CACL-RandLA-Net model, which is suitable for large-scale point cloud segmentation, for tree trunk detection. Compared to other deep learning models [23–27], the RandLA-Net model employs a random sampling (RS) method that can process one million points at a time, and through the LFA structural design, it achieves aggregation of local features. This overcomes the problem of insufficient sparse keypoints information caused by RS. The study findings reveal that KPConv is marginally inferior to RandLA-Net by approximately 2%, with PointNet++ falling short in terms of trunk segmentation precision. These observations resonate with the findings of Wang et al. [20] and Hu et al. [28,34], who found RandLA-Net's segmentation prowess outshone other leading semantic segmentation methods. By introducing the WCF and

CACL, the WCF-CACL-RandLA-Net model demonstrates better performance in tree trunk detection tasks compared to the RandLA-Net model. Firstly, the introduction of the WCF enables the model to learn trunk features more effectively, which helps eliminate the impact of point cloud noise and occlusion and improves tree trunk detection accuracy. Secondly, the introduction of the CACL helps the model adaptively adjust the learning rate during training, ensuring a suitable learning speed at different stages. This helps to avoid the model falling into local optima and enhances its generalization performance.

Traditional trunk detection methodologies, such as those reliant on vector features (VF) [18,19], geometric circle fitting [35], cylinder fitting [36], and cross-section detection [37], are excessively sensitive to incomplete trunk point clouds and lack robustness. In contrast, deep learning algorithms demonstrate notable superiority in detection accuracy. For instance, our framework method outperforms VF in trunk point cloud recognition precision by 48%. Moreover, our approach exhibits greater robustness when handling large-scale forest data without necessitating point cloud-raster data conversion [37] or manual elimination of non-target point clouds. In addition, compared to other deep learning-based improved methods, our model shows higher accuracy and stability in tree trunk detection tasks, with an *IOU* segmentation accuracy of 0.88 for the hierarchically coupled data layer, which is superior to the segmentation performance of RandLA-Net on other datasets [29,37]. In future research, we may consider further enhancing the model's performance in handling large-scale tree trunk detection tasks by improving the model structure or employing distributed computing techniques.

Of course, the fitting method for DBH also affects the accuracy of DBH estimation. In previous studies [12–14,38], circle fitting-based DBH estimation has developed into a mature approach. Different methods have their respective strengths and weaknesses. Least squares (LS) [11,12] can achieve satisfactory fitting results when processing simple and regular point clouds, but its ability to handle outliers is limited. Hough transform (HT) [14] has a high computational complexity, is sensitive to parameters, and requires manual selection of appropriate parameters, resulting in limited robustness. The convex hull algorithm (CHA) [11] may perform poorly when processing data from densely populated forest areas and is susceptible to noise and occlusion. RANSAC [3] has certain advantages in handling noise and occlusion issues but requires manual parameter settings, such as inlier threshold and iteration count, and is sensitive to parameter selection. In view of these shortcomings, this research proposes a LSA-RANCAC method. The experimental results show that this method combines the fitting accuracy of the least squares method with the robustness of the RANSAC method, providing a strong outlier processing capability, adaptive parameter settings, and high fitting accuracy, which is significantly better than RANSAC, LS, and CHA fitting methods. Of course, there is still room for improvement in the LSA-RANSAC method used in this study. For example, Duanmu, J. et al. [2] proposed an adjacent point distribution analysis (ANPDA) method to reduce noisy point clouds, which could be an effective way to further enhance the LSA-RANSAC method.

In this study, the forest stand density of our test area was set at 394 stems per hectare, and 1270 trees were detected using our method. When the LiDAR scanning distance level was set at 20 m, the average *ER* was 90.03%, significantly higher than the 17.4–32% reported in other studies [2,11]. Considering the three forest types established in this study—mixed coniferous, mixed broadleaf, and mixed coniferous–broadleaf forests—the forest type also has an impact on the accuracy on BLS-based DBH estimation. The trunk extraction and fitting accuracy of mixed coniferous and mixed broad-leaved forests are higher than that of mixed coniferous and broad-leaved forests, possibly due to higher spatial clustering and wider diameter distribution in mixed coniferous–broadleaf forests. If the accuracy of individual tree trunk detection is guaranteed, BLS will be the most suitable solution for estimating mountain forest parameters. This is because BLS offers greater portability and excellent penetration capabilities, with the highest accuracy in the main tree trunk detection and DBH estimation through a single scan [5,39].

5. Conclusions

This study presented a novel framework for forest tree trunk detection and DBH estimation based on BLS (BSTDF), aimed at addressing the difficulty and low accuracy of obtaining large-scale forest DBH information. In BSTDF, a stratified coupling method using BLS was employed to construct a tree trunk detection dataset, significantly reducing the dataset size, improving computational efficiency, and enhancing the generalization capability of deep learning models. BSTDF improved the RandLA-Net deep learning algorithm by introducing WCF and CACL, effectively mitigating class imbalance issues in the tree trunk detection process, accelerating model convergence, and enhancing model stability and accuracy. Furthermore, BSTDF estimated DBH by using the cylindrical fitting method of LSA-RANSAC, fully leveraging the parameter search capabilities of adaptive methods and least squares optimization algorithm, and model error convergence, making the entire DBH fitting process more accurate. The experimental results showed that the *OA*, *IoU*, and F_1 of the WCF-CACL-RandLA-Net model were 0.98, 0.88, and 0.92, respectively. Better than segmentation algorithms such as KPConv, PointNet++, and VF. Compared to RandLA-Net, the *IoU* increased from 0.85 to 0.88, and the F_1 increased from 0.90 to 0.92. Particularly, the segmentation accuracy of the trunk category improved by 6%, and the F_1 increased by 3%, effectively addressing the class accuracy imbalance issue when RandLA-Net segmented trunks. The overall tree trunk detection rate was 70.41%. *RMSE* and R^2 for the overall DBH estimation accuracy of LSA-RANSAC were 6.50 cm and 0.75, respectively. Compared to RANSAC, *RMSE* decreased by 1.08 cm, and R^2 increased by 14%, indicating that LSA-RANSAC effectively mitigated RANSAC's underfitting problem and improved DBH estimation accuracy. Both *ER* and DBH estimation accuracy were negatively correlated with the LiDAR scanning distance (bandwidth). Considering the accuracy and efficiency, the optimal distance level for single-path BLS data collection was 20 m, and the optimal scanning bandwidth for multiple parallel paths should not exceed 40 m. Under these conditions, the average *ER* of BSTDF was 90.03%, with a DBH estimation accuracy of *RMSE* 4.41 cm and R^2 0.87.

In conclusion, the BSTDF provided effective technical support for large-scale forest trunk identification and DBH estimation. We recommended applying the proposed method in other forests to evaluate its performance and limitations when encountering different environments. Additionally, by testing under various terrain conditions, BSTDF could be further optimized and improved, making it more versatile and adaptable.

Author Contributions: Conceptualization, H.Z. (Huacong Zhang) and H.Z. (Huaiqing Zhang); methodology, H.Z. (Huacong Zhang) and H.Z. (Huaiqing Zhang); software, H.Z. (Huacong Zhang); validation, H.Z. (Huacong Zhang); formal analysis, H.Z. (Huacong Zhang); investigation, K.X., Y.L., and L.W.; resources, H.Q. and L.Y.; data curation, H.Z. (Huacong Zhang); writing—original draft preparation, H.Z. (Huacong Zhang); writing—review and editing, H.Z. (Huaiqing Zhang); visualization, K.X. and R.L.; supervision, H.Z. (Huaiqing Zhang) and Y.L.; project administration, H.Z. (Huaiqing Zhang); funding acquisition, H.Z. (Huaiqing Zhang). All authors have read and agreed to the published version of the manuscript.

Funding: This research was funded by the Fundamental Research Funds of Chinese Academy of Forestry (CAF), grant number CAFYBB2021ZE005, CAFYBB2019SZ004.

Data Availability Statement: Not applicable.

Acknowledgments: This work is supported by Artificial Intelligence and Visualization Team in IRIFIT, CAF. Special thanks to the teammates at the Experimental Center of Subtropical Forestry for their assistance in collecting and processing field data.

Conflicts of Interest: The authors declare that there are no conflict of interest.

References

1. Liu, C.; Xing, Y.; Duanmu, J.; Tian, X. Evaluating Different Methods for Estimating Diameter at Breast Height from Terrestrial Laser Scanning. *Remote Sens.* **2018**, *10*, 513. [CrossRef]
2. Duanmu, J.; Xing, Y. Annular Neighboring Points Distribution Analysis: A Novel PLS Stem Point Cloud Preprocessing Algorithm for DBH Estimation. *Remote Sens.* **2020**, *12*, 808. [CrossRef]
3. Olofsson, K.; Holmgren, J.; Olsson, H. Tree Stem and Height Measurements using Terrestrial Laser Scanning and the RANSAC Algorithm. *Remote Sens.* **2014**, *6*, 4323–4344. [CrossRef]
4. Brede, B.; Lau, A.; Bartholomeus, H.M.; Kooistra, L. Comparing RIEGL RiCOPTER UAV LiDAR Derived Canopy Height and DBH with Terrestrial LiDAR. *Sensors* **2017**, *17*, 2371. [CrossRef]
5. Oveland, I.; Hauglin, M.; Giannetti, F.; Schipper Kjørsvik, N.; Gobakken, T. Comparing Three Different Ground Based Laser Scanning Methods for Tree Stem Detection. *Remote Sens.* **2018**, *10*, 538. [CrossRef]
6. Wang, Z.; Lu, X.; An, F.; Zhou, L.; Wang, X.; Wang, Z.; Zhang, H.; Yun, T. Integrating Real Tree Skeleton Reconstruction Based on Partial Computational Virtual Measurement (CVM) with Actual Forest Scenario Rendering: A Solid Step Forward for the Realization of the Digital Twins of Trees and Forests. *Remote Sens.* **2022**, *14*, 6041. [CrossRef]
7. Van Leeuwen, M.; Nieuwenhuis, M. Retrieval of forest structural parameters using LiDAR remote sensing. *Eur. J. For. Res.* **2010**, *129*, 749–770. [CrossRef]
8. Liang, X.; Kukko, A.; Kaartinen, H.; Hyyppä, J.; Yu, X.; Jaakkola, A.; Wang, Y. Possibilities of a Personal Laser Scanning System for Forest Mapping and Ecosystem Services. *Sensors* **2014**, *14*, 1228–1248. [CrossRef]
9. Bauwens, S.; Bartholomeus, H.; Calders, K.; Lejeune, P. Forest Inventory with Terrestrial LiDAR: A Comparison of Static and Hand-Held Mobile Laser Scanning. *Forests* **2016**, *7*, 127. [CrossRef]
10. Chen, S.; Liu, H.; Feng, Z.; Shen, C.; Chen, P. Applicability of personal laser scanning in forestry inventory. *PLoS ONE* **2019**, *14*, e0211392. [CrossRef]
11. Pueschel, P.; Newnham, G.; Rock, G.; Udelhoven, T.; Werner, W.; Hill, J. The influence of scan mode and circle fitting on tree stem detection, stem diameter and volume extraction from terrestrial laser scans. *ISPRS J. Photogramm. Remote Sens.* **2013**, *77*, 44–56. [CrossRef]
12. Calders, K.; Newnham, G.; Burt, A.; Murphy, S.; Raumonen, P.; Herold, M.; Culvenor, D.; Avitabile, V.; Disney, M.; Armston, J.; et al. Nondestructive estimates of above-ground biomass using terrestrial laser scanning. *Methods Ecol. Evol.* **2015**, *6*, 198–208. [CrossRef]
13. Trochta, J.; Krucek, M.; Vrska, T.; Kral, K. 3D forest: An application for descriptions of three-dimensional forest structures using terrestrial lidar. *PLoS ONE* **2017**, *12*, e0176871. [CrossRef]
14. Sun, H.; Wang, G.; Lin, H.; Li, J.; Zhang, H.; Ju, H. Retrieval and accuracy assessment of tree and stand parameters for chinese fir plantation using terrestrial laser scanning. *IEEE Geosci. Remote Sens. Lett.* **2015**, *12*, 1993–1997. [CrossRef]
15. Del Perugia, B.; Giannetti, F.; Chirici, G.; Travaglini, D. Influence of Scan Density on the Estimation of Single-Tree Attributes by Hand-Held Mobile Laser Scanning. *Forests* **2019**, *10*, 277. [CrossRef]
16. Angermueller, C.; Pärnamaa, T.; Parts, L.; Stegle, O. Deep learning for computational biology. *Mol. Syst. Biol.* **2016**, *12*, 878. [CrossRef]
17. Olofsson, K.; Holmgren, J. Single Tree Stem Profile Detection Using Terrestrial Laser Scanner Data, Flatness Saliency Features and Curvature Properties. *Forests* **2016**, *7*, 207. [CrossRef]
18. Wang, X.; Yang, Z.; Cheng, X.; Stoter, J.; Xu, Z.; Wu, Z.; Nan, L. GlobalMatch: Registration of forest terrestrial point clouds by global matching of relative stem positions. *ISPRS J. Photogramm. Remote Sens.* **2023**, *197*, 71–86. [CrossRef]
19. Li, J.; Cheng, X.; Xiao, Z. A branch-trunk-constrained hierarchical clustering method for street trees individual extraction from mobile laser scanning point clouds. *Measurement* **2022**, *189*, 110440. [CrossRef]
20. Wang, P.; Tang, Y.; Liao, Z.; Yan, Y.; Dai, L.; Liu, S.; Jiang, T. Road-Side Individual Tree Segmentation from Urban MLS Point Clouds Using Metric Learning. *Remote Sens.* **2023**, *15*, 1992. [CrossRef]
21. Liu, Y.; Zhang, H.; Cui, Z.; Lei, K.; Zuo, Y.; Wang, J.; Hu, X.; Qiu, H. Very High Resolution Images and Superpixel-Enhanced Deep Neural Forest Promote Urban Tree Canopy Detection. *Remote Sens.* **2023**, *15*, 519. [CrossRef]
22. Armitage, S.; Awty-Carroll, K.; Clewley, D.; Martinez-Vicente, V. Detection and Classification of Floating Plastic Litter Using a Vessel-Mounted Video Camera and Deep Learning. *Remote Sens.* **2022**, *14*, 3425. [CrossRef]
23. Ning, X.; Ma, Y.; Hou, Y.; Lv, Z.; Jin, H.; Wang, Y. Semantic Segmentation Guided Coarse-to-Fine Detection of Individual Trees from MLS Point Clouds Based on Treetop Points Extraction and Radius Expansion. *Remote Sens.* **2022**, *14*, 4926. [CrossRef]
24. Chen, X.; Jiang, K.; Zhu, Y.; Wang, X.; Yun, T. Individual Tree Crown Segmentation Directly from UAV-Borne LiDAR Data Using the PointNet of Deep Learning. *Forests* **2021**, *12*, 131. [CrossRef]
25. Thomas, H.; Qi, R.; Deschaud, E.; Marcotegui, B.; Goulette, F.; Guibas, L. Kpconv: Flexible and deformable convolution for point clouds. In Proceedings of the IEEE/CVF International Conference on Computer Vision, Seoul, Republic of Korea, 27 October–2 November 2019; pp. 6411–6420.
26. Aloysius, N.; Geetha, M. A Review on Deep Convolutional Neural Networks. In Proceedings of the 2017 International Conference on Communication and Signal Processing (ICCSP 2017), Chennai, India, 6–8 April 2017; pp. 588–592. [CrossRef]
27. Lee, H.; Slatton, K.C.; Roth, B.E.; Cropper, W.P. Adaptive clustering of airborne LiDAR data to segment individual tree crowns in managed pine forests. *Int. J. Remote Sens.* **2010**, *31*, 117–139. [CrossRef]

28. Hu, Q.; Yang, B.; Xie, L.; Rosa, S.; Guo, Y.; Wang, Z.; Trigoni, N.; Markham, A. RandLA-Net: Efficient Semantic Segmentation of Large-Scale Point Clouds. In Proceedings of the 2020 IEEE/CVF Conference on Computer Vision and Pattern Recognition (CVPR), Seattle, WA, USA, 13–19 June 2020; pp. 11105–11114.
29. Ester, M.; Kriegel, P.; Sander, J.; Xu, X. *A Density-Based Algorithm for Discovering Clusters in Large Spatial Databases with Noise*; AAAI Press: Washington, DC, USA, 1996; Volume 34, pp. 226–231. [CrossRef]
30. Qi, C.R.; Yi, L.; Su, H.; Guibas, L.J. PointNet++: Deep hierarchical feature learning on point sets in a metric space. In Proceedings of the Advances in Neural Information Processing Systems, Long Beach, CA, USA, 4–9 December 2017; pp. 5099–5108.
31. Demol, M.; Calders, K.; Krishna Moorthy, S.M.; Van den Bulcke, J.; Verbeeck, H.; Gielen, B. Consequences of vertical basic wood density variation on the estimation of aboveground biomass with terrestrial laser scanning. *Trees* **2021**, *35*, 671–684. [CrossRef]
32. Xu, D.; Chen, G.; Jing, W. A Single-Tree Point Cloud Completion Approach of Feature Fusion for Agricultural Robots. *Electronics* **2023**, *12*, 1296. [CrossRef]
33. Luo, H.; Khoshelham, K.; Chen, C.; He, H. Individual tree extraction from urban mobile laser scanning point clouds using deep pointwise direction embedding. *ISPRS J. Photogramm. Remote Sens.* **2021**, *175*, 326–339. [CrossRef]
34. Hu, Q.; Yang, B.; Khalid, S.; Xiao, W.; Trigoni, N.; Markham, A. Towards Semantic Segmentation of Urban-Scale 3D Point Clouds: A Dataset, Benchmarks and Challenges. In Proceedings of the 2021 IEEE/CVF Conference on Computer Vision and Pattern Recognition (CVPR), Nashville, TN, USA, 20–25 June 2021; pp. 4977–4987.
35. Liang, X.; Litkey, P.; Hyyppä, J.; Matikainen, L. Automatic stem mapping using single-scan terrestrial laser scanning. *IEEE Trans. Geosci. Remote Sens.* **2012**, *50*, 661–670. [CrossRef]
36. Dassot, M.; Constant, T.; Fournier, M. The use of terrestrial LiDAR technology in forest science: Application fields, benefits and challenges. *Ann. For. Sci.* **2011**, *68*, 959–974. [CrossRef]
37. Zhang, Y.; Tan, Y.; Onda, Y.; Hashimoto, A.; Gomi, T.; Chiu, C.; Inokoshi, S. A tree detection method based on trunk point cloud section in dense plantation forest using drone LiDAR data. *For. Ecosyst.* **2023**, *10*, 100088. [CrossRef]
38. Hackel, T.; Savinov, N.; Ladicky, L.; Wegner, J.D.; Schindler, K.; Pollefeys, M. Semantic3D.Net: A New Large-Scale Point Cloud Classification Benchmark. *arXiv* **2017**, arXiv:1704.03847. [CrossRef]
39. Giannetti, F.; Puletti, N.; Quatrini, V.; Travaglini, D.; Bottalico, F.; Corona, P.; Chirici, G. Integrating terrestrial and airborne laser scanning for the assessment of single-tree attributes in Mediterranean forest stands. *Eur. J. Remote Sens.* **2018**, *51*, 795–807. [CrossRef]

forests

MDPI

Article

A Trunk Detection Method for *Camellia oleifera* Fruit Harvesting Robot Based on Improved YOLOv7

Yang Liu [1,2], Haorui Wang [1], Yinhui Liu [3], Yuanyin Luo [1], Haiying Li [1], Haifei Chen [1], Kai Liao [1] and Lijun Li [1,*]

[1] School of Mechanical and Electrical Engineering, Central South University of Forestry and Technology, Changsha 410004, China; yangliu@csuft.edu.cn (Y.L.)

[2] Hunan Automotive Engineering Vocational College, Zhuzhou 412001, China

[3] Zhongqing Changtai (Changsha) Intelligent Technology Co., Ltd., Changsha 410116, China

* Correspondence: junlili1122@163.com; Tel.: +86-13974841598

Abstract: Trunk recognition is a critical technology for *Camellia oleifera* fruit harvesting robots, as it enables accurate and efficient detection and localization of vibration or picking points in unstructured natural environments. Traditional trunk detection methods heavily rely on the visual judgment of robot operators, resulting in significant errors and incorrect vibration point identification. In this paper, we propose a new method based on an improved YOLOv7 network for *Camellia oleifera* trunk detection. Firstly, we integrate an attention mechanism into the backbone and head layers of YOLOv7, enhancing feature extraction for trunks and enabling the network to focus on relevant target objects. Secondly, we design a weighted confidence loss function based on Facol-EIoU to replace the original loss function in the improved YOLOv7 network. This modification aims to enhance the detection performance specifically for *Camellia oleifera* trunks. Finally, trunk detection experiments and comparative analyses were conducted with YOLOv3, YOLOv4, YOLOv5, YOLOv7 and improved YOLOv7 models. The experimental results demonstrate that our proposed method achieves an mAP of 89.2% , Recall Rate of 0.94, F1 score of 0.87 and Average Detection Speed of 0.018s/pic that surpass those of YOLOv3, YOLOv4, YOLOv5 and YOLOv7 models. The improved YOLOv7 model exhibits excellent trunk detection accuracy, enabling *Camellia oleifera* fruit harvesting robots to effectively detect trunks in unstructured orchards.

Keywords: trunk detection; *Camellia oleifera;* attention mechanism; CBAM; Facol-EIoU; improved YOLOv7

1. Introduction

Camellia oleifera an important oil-bearing crop, is cultivated extensively in the south of China [1]. The morphology of *Camellia oleifera* is characterized by evergreen shrubs or medium-sized trees and its leaves are elliptical, oblong, or inversely ovate in shape [2]. The harvesting of *Camellia oleifera* is the most expensive stage, demanding significant human and financial resources. The harvesting of *Camellia oleifera* fruit is particularly challenging and hazardous due to the complex growth environments, which are often found in hilly and mountainous areas [3]. Consequently, there is a pressing need to develop a mechanized and intelligent picking method that can reduce labor requirements and enhance harvesting efficiency for *Camellia oleifera*. The development of the *Camellia oleifera* fruit harvesting industry has witnessed the design and application of numerous mechanized fruit-picking machines [4]. In our previous work, we demonstrated that vibration-based methods are the most effective approach for mechanized harvesting of *Camellia oleifera*, leading us to design a *Camellia oleifera* fruit harvesting robot that utilizes vibration to pick fruits, thereby achieving mechanized harvesting [5]. However, the current robot still requires manual intervention to select the vibration points on the trunk, which limits its level of intelligence. Therefore, the rapid and accurate detection of *Camellia oleifera* trunks using computer vision methods has emerged as a crucial technology for harvesting robots and has become the primary focus of research.

Exiting methods for trunk and crop detection mainly rely on imaging process technology [6–8]. Several methods have been proposed and improved for detecting various crops, including *pineapple* [9], *grape* [10,11], *apple* [12–15], *kiwifruit* [16,17] and others. While there has been some research on the detection of *Camellia oleifera* fruit and flowers [3,8], there is limited study on the detection of *Camellia oleifera* trunks. In order to assist the *Camellia oleifera* harvesting robot in locating the vibration point, trunk detection is the crucial initial step. There are two main learning methods of crop detection, traditional machine image processing and deep learning image processing, which aim to identify the class and determine the position of targets in the image [18]. Traditional methods, such as support-vector-machine-based (SVM) [19], histogram of oriented gradient (HOG) [20], deformable parts model (DPM) [21] and others, have commonly been employed for image detection and segmentation. These methods can mitigate the impact of illumination changes on image detection. However, traditional detection methods suffer from the drawback of being time-consuming.

Deep learning has gained widespread application in the field of computer vision, particularly in target detection, where it has achieved remarkable success. Convolutional neural networks (CNN) have been applied to detect various fruits in images. Models such as AlexNet [22], VGGNet [23] and Inception [24] have been updated to enhance recognition accuracy for harvesting robots in field experiments. With advancements in learning algorithms, learning-based object detection methods can be categorized into two types: the one-stage method and two-stage method. R-CNN, Fast R-CNN and Faster R-CNN [25] use the two-stage method. While these methods improve recognition accuracy, they suffer from slow detection speed, making them unsuitable for real-time applications. SSD [26] and the YOLO series [27] are the typical one-stage object-detection methods, which divide the image into regions and promptly determine the object boundaries and classification probabilities of for each object. They offer faster detection speeds but may sacrifice a slight decrease in recognition accuracy compared to two-stage methods.

In terms of balancing detection accuracy and recognition speed, one-stage object detection algorithms are more suitable for vision systems in harvesting robots. The YOLO serial algorithms propose the use of an end-to-end neural network that simultaneously predicts bounding boxes and class probabilities. Wu et al. [8] applied the YOLOv7 model to recognize and locate the *Camellia oleifera* fruit and improve the detection accuracy. YOLOv7 [28] is the latest detector of the YOLO series, designed with a trainable bag-of-freebies. This design enables real-time detectors to significantly enhance accuracy without increasing the inference cost. It incorporates extended compound scaling techniques to effectively reduce the number of parameters and calculations in the target detector, thereby greatly improving detection speed. Therefore, we have applied YOLOv7 to detect *Camellia oleifera* trunks in our work.

The primary motivation of this study is to develop a robust and reliable approach to detect *Camellia oleifera* trunks using an improved YOLOv7 model. This improvement aims to enhance trunk detection performance in unstructured environments. The contributions of this study are summarized as follows:

(1) The creation of a dataset comprising manually annotated visible images of *Camellia oleifera* trunks and fruits captured in *Camellia oleifera* orchards.
(2) An algorithm for trunk detection was proposed based on the improved YOLOv7 model using monocular vision images. This method enables *Camellia oleifera* harvesting robots to identify and detect trunks. An attention mechanism, specifically, a CBAM (Convolutional block attention module) module, is incorporated in the backbone of YOLOv7 to enhance the detection accuracy of *Camellia oleifera* trunks.
(3) The application of the Facol-EIoU loss function to replace the loss function in the improved YOLOv7 network, further enhancing the detection of *Camellia oleifera* trunks. A comparison is made between the Precision (P), Recall(R), F1 and Detection Speed of

the improved YOLOv7 and other algorithms, including YOLOv3, YOLOv4, YOLOv5 and YOLOv7.

2. Materials and Methods

2.1. *Camellia oleifera* Trunk Image Acquisition

The images of *Camellia oleifera* trunks used in this study were obtained from the *Camellia oleifera* base located in Zhentou Village, Liuyang county, Changsha city, Hunan Province. The *Camellia oleifera* in the base were planted at an approximate distance of three meters from each other, which creates a suitable environment for the robot to perform harvesting tasks. All pictures were captured using a monocular camera (Micovision Co., Ltd., Washington, DC, USA) with a resolution of 4096 × 3072 pixels. The camera was mounted over the *Camellia oleifera* trunks at a fixed distance of 130 cm and a height of 100 cm, as shown in Figure 1, in order to capture clear trunk data. A total of 1500 pictures of *Camellia oleifera* trunks were collected under different conditions, including single trunks with front-light, single trunks with back-light, clusters of trunks with front-light and clusters of trunks with back-light, as shown in Figure 2. The resolution of the trunk images was reduced to 640 × 640 pixels to meet the requirements of the YOLOv7 model.

2.2. Data Annotation

The *Camellia oleifera* trunks in the images were individually annotated as a single class. Data labeling involves the process of adding tags or labels to raw data. In this study, the datasets were labeled using LabelImage tool and followed the Pascal Visual Object Classes format [29]. The annotation samples of labeled *Camellia oleifera* trunk images are shown in Figure 3. For each image, the image name, object classification, and position of the trunk were recorded in .xml file format. All the generated XML files were saved and converted to TXT files.

Figure 1. View of the images acquistion.

Figure 2. Images of *Camellia oleifera* trunk under different conditions. (**a**) Single trunk with front-light. (**b**) Single trunk with back-light. (**c**) Cluster of trunks with front-light. (**d**) Cluster of trunks with back-light.

Figure 3. Annotation of *Camellia oleifera* trunks in images. (**a**) Single trunk. (**b**) Cluster of trunks.

2.3. Data Augmentation

Data augmentation plays a crucial role in deep-learning-based object detection approaches as it enhances data diversity and regularizes the mode [30]. To better extract trunk features and ensure recognition accuracy while avoiding overfitting during training, various modification methods were employed for data augmentation. The data augmentation tool and OpenCV were utilized to expand the dataset through operations such as image brightness enhancement, image rotation, image mirror flipping, image random clipping, image noise increases and mosaic. Table 1 provides an explanation of six augmentation operations, and the resulting augmented images are shown in Figure 4. Following the data augmentation process, the *Camellia oleifera* trunk dataset consisted of 9000 (5 × 1500 + 1500) images. From this dataset, 6000 images were randomly selected for the training dataset, 2500 images for test dataset, and 500 images for verification set.

Table 1. Data Augmentation applied to the *Camellia oleifera* trunk original images.

Operation	Value	Description	The Percentage of Total Dataset(%)
Hue, Saturation and Value	Random	Enhance and reduce image's hue, saturation and value	10
Mirror	Random	Horizontal and vertical mirroring	10
Noise	Random	Add Gaussian noise	25
Mosaic	Random	Image Mosaic	10
Rotation	90°, 180°, and 270°	Image Rotation	10
Scale	Random	Image Scale	20

Figure 4. Image augmentation results. (**a**) Original image. (**b**) Image noise. (**c**) Image flipping. (**d**) Image mirror. (**e**) Image rotation. (**f**) Image mosaic.

2.4. *Camellia oleifera* Trunk Detection Algorithm Based on the Improved YOLOv7

2.4.1. YOLOv7

The current state-of-the-art real-time one-stage object detectors primarily rely on the YOLO series, which is capable of quickly and accurately detecting and classifying targets. This makes it well-suited for meeting the requirements of real-time detection in harvesting robots. In this study, a detection algorithm based on the latest YOLOv7 model is proposed for object detection of *Camellia oleifera* trunks. The YOLOv7 model, proposed by Alexey Bochkovskiy et al. [28], utilizes a re-parameterized approach to replace the original modules and expands the aggregation network. This approach effectively reduces the computational cost of real-time object detection.

The architecture of the YOLOv7 network consists of three main parts, the input network, backbone network and head network, as illustrated in Figure 5. Comparing with YOLOv5, the neck layer and head layer in YOLOv7 are referred to as the head network, and the use of mosaic data augmentation in YOLOv7 is particularly suitable for small object detection [31]. The backbone network is responsible for extracting features, while the head network is used for prediction. Prior to entering the backbone layer, the resolution of each image is preprocessed and resized to 640 × 640 × 3, which is the standard training image size for YOLOv7). Then images are fed into the backbone network. In the backbone layer, the Efficient Layer Aggregation Networks (ELAN) module is employed to replace the CSPDarknet53 network. This module allows the model to learn more features and enhances its robustness. DownC module is utilized for downsampling and the SPPCSP module is redesigned based on CSP [32] to enable richer combinations of gradients and reduce computation. The head network utilizes the outputs from the three layers of the backbone network to generate three different-sized feature maps. Finally, RepVGG blocks and convolutions are used to perform image detection and output the detection results [33].

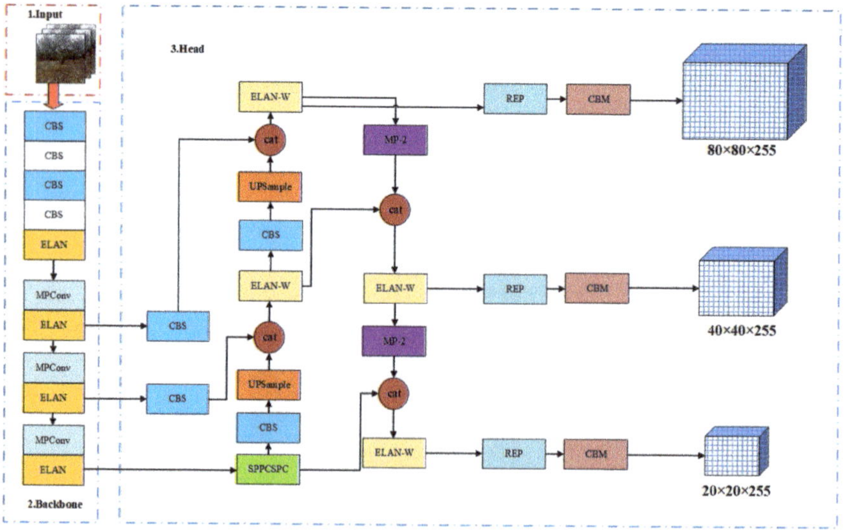

Figure 5. The architecture of YOLOv7 network.

2.4.2. Improvement of the YOLOv7 Network

The complex and irregular environment of the *Camellia oleifera* orchard poses challenges for accurate trunk identification. To assist the harvesting robot in detecting and locating the vibration point, it is imperative to improve the detection accuracy by optimizing the network based on YOLOv7.

In order to enhance feature extraction and focus on relevant target objects, attention mechanisms were incorporated into both the backbone and head layers. Attention mechanisms are data processing methods commonly applied in deep learning tasks [34]. They effectively highlight the important and relevant feature information, including channel attention mechanisms [35], spatial attention mechanisms [36], mixed attention mechanisms [37] and so on. These attention mechanisms play a significant role in directing the model's attention to specific regions of interest and improving the overall detection performance.

The CBAM (Convolutional Block Attention Module) block integrates attention maps along two separate channels and spatial dimensions in series for adaptive feature refinement, which emphasizes useful channels as well as enhancing informative local regions [38], as shown in Figure 6. In our study, we applied the CBAM attention model, which incorporates both channel attention model and spatial attention model attention operations. The channel attention model pays more attention to the foreground objects and the meaningful area within the input image, while the spatial attention model focuses on the position information and contextual information across the entire image [39].

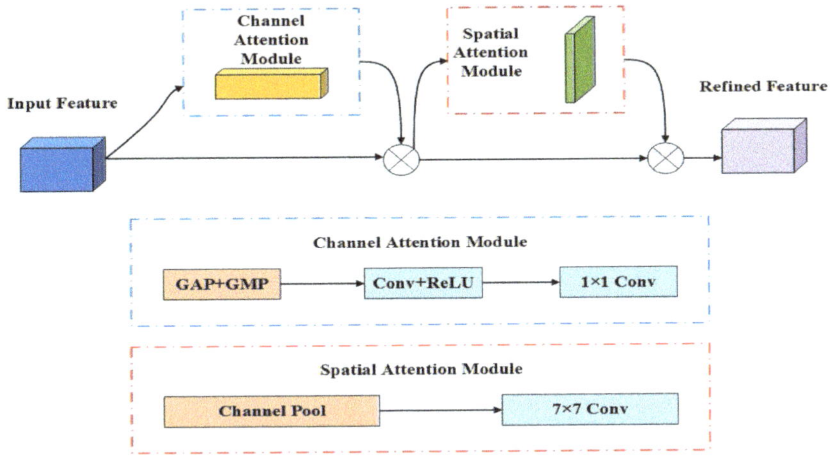

Figure 6. The main structure of CBAM attention module.

To preserve the original weights of the backbone network, the CBAM block was added at the beginning of both the backbone and head layers of the YOLOv7 model. The improved YOLOv7 network structure is shown in Figure 7. By incorporating the CBAM model, the feature extraction capabilities of the backbone network are enhanced. This method enables the network to pay more attention to detect key objects and mitigate the impact of interfering elements in complex orchard environments.

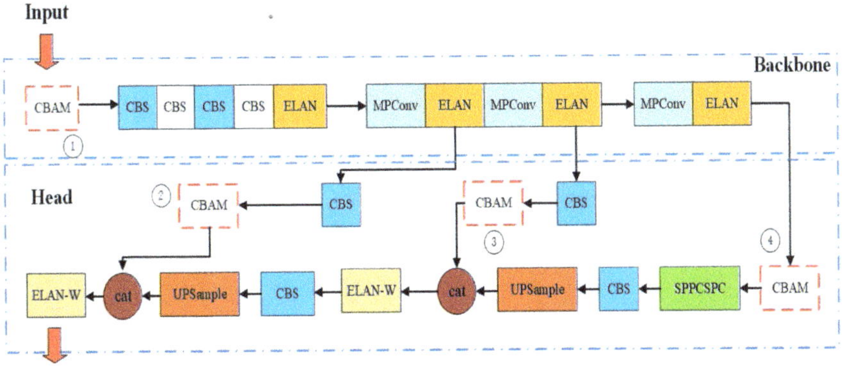

Figure 7. The structure of improved YOLOv7.

In the improved backbone network of YOLOv7, the pre-processed image with a size of 640 × 640 × 3 is fed into the backbone. The input feature map undergoes global max pooling (GMP) and global average pooling (GAP) operations, resulting in two feature maps. These two feature maps are then passed through a two-layer multilayer perceptron and activated using the ReLU activation function. The ReLU activation function generates the channel attention feature which is multiplied with the original input feature map to obtain the input feature of the spatial attention module [36]. Finally, the outputs of the channel attention module and the spatial attention module are multiplied together to obtain the output feature map of CBAM. The feature maps from the CBAM are fed into the CBS module in the original backbone, and the final predictions are generated to implement object detection by the model.

2.4.3. Loss Function

The loss function of the improved YOLOv7 in our method consists of confidence loss and location loss. The confidence loss is used to gauge the probability that the predicted bounding box contains the actual target and assess the model's performance and identify areas for improvement. Binary cross entropy was employed to calculate the confidence loss in this study; as shown in Equation (1), IoU represents the value of the IoU loss, Ci denotes the prediction confidence, and N refers to the total number of samples.

$$L_{conf} = -\frac{\sum_{i \in N}(IoU \times \ ln((C_i'))) + (1 - IoU) \times \ln(1 - (C_i'))}{N} \tag{1}$$

$$C_i' = sigmoid(C_i) \tag{2}$$

The location loss is responsible for evaluating the discrepancy between the predicted bounding box and the ground-truth box. In this study, we replaced the original location loss function of YOLOv7 with the Focal-EIoU loss [40]. Focal-EIoU loss is the combination of focal loss and EIoU loss. The focal loss [41] dynamically scales the cross-entropy loss to address the significant class imbalance between foreground and background classes during training, thereby enabling high-accuracy detection in one-stage object detection scenarios. The EIoU loss function directly measures the overlap area, central point and side length of targets, and anchors to ensure convergence speed and localization accuracy [39]. The Focal-EIoU loss, as shown in Equation (3), incorporates both the focal loss and the EIoU loss. Here b and b^{gt} represent the central points of the predicted box and ground-truth box, while w and h are the length and width of ground-truth box, and w^{gt} and h^{gt} represent the length and width of the predicted bounding box. Furthermore, ρ is the Euclidean distance between the predicted box and the ground-truth box, w^c and h^c donate the width and height of the smallest enclosing box that covers the two boxes, and W_i is a parameter that controls the degree of suppression of outliers. Facol-EIoU is designed to address the limitations of CIoU by minimizing the difference between the width and height of the target box and the anchor, resulting in faster convergence and better localization results.

$$L_{EIoU} = 1 - IoU + \frac{\rho^2(b, b^{gt})}{(w^c)^2 + (h^c)^2} + \frac{\rho^2(w, w^{gt})}{(w^c)^2} + \frac{\rho^2(h, h^{gt})}{(h^c)^2} \tag{3}$$

$$L_{loc} = \frac{\sum_{i=1}^{n} W_i \times L_{EIoU}}{\sum_{i=1}^{n} W_i} \tag{4}$$

Combined with location loss and confidence loss, the loss function of the improved YOLOv7 in this study is shown in Equation (5).

$$Loss = L_{conf} + L_{loc} \tag{5}$$

2.5. Model Training

2.5.1. Training Platforms and Parameter Settings

In this study, the target detection model was implemented using the PyTorch deep learning framework. The experiments were conducted on a system equipped with an Intel Core i7 CPU and an NVIDIA GeForce RTX2070 GPU with 12 GB of video memory, running on the Windows 10 operating system. Python 3.8 served as the programming language, and the training and testing of the model were carried out using CUDA 11.3, CUDNN 8.2, OpenCV 3.4.5 and Pycharm2016. The training configuration consisted of using the stochastic gradient descent (SGD) optimizer with an initial learning rate of 1×10^{-2}, a momentum of 0.937, and a weight decay of 0.0005.

The batch size was set to 16 and training was 600 epochs. The input size of images was set to 640 × 640 pixels. The momentum decay and weight decay were both set to 0.937 and 0.0005. During training, the initial vector was set to 0.01 and the IOU threshold was set to 0.5. Data augmentation was applied with the coefficient of hue (H), saturation (S) and lightness (V) set to 0.2, 0.5 and 0.4. Throughout the training process, the training data, loss values and model weights were saved at each epoch. The performance of the model was evaluated using the test set. Six detection algorithms were trained and compared in this study: YOLOv3, YOLOv4, YOLOv5, YOLOv7 and the improved YOLOv7. The training process used the Adam optimizer with 600 training epochs. To prevent overfitting, the training would automatically stop if there was no improvement in accuracy over the last 50 training epochs. The training parameters used in the experiments are summarized in Table 2.

Table 2. Training Parameter.

Parameter	Value	Parameter	Value
Ephochs	600	Batch Size	16
Learning Rate	0.01	Weight Decay	0.0005
Image Size	640 × 640	Momentum	0.937

2.5.2. Evaluation Indicators of the Model

In this paper, the performance of the detection model was evaluated using Precision (*P*), Recall (*R*), Mean Average Precision (*mAP*) and *F1* [41] score. A higher mAP score indicates better performance. The mAP50 represents the mAP at an IoU threshold of 0.5, while mAP75 refers to the mAP at an IoU threshold of 0.75. The overall mAP is calculated as the average of mAP values across IoU thresholds ranging from 0.5 to 0.95, with an interval of 0.05. The evaluation metrics rely on the computation of true positives (*TP*), false positives (*FP*), and false negatives (*FN*), as shown in Equation (6). Precision measures the ratio of correctly detected targets to the total number of detected targets and serves as an intuitive index for detection evaluation. The (*mAP*) Recall and F1 score were introduced to provide a comprehensive evaluation.

$$P = \frac{TP}{TP + FP} \tag{6}$$

$$R = \frac{TP}{TP + FN} \tag{7}$$

$$AP = \int_0^1 P(R)dR \tag{8}$$

$$mAP = \frac{1}{n}\sum_{i=1}^{n} AP_i \tag{9}$$

$$F1 = 2 \times \frac{P \times R}{P + R} \tag{10}$$

In this study, the detection task focused on a single category, which is the *Camellia oleifera* trunk. Therefore, the *AP* and *mAP* values are equal since there is only one category involved in the detection process. *TP* is the number of *Camellia oleifera* trunks that are detected correctly. *FP* is the number of negative objects that were detected as the *Camellia oleifera* trunk, and *FN* is the number of positive samples incorrectly detected.

3. Results

In order to evaluate the performance of our model in *Camellia oleifera* trunk detection, this section provides a detailed description of the experiments conducted, including the training process, experimental results and ablation experimental results. A comparison is made between different detection algorithms to assess the contributions of the methods proposed in this paper. The test set consists of 2500 images, categorized as follows: 600 images of single trunks with front-light, 600 images of single trunks with back-light, 600 images of clusters of trunks with front-light, and 700 images of clusters of trunks with back-light. Our model is compared with other networks such as YOLOv3, YOLOv4, YOLOv5 and the original YOLOv7, using the images from the test set. The workflow of the proposed study is shown in Figure 8.

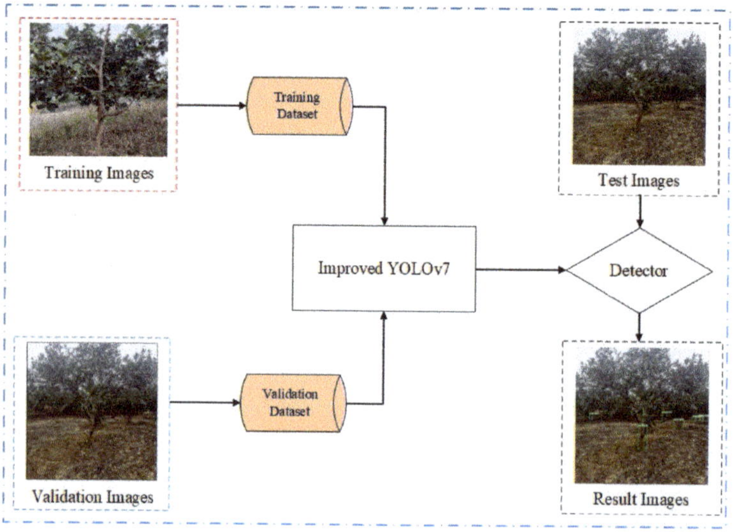

Figure 8. Workflow of the proposed study.

3.1. Training Results

The loss curves of the training and validation during the training process are shown in Figure 9. The curves demonstrate a rapid decrease in loss value during the initial 150 epochs, followed by a steady decline beyond 600 epochs. Notably, there was no evidence of overfitting, and the results were satisfactory. Based on these observations, the model trained for 600 epochs was deemed suitable for detecting *Camellia oleifera* trunks, thus serving the purpose of the *Camellia oleifera* fruit harvesting robot.

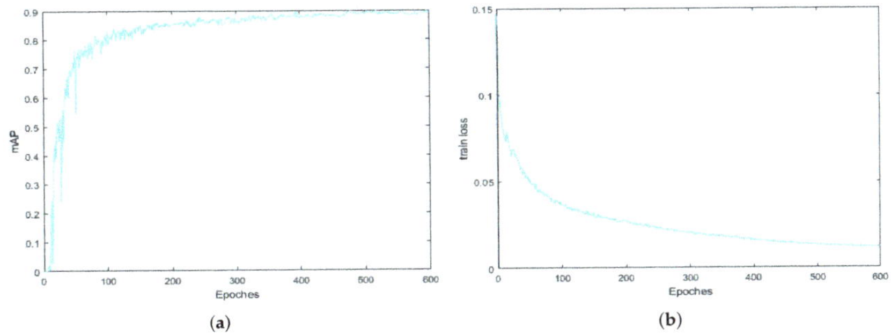

Figure 9. Training Results. (**a**) mAP curve. (**b**) Loss curve.

3.2. Ablation Experiments

Ablation experiments were conducted to assess the impact of the CBAM and Focal-EIoU loss modules on the detection performance. Table 3 presents the results of different combinations and their respective detection metrics. Four models were compared: YOLOv7 without using CBAM and Focal-EIoU loss function, YOLOv7 improved with CBAM only, YOLOv7 improved with Focal-EIoU loss function only, and YOLOv7 improved with CBAM and Focal-EIoU loss function. The performance metrics of these models were evaluated and compared.

Table 3. Results of ablation experiments.

CBAM	Focal-EIoU Loss Function	mAP(%)	Recall	F1
		84.2	0.89	0.84
√		86	0.92	0.84
	√	85.6	0.9	0.84
√	√	89.2	0.94	0.87

Results in Table 3 demonstrate that the accuracy of the models varies depending on the combination of the CBAM and Focal-EIoU loss function. In terms of mAP metric, the introduction of CBAM improves the YOLOv7 model by 2.1%. Additionally, the introduction of the Focal-EIoU loss function improves the YOLOv7 model by 1.7% , which means the CBAM enhances the model to focus on detection targets and the Focal-EIoU loss function improves the accuracy of the detection. When both CBAM and Focal-EIoU loss function are applied in YOLOv7, The mAP of model improves by 5.9% and the detection results achieve the optimal detection performance.

3.3. Experiment Results

To verify the effectiveness of the improved detection network model for *Camellia oleifera* trunks, test datasets consisting of trunk images were applied in the experiments. The detection results for single trunks with front-light, single trunks with back-light, clusters of trunks with front-light and clusters of trunks with back-light were analyzed. As shown in Figure 10, the majority of trunks in the orchard were identified even in the presence of occlusions and different light conditions. The detection results obtained from our model were found to be excellent.

(a) (b) (c) (d)

Figure 10. Experimental results of trunk detection under different conditions. (**a**) Single trunk with front-light. (**b**) Single trunk with back-light. (**c**) Cluster of trunks with front-light. (**d**) Cluster of trunks with back-light.

3.4. Comparison of Detection Algorithms

To demonstrate the recognition performance of the improved YOLOv7 network for *Camellia oleifera* trunks, we compared models proposed in our study with the YOLOv3, YOLOv4, YOLOv5 and YOLOv7 networks using the test dataset. The metrics used for comparison were mAP, Recall, F1 and Average Detection Speed. All object detection algorithms were trained using the same dataset as described in this study.

We compared the effectiveness of detecting *Camellia oleifera* trunks with different models, and the results are presented in Table 4 and Figure 11. It can be observed that our model achieves the highest mAP, which is 1.05 times higher than YOLOv3, 1.1 times higher than YOLOv4, 1.06 times higher than YOLOv5 and 1.05 times higher than YOLOv7. The Average Detection Speed of our model is 0.018 s/pic, which is 0.5 times, 0.4 times, 0.43 times and 0.72 times faster than those of the YOLOv3, YOLOv4, YOLOv5 and YOLOv7 networks. These results indicate our model performed better in detecting *Camellia oleifera* trunks and could meet the requirements of *Camellia oleifera* fruit harvesting robots for trunk recognition.

Table 4. Comparison of different models of *Camellia oleifera* trunks detection.

Object Detection Networks	mAP (%)	Recall	F1	Average Detection Speed (s/pic)
YOLOv3	84.9	0.83	0.84	0.032
YOLOv4	80.9	0.82	0.81	0.045
YOLOv5	83.9	0.88	0.83	0.041
YOLOv7	84.2	0.89	0.84	0.025
Our model	89.2	0.94	0.87	0.018

The comparison results demonstrate that our model not only ensures high detection accuracy but also meets the detection speed requirements of the harvesting robot. The improved YOLOv7 proposed in this study achieves excellent recognition results under different light conditions. Our model exhibits faster detection speed compared to YOLOv3, YOLOv4, YOLOv5 and YOLOv7. As shown in Figures 12 and 13, our model is capable of recognizing more trunks that are out of focus in the camera, indicating its robustness. Particularly, under back-light conditions, YOLOv3, YOLOv4 and YOLOv5 exhibit insensitivity in detecting small objects, resulting in the failure to detect small trunks on the left side of the image (as shown in Figure 13b,d,f,h,j). This is likely due to the lack of sufficient color features of *Camellia oleifera* trunks in back-light conditions. The improved YOLOv7 and Focal-EIoU loss function generate better trunk features, effectively enhancing the performance and detection speed of our model.

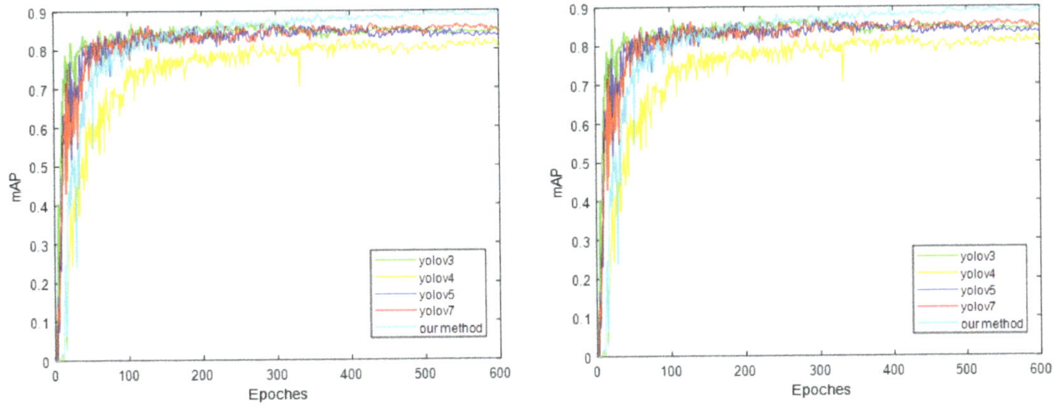

(a) (b)

Figure 11. Comparison of different detection models. (**a**) mAP curve. (**b**) Loss curve.

(a) (b) (c) (d) (e)

(f) (g) (h) (i) (j)

Figure 12. Comparison results of *Camellia oleifera* trunk detection with front-light by different models. (**a**)YOLOv3 (Single trunk). (**b**) YOLOv4 (Single trunk). (**c**) YOLOv5 (Single trunk). (**d**) YOLOv7 (Single trunk). (**e**) Our model (Single trunk). (**f**) YOLOv3 (Cluster of trunks). (**g**) YOLOv4 (Cluster of trunks). (**h**) YOLOv5 (Cluster of trunks). (**i**) YOLOv7 (Cluster of trunks). (**j**) Our model (Cluster of trunks).

Figure 13. Comparison results of *Camellia oleifera* trunks detection with back-light by different models. (**a**) YOLOv3 (Single trunk). (**b**)YOLOv4 (Single trunk). (**c**) YOLOv5 (Single trunk). (**d**) YOLOv7 (Single trunk). (**e**) Our model (Single trunk). (**f**) YOLOv3 (Cluster of trunks). (**g**) YOLOv4 (Cluster of trunks). (**h**) YOLOv5 (Cluster of trunks). (**i**) YOLOv7 (Cluster of trunks). (**j**) Our model (Cluster of trunks).

4. Discussion

In a forest crop harvesting application, trunk detection can help the harvesting robot to locate the trunk. However, due to the complex environment of orchards, the accuracy of tree trunk detection is affected by different lighting and shadows. Our research contributes to trunk detection with a comparison of different detection methods. Su, F et al. [14] proposed a tree trunk and obstacle detection method in a semi-structured apple orchard environment based on improved YOLOv5s, without considering different light conditions. Yang et al. [42] proposed a method of Y3TM to detect trunks, streetlight and telephone poles in the forest. The presented research demonstrates that the deep-learning-based detection algorithm has excellent performance in an unconstructed environment and provides the basis for harvesting robots.

Compared with relevant studies, the *Camellia oleifera* trunk detection method based on an improved YOLOv7 proposed in this paper has better recognition accuracy and faster speed. The application of the attention mechanism CBAM in the YOLOv7 backbone and the replacement of the original loss function with the Focal-EIoU loss function have significantly contributed to enhancing the intelligence of the *Camellia oleifera* fruit harvesting robot, enabling it to determine vibration points using the vision sensing system. The training results demonstrate that the improved YOLOv7 model can effectively prevent overfitting and achieve faster convergence during the training process. The ablation experiment results indicate that our model outperforms the other recognition methods in terms of both detection speed and accuracy for *Camellia oleifera* trunk recognition. The results of comparison experiments show that our model can achieve the best detection results over the mainstream detection algorithms under different light conditions. In future research, the improved YOLOv7 model can also applied to detect *Camellia oleifera* fruit,

Camellia oleifera flowers and other types of forest crops. Thus, the improved YOLOv7 model presented in this paper has the potential to not only contribute to the fields of *Camellia oleifera* trunks detection, but also advanced crop detection in smart forestry applications.

5. Conclusions

The detection and recognition of *Camellia oleifera* trunks in unstructured environments is a crucial technology for the advancement of *Camellia oleifera* fruit harvesting robots. In this paper, we propose a highly accurate and efficient method for detecting and recognizing*Camellia oleifera* trunks. Our approach utilizes an improved YOLOv7 network as the detection model. By integrating the CBAM attention mechanism into the YOLOv7 backbone, we enhance the model's detection accuracy. Additionally, we adopt the Facol-EIoU loss function, which replaces the original YOLOv7 location loss function, improving the model's robustness. Various well-known detection networks are compared and analyzed for their performance in *Camellia oleifera* trunk detection. Experimental results demonstrate that our proposed detection model outperforms other methods in terms of accuracy and speed when applied to the *Camellia oleifera* trunk dataset. The development of intelligent *Camellia oleifera* fruit harvesting robots holds profound significance for the *Camellia oleifera* industry. This study offers an effective vision system solution for *Camellia oleifera* fruit harvesting robots. In the future, we will continue to advance object detection technology in harvesting robots and update the detection method to enable intelligent harvesting.

Author Contributions: Y.L. (Yang Liu) and L.L. proposed the trunk detection method and designed the research; H.W. and Y.L. (Yang Liu) performed the experiments; H.W., Y.L. (Yuanyin Luo), H.L., Y.L. (Yinhui Liu), H.C. and K.L. collected and analyzed the experiment data; Y.L. (Yang Liu) and H.W. wrote the manuscript. All authors have read and agreed to the published version of the manuscript.

Funding: This research was supported by National Key Research and Development Program (2022YFD2202103), Scientific Innovation Fund for Post-graduates of Central South University of Forestry and Technology (2023CX01025).

Data Availability Statement: The data presented in this study are available on request from the corresponding author.

Conflicts of Interest: The authors declare that there are no conflict of interest.

References

1. Deng, Q.; Li, J.; Gao, C.; Cheng, J.; Deng, X.; Jiang, D.; Li, L.; Yan, P. New perspective for evaluating the main *Camellia oleifera* cultivars in China. *Sci. Rep.* **2020**, *10*, 20676. [CrossRef] [PubMed]
2. Gao, S.; Wang, B.; Liu, F.; Zhao, J.; Yuan, J.; Xiao, S.; Masabni, J.; Zou, F.; Yuan, D. Variation in Fruit Morphology and Seed Oil Fatty Acid Composition of *Camellia oleifera* Collected from Diverse Regions in Southern China. *Horticulturae* **2022**, *8*, 818. [CrossRef]
3. Zhou, Y.; Tang, Y.; Zou, X.; Wu, M.; Tang, W.; Meng, F.; Zhang, Y.; Kang, H. Adaptive Active Positioning of *Camellia oleifera* Fruit Picking Points: Classical Image Processing and YOLOv7 Fusion Algorithm. *Appl. Sci.* **2022**, *12*, 12959. [CrossRef]
4. Wu, D.; Zhao, E.; Fang, D.; Jiang, S.; Wu, C.; Wang, W.; Wang, R. Determination of Vibration Picking Parameters of *Camellia oleifera* Fruit Based on Acceleration and Strain Response of Branches. *Agriculture* **2022**, *12*, 1222. [CrossRef]
5. Zicheng, G.; Lijun, L.; Penghui, W.; et al. Design and Analysis of Picking Head of Camellia Fruit Picking Robot. *J. Cent. South Univ. For. Technol. Pap.* **2016**, *36*, 114–118.
6. Liang, C.; Xiong, J.; Zheng, Z.; Zhong, Z.; Li, Z.; Chen, S.; Yang, Z. A visual detection method for nighttime litchi fruits and fruiting stems. *Comput. Electron. Agric.* **2020**, *169*, 105192. [CrossRef]
7. Mirhaji, H.; Soleymani, M.; Asakereh, A.; Mehdizadeh, S.A. Fruit detection and load estimation of an orange orchard using the YOLO models through simple approaches in different imaging and illumination conditions. *Comput. Electron. Agric.* **2021**, *191*, 106533. [CrossRef]
8. Wu, D.; Jiang, S.; Zhao, E.; Liu, Y.; Zhu, H.; Wang, W.; Wang, R. Detection of *Camellia oleifera* Fruit in Complex Scenes by Using YOLOv7 and Data Augmentation. *Appl. Sci.* **2022**, *12*, 11318. [CrossRef]
9. Lai, Y.; Ma, R.; Chen, Y.; Wan, T.; Jiao, R.; He, H. A Pineapple Target Detection Method in a Field Environment Based on Improved YOLOv7. *Appl. Sci.* **2023**, *13*, 2691. [CrossRef]
10. Jin, Y.; Yu, C.; Yin, J.; Yang, S.X. Detection method for table grape ears and stems based on a far-close-range combined vision system and hand-eye-coordinated picking test. *Comput. Electron. Agric.* **2022**, *202*, 107364. [CrossRef]

11. Qiu, C.; Tian, G.; Zhao, J.; Liu, Q.; Xie, S.; Zheng, K. Grape Maturity Detection and Visual Pre-Positioning Based on Improved YOLOv4. *Electronics* **2022**, *11*, 2677. [CrossRef]
12. Ji, W.; Pan, Y.; Xu, B.; Wang, J. A real-time Apple targets detection method for picking robot based on ShufflenetV2-YOLOX *Agriculture* **2022**, *12*, 856. [CrossRef]
13. Zhang, C.; Kang, F.; Wang, Y. An Improved Apple Object Detection Method Based on Lightweight YOLOv4 in Complex Backgrounds. *Remote Sens.* **2022**, *14*, 4150. [CrossRef]
14. Su, F.; Zhao, Y.; Shi, Y.; Zhao, D.; Wang, G.; Yan, Y.; Zu, L.; Chang, S. Tree Trunk and Obstacle Detection in Apple Orchard Based on Improved YOLOv5s Model. *Agronomy* **2022**, *12*, 2427. [CrossRef]
15. Yan, B.; Fan, P.; Lei, X.; Liu, Z.; Yang, F. A real-time apple targets detection method for picking robot based on improved YOLOv5 *Remote Sens.* **2021**, *13*, 1619. [CrossRef]
16. Xia, Y.; Nguyen, M.; Yan, W.Q. A Real-Time Kiwifruit Detection Based on Improved YOLOv7. In *Image and Vision Computing, Proceedings of the 37th International Conference, IVCNZ 2022, Auckland, New Zealand, 24–25 November 2022, Revised Selected Papers*; Springer: Berlin/Heidelberg, Germany, 2023; pp. 48–61.
17. Suo, R.; Gao, F.; Zhou, Z.; Fu, L.; Song, Z.; Dhupia, J.; Li, R.; Cui, Y. Improved multi-classes kiwifruit detection in orchard to avoid collisions during robotic picking. *Comput. Electron. Agric.* **2021**, *182*, 106052. [CrossRef]
18. Wang, K.; Meng, Z.; Wu, Z. Deep Learning-Based Ground Target Detection and Tracking for Aerial Photography from UAVs. *Appl. Sci.* **2021**, *11*, 8434. [CrossRef]
19. Lin, G.; Tang, Y.; Zou, X.; Li, J.; Xiong, J. In-field citrus detection and localisation based on RGB-D image analysis. *Biosyst. Eng.* **2019**, *186*, 34–44. [CrossRef]
20. Pang, Y.; Yuan, Y.; Li, X.; Pan, J. Efficient HOG human detection. *Signal Process.* **2011**, *91*, 773–781. [CrossRef]
21. Devietti, J.; Lucia, B.; Ceze, L.; Oskin, M. DMP: Deterministic shared memory multiprocessing. In Proceedings of the 14th international Conference on Architectural Support for Programming Languages and Operating Systems, Washington, DC, USA, 7–11 March 2009; pp. 85–96.
22. Alom, M.Z.; Taha, T.M.; Yakopcic, C.; Westberg, S.; Sidike, P.; Nasrin, M.S.; Van Esesn, B.C.; Awwal, A.A.S.; Asari, V.K. The history began from alexnet: A comprehensive survey on deep learning approaches. *arXiv* **2018**, arXiv:1803.01164.
23. Wang, L.; Guo, S.; Huang, W.; Qiao, Y. Places205-vggnet models for scene recognition. *arXiv* **2015**, arXiv:1508.01667.
24. Szegedy, C.; Ioffe, S.; Vanhoucke, V.; Alemi, A. Inception-v4, inception-resnet and the impact of residual connections on learning. In Proceedings of the AAAI Conference on Artificial Intelligence, San Francisco, CA, USA, 4–9 February 2017; Volume 31.
25. Williams, H.A.; Jones, M.H.; Nejati, M.; Seabright, M.J.; Bell, J.; Penhall, N.D.; Barnett, J.J.; Duke, M.D.; Scarfe, A.J.; Ahn, H.S.; et al. Robotic kiwifruit harvesting using machine vision, convolutional neural networks, and robotic arms. *Biosyst. Eng.* **2019**, *181*, 140–156. [CrossRef]
26. Zheng, Y.; Izzat, I.H.; Ziaee, S. GFD-SSD: Gated fusion double SSD for multispectral pedestrian detection. *arXiv* **2019**, arXiv:1903.06999.
27. Qiu, Y.; Lu, Y.; Wang, Y.; Jiang, H. IDOD-YOLOV7: Image-Dehazing YOLOV7 for Object Detection in Low-Light Foggy Traffic Environments. *Sensors* **2023**, *23*, 1347. [CrossRef] [PubMed]
28. Wang, C.Y.; Bochkovskiy, A.; Liao, H.Y.M. YOLOv7: Trainable bag-of-freebies sets new state-of-the-art for real-time object detectors. In Proceedings of the IEEE/CVF Conference on Computer Vision and Pattern Recognition, Vancouver, BC, Canada, 18–22 June 2023; pp. 7464–7475.
29. Everingham, M.; Van Gool, L.; Williams, C.K.; Winn, J.; Zisserman, A. The pascal visual object classes (voc) challenge. *Int. J. Comput. Vis.* **2010**, *88*, 303–338. [CrossRef]
30. Dai, J.; Qi, H.; Xiong, Y.; Li, Y.; Zhang, G.; Hu, H.; Wei, Y. Deformable convolutional networks. In Proceedings of the IEEE International Conference on Computer Vision, Venice, Italy, 22–29 October 2017; pp. 764–773.
31. Jiang, K.; Xie, T.; Yan, R.; Wen, X.; Li, D.; Jiang, H.; Jiang, N.; Feng, L.; Duan, X.; Wang, J. An Attention Mechanism-Improved YOLOv7 Object Detection Algorithm for Hemp Duck Count Estimation. *Agriculture* **2022**, *12*, 1659. [CrossRef]
32. Wang, C.Y.; Liao, H.Y.M.; Wu, Y.H.; Chen, P.Y.; Hsieh, J.W.; Yeh, I.H. CSPNet: A new backbone that can enhance learning capability of CNN. In Proceedings of the IEEE/CVF Conference on Computer Vision and Pattern Recognition Workshops, Seattle, WA, USA, 14–19 June 2020; pp. 390–391.
33. Ding, X.; Zhang, X.; Ma, N.; Han, J.; Ding, G.; Sun, J. Repvgg: Making vgg-style convnets great again. In Proceedings of the IEEE/CVF Conference on Computer Vision and Pattern Recognition, Virtual, 19–25 June 2021; pp. 13733–13742.
34. Niu, Z.; Zhong, G.; Yu, H. A review on the attention mechanism of deep learning. *Neurocomputing* **2021**, *452*, 48–62. [CrossRef]
35. Li, C.; Zhao, G.; Gu, D.; Wang, Z. Improved Lightweight YOLOv5 Using Attention Mechanism for Satellite Components Recognition. *IEEE Sensors J.* **2022**, *23*, 514–526. [CrossRef]
36. Ma, X.; Guo, J.; Sansom, A.; McGuire, M.; Kalaani, A.; Chen, Q.; Tang, S.; Yang, Q.; Fu, S. Spatial pyramid attention for deep convolutional neural networks. *IEEE Trans. Multimed.* **2021**, *23*, 3048–3058. [CrossRef]
37. Jiang, G.; Jiang, X.; Fang, Z.; Chen, S. An efficient attention module for 3d convolutional neural networks in action recognition. *Appl. Intell.* **2021**, *51*, 7043–7057. [CrossRef]
38. Vaswani, A.; Shazeer, N.; Parmar, N.; Uszkoreit, J.; Jones, L.; Gomez, A.N.; Kaiser, Ł.; Polosukhin, I. Attention is all you need. *Adv. Neural Inf. Process. Syst.* **2017**, *30*.

39. Ying, X.; Wang, Y.; Wang, L.; Sheng, W.; An, W.; Guo, Y. A stereo attention module for stereo image super-resolution. *IEEE Signal Process. Lett.* **2020**, *27*, 496–500. [CrossRef]
40. Zhang, Y.F.; Ren, W.; Zhang, Z.; Jia, Z.; Wang, L.; Tan, T. Focal and efficient IOU loss for accurate bounding box regression. *Neurocomputing* **2022**, *506*, 146–157. [CrossRef]
41. Lin, T.Y.; Goyal, P.; Girshick, R.; He, K.; Dollár, P. Focal loss for dense object detection. In Proceedings of the IEEE international Conference on Computer Vision, Venice, Italy, 22–29 October 2017; pp. 2980–2988.
42. Yang, T.T.; Zhou, S.Y.; Ai-Jun, X.U. Rapid Image Detection of Tree Trunks Using a Convolutional Neural Network and Transfer Learning. *IAENG Int. J. Comput. Sci.* **2021**, *48*, 257–265.

forests

MDPI

Article

A Novel Strategy for Constructing Large-Scale Forest Scene: Integrating Forest Hierarchical Models and Tree Growth Models to Improve the Efficiency and Stability of Forest Polymorphism Simulation

Kexin Lei [1,2], Huaiqing Zhang [1,2,*], Hanqing Qiu [1,2], Tingdong Yang [1,2], Yang Liu [1,2], Jing Zhang [1,2], Xingtao Hu [1,2] and Zeyu Cui [1,2]

1 Institute of Forest Resource Information Techniques, Chinese Academy of Forestry, Beijing 100091, China; leikx@ifrit.ac.cn (K.L.); qiuhanqing123@ifrit.ac.cn (H.Q.); yangtd@ifrit.ac.cn (T.Y.); liuyang@ifrit.ac.cn (Y.L.); zhangjing@ifrit.ac.cn (J.Z.); 20010090319@gznu.edu.cn (X.H.); cuizeyu@ifrit.ac.cn (Z.C.)

2 Key Laboratory of Forestry Remote Sensing and Information System, The National Forestry and Grassland Administration, Beijing 100091, China

* Correspondence: zhang@ifrit.ac.cn; Tel.: +86-1-365-116-6672

Abstract: Modeling large-scale scenarios of diversity in real forests is a hot topic in forestry research. At present, there is a common problem of simple and poor model scalability in large-scale forest scenes. Forest growth is often carried out using a holistic scaling approach, which does not reflect the diversity of trees in nature. To solve this problem, we propose a method for constructing large-scale forest scenes based on forest hierarchical models, which can improve the dynamic visual effect of large-scale forest landscape polymorphism. In this study, we constructed tree hierarchical models of corresponding sizes using the detail attribute data of 29 subplots in the Shanxia Experimental Forest Farm in Jiangxi Province. The growth values of trees of different ages were calculated according to the hierarchical growth model of trees, and the growth dynamic simulation of large-scale forest scenes constructed by the integrated model and hierarchical model was carried out using three-dimensional visualization technology. The results indicated that the runtime frame rate of the scene constructed by the hierarchical model was 30.63 fps and the frame rate after growth was 29.68 fps, which met the operational requirements. Compared with the traditional integrated model, the fluctuation value of the frame rate of the hierarchical model was 0.036 less than that of the integrated model, and the scene ran stably. The positive feedback rate of personnel evaluation reached 95%. In this study, the main conclusion is that our proposed method achieves polymorphism in large-scale forest scene construction and ensures the stability of large-scale scene operation.

Keywords: large-scale forest scenes; stand structure; forest hierarchical model; dynamic visualization simulation of growth

1. Introduction

Forest ecosystems are highly complex and dynamic assemblages of biotic and abiotic webs that sustain human survival and development [1,2]; they play a crucial role in biodiversity conservation [3], wildlife habitat protection [4], forest products, and entertainment [5]. Visualization technology for tree growth utilizes three-dimensional simulation, enabling forest managers to intuitively perceive the individual tree growth states and forest resource distribution they are studying, which contributes to forest planning, forest decision-making, and forest resource management [5–8]. Therefore, large-scale three-dimensional forest simulation is vital for sustainable forest resource management. With the development of computer software, hardware, and graphic acceleration algorithms, many scholars at home and abroad have conducted relevant research on large-scale forest scene construction.

At present, the commonly used forest scene construction methods primarily include Polygon-Based Rendering [9,10], Image-Based Rendering [11,12], and Polygon and Image Hybrid Rendering [13]. The polygon-Based modeling method could fully exhibit individual tree polymorphisms. Therefore, heuristic multiresolution-based models [14], viewpoint multiresolution-based models [15], and other models were successively developed by grouping tree models, constructing the Level of Details (LOD) of different tree parts (trunk LOD, branch LOD, and leaf LOD) [16], and replacing tree geometric modeling that exceeded a certain distance by impostor texture replacement technology, which is considered an effective method for accelerating tree rendering [17]. Nowadays, some scholars use complex polygons to represent large objects and simple quadrilateral representations for small objects to accomplish large-scale scene rendering. Although geometric-based models have abundant structure details, the rendering speed is relatively slow and difficult in the application of large-scale forest scenes [18]. Due to the inability of geometric-based methods to meet people's needs, many scholars have focused on image-based methods. In contrast to simple geometric elements, images can present different aspects of complicated objects. Therefore, in the construction of large-scale forest scenes, image-based methods are the preferred selection technology. Some scholars have applied highly complex tree photos as a texture to a simple polygon to present the tree model information [19] and construct a forest landscape containing 100,000 trees using virtual reality modeling language (VRML) [20]. Volume texture was used to implement the individual tree visualization expression [21]. The trees generated offer multiple advantages; for example, they are more physically realistic and less time-consuming to create. However, they are essentially two-dimensional static images; therefore, they cannot fully exhibit the three-dimensional geometric structure of real trees, tree growth, and competition characteristics [22,23]. Although the speed satisfies the needs of large-scale rendering, the entire forest is relatively simple and cannot reflect the differences between individual trees. Geometric-based simplification methods can effectively simplify tree models, but the simplified models still contain a large number of triangular patches. Image-based simplification methods can greatly simplify tree models, but the close-up view is not ideal. Therefore, researchers have made attempts to mix geometric- and image-based methods in the simplification process of tree models to find a balance between magnitude and quality simplification [24]. SpeedTree software 9.0.0 (https://store.speedtree.com/, accessed on 20 May 2023) uses triangular patches to express the main branch structure of trees and a series of billboards to represent clusters of neighboring twigs and leaves. By introducing detailed textures, it generates a tree model that has a geometric structure and is highly physically realistic [13,25]. The tree models constructed with SpeedTree software are realistic, but large-scale forest scenes typically consist of tens of thousands of trees with variable shapes and morphologies. SpeedTree software requires numerous tree structural parameters to construct models with different shapes, which is time-consuming and laborious [26].

In addition, forest growth visualization simulation is an essential component of forest management and has great guiding significance for forestry production practice. Trees usually have diverse morphological characteristics [27,28]. In order to simulate tree growth status and spatial structure and predict the forest development succession more intuitively, vividly, and realistically, virtual forest construction methods based on stand growth laws have emerged [29]. Among them, some methods use individual tree growth models to calculate the growth parameters of forest morphological structures (diameter at breast height, tree height, and crown width) and use three-dimensional programming tools to accomplish stand growth visualization [30,31]. Some studies have developed distance-independent tree spatial growth models, which can better estimate stand development [32,33]. The distance-dependent models have the advantage that, in contrast to distance-independent models, their competition indices are optimized [34,35]. Some researchers have developed distance-dependent growth models from irregularly measured sample plot data; these allow model-based analyses of spatial problems such as systematic row thinning and different tree planting patterns [35–37]. By combining tree growth constraints, they also use three-dimensional visualization and artificial intelligence (AI) algorithms to simulate the growth process of trees [7,38,39]. Although tree growth simulation has been achieved, many existing studies use discrete static simulation methods with overall scaling, but the polymorphism in the tree growth process cannot be depicted fully [40]. Existing studies on tree growth simulation generally follow the same pattern. In forestry, a tree growth model that conforms to the tree growth laws can be established on the basis of field-measured individual tree scale data. The growth parameters of various parts of individual trees can be calculated on the basis of the growth model, and three-dimensional visualization technology can be used to achieve dynamic visualization simulation of tree growth [41–43]. However, for large-scale forest scenes, the best way to model different characteristics of trees and dynamically simulate the polymorphism of the tree growth process problem remains unclear.

In summary, recent studies on forest scene dynamic visualization simulation mainly have focused on the construction and rendering optimization of small-scale stand scenes or middle-scale forest landscapes. However, fewer studies have been carried out on the construction of large-scale forest scenes combined with field-measured data, so large-scale forest scenes often exhibit the characteristics of a single form. Therefore, in order to address these issues, this study selected Shanxia Experimental Forest Farm as the study area to develop a large-scale forest scene with a tree hierarchical growth model based on data from 29 subplots. This study aimed to (1) develop a method for constructing a diverse large-scale forest scene using a tree hierarchical growth model; (2) conduct a comparative analysis on the construction of scenes with 100, 1000, 10,000, and 100,000 trees, respectively, to quantify the advantages of this research method compared with previous research methods; and (3) conduct a large-scale forest growth simulation using Unreal Engine 4 and tree hierarchical models and optimize the scene's rendering efficiency.

2. Materials and Methods

2.1. Study Areas

The study area was located in Shanxia Experimental Forest Farm (27°40′–27°45′ N, 114°35′–114°40′ E) in the southwest of Fenyi County, Xinyu City, Jiangxi Province, China (Figure 1). The forest farm has rich tree species resources, with 115,000 cubic meters of standing volume of living trees. The forest coverage is 95.6%. There are 115 precious broad-leaved tree species preserved, making it a relatively complete experimental forest in a subtropical region. The dominant tree species are *Cunninghamia lanceolata*, *Schima superba*, and *Machilus pauhoi*. It belongs to the subtropical monsoon climate with abundant rainfall and sunshine. The frost-free period is 270 days. It has abundant rainfall, with an annual precipitation of 1600 mm. The mean annual temperature is 15~17 °C. (Figure 1).

2.2. Data Collection

In May 2022, a sampling survey was conducted on 29 subplots in Shanyia Experimental Forest Farm, Fenyi County, and Jiangxi Province. Subplots are the fundamental management units within the forest farm and serve as the basic units for forest resource inventory, statistical calculations, and management. They are delineated according to ownership, land type, tree species, dominant tree species, and other factors. Considering the rich species diversity and high biodiversity of subplot 1, which makes it representative, all the trees in subplot 1 were measured. For other subplots with relatively homogeneous species composition, 2–3 sample plots with a size of 20 × 30, selected on the basis of the abundance of tree species and the good spatial structure of the forest stand, were used for the sampling survey. Tree species were identified by tree species experts. The relative coordinates (x, y, z) of the roots of each tree were measured using Topcon Total station made in Japan. The diameter at breast height (DBH) of each tree was measured using a DBH tape measure. A Leica D210 laser rangefinder made in Switzerland was used to measure the tree height (H) and the height under the living branches (UBH) of the trees. The crown width (CW) of the trees was measured using a tape measure. The crown height (CH) was obtained by subtracting the height under the branches (UBH) from the tree height (H). The remaining 28 subplots were sampled and surveyed to obtain the main tree species, number of trees, average DBH, average H, average CW, average UBH, and average CH within the subplots. To build three-dimensional models of the trees, we used a Sony digital camera made in Japan to photograph the bark and leaf texture of the trees, shrubs, and herbs. The attribute data of 29 subplots are shown in Table 1.

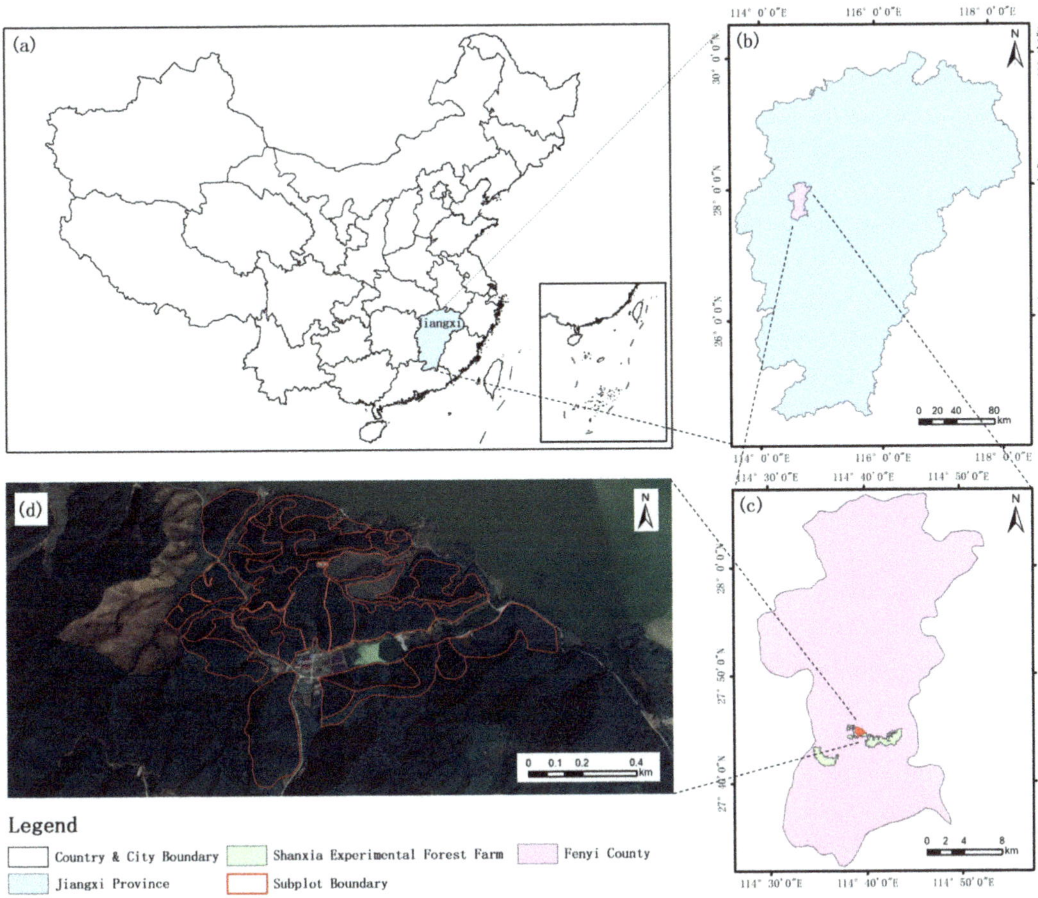

Figure 1. The location of the study area. (**a**) China's administrative boundary; (**b**) Jiangxi Province; (**c**) Fenyi County; (**d**) Shanxia Experimental Forest Farm. The red lines represent the subplot boundaries.

Table 1. Main forest stand attribute information of 29 subplots in Shanxia Experimental Forest Farm.

Subplot Number	Tree Species	Average DBH (cm)	Maximum Value	Minimum Value	Standard Deviation	Average H (m)	Maximum Value	Minimum Value	Standard Deviation	Average CW (m)	Number of Trees (Trees)	Canopy Density (%)	Stand Density (Trees/ha)	Area (ha)
1	*Cunninghamia lanceolata*, *Schima superba*, *Machilus pauhoi*	18.22	37.9	1.7	10.22	21.8	22.3	1.6	5.66	3.75	4092	70	54.25	5.03
2	*Cunninghamia lanceolata*, *Schima superba*, *Machilus pauhoi*	19.3	35.4	0.8	9.48	23.4	23.2	1.7	5.51	3.83	1896	70	52.79	2.39
3	*Cunninghamia lanceolata*, *Schima superba*, *Machilus pauhoi*	22.8	43.8	0.8	10.73	24.1	24.5	1.5	6.33	3.4	4042	70	71	3.80
4	*Cunninghamia lanceolata*, *Schima superba*, *Machilus pauhoi*	20.3	35.6	1.1	10.1	16.7	23.3	1.8	5.32	3.1	1156	70	65	1.19
5	*Cunninghamia lanceolata*	24.5	52.9	0.8	11.06	18.5	26	1.5	6.75	3.33	3729	70	53	4.69
6	*Schima superba*	31.5	35.5	1	10.21	31.9	22.8	1.9	5.62	8.5	978	80	35	1.86
7	*Cunninghamia lanceolata*	6.2	42.3	1	12.64	3.1	23.6	1.8	6.74	4.2	4161	70	50	5.55
8	*Cunninghamia lanceolata*	16.4	39	0.6	10.82	17.5	25.7	1.6	7.18	2.6	929	70	78	0.79
9	*Schima superba*, *Sassafras tzumu*, *Cinnamomum camphora*	21.2	39.3	2.3	9.67	20.2	22	3.1	4.6	7.1	199	75	29	0.46

Table 1. *Cont.*

Subplot Number	Tree Species	Average DBH (cm)	Maximum Value	Minimum Value	Standard Deviation	Average H (m)	Maximum Value	Minimum Value	Standard Deviation	Average CW (m)	Number of Trees (Trees)	Canopy Density (%)	Stand Density (Trees/ha)	Area (ha)
10	*Schima superba, Sassafras tzumu*	24.5	46.5	0.6	10.95	21.5	24.5	1.6	6.73	6.74	685	70	43	1.06
11	*Cunninghamia lanceolata*	22.5	36.9	0.6	10.36	22	25.8	1.8	6.42	3.33	7238	75	78	6.19
12	*Cunninghamia lanceolata, Schima superba*	20.3	45.1	1.1	11.66	18.1	24.4	1.9	7.48	4.65	4458	75	77	3.86
13	*Schima superba, Castanopsis sclerophylla, Cinnamomum camphora, Sassafras tzumu*	26.2	35.4	2.6	8.39	14.6	21.9	2.9	4.4	5.4	481	60	33	0.97
14	*Cunninghamia lanceolata*	22.8	46.5	0.3	11.39	18.6	24.1	1.8	6.18	3.65	2029	70	74	1.83
15	*Pinus massoniana, Cunninghamia lanceolata, Vernicia fordii*	21.5	36.7	1.3	7.36	15.9	23	3.7	3.68	5.31	7674	85	85	6.02
16	*Cunninghamia lanceolata*	24.1	39.6	2.2	9.43	20.2	22.8	4	4.8	4.05	5811	70	78	4.97
17	*Schima superba, Vernicia fordii*	29.4	42.4	1	11.02	19.5	21.5	1.5	4.85	8.5	3896	70	40	6.49
18	*Cunninghamia lanceolata*	13.8	42.1	1.2	11.52	12.2	24.3	2.2	6.84	2.54	11629	95	162	4.79
19	*Cunninghamia lanceolata*	27.5	46.1	3.3	9.88	21.1	23.1	0.7	5.11	3.62	1656	70	82	1.35

Table 1. *Cont.*

Subplot Number	Tree Species	Average DBH (cm)	Maximum Value	Minimum Value	Standard Deviation	Average H (m)	Maximum Value	Minimum Value	Standard Deviation	Average CW (m)	Number of Trees (Trees)	Canopy Density (%)	Stand Density (Trees/ha)	Area (ha)
20	*Castanea mollissima, Paulownia fortunei, Cunninghamia lanceolata*	17.6	45	2	11.29	12.7	25.4	2.7	6.23	7.21	921	70	48	1.28
21	*Michelia platypetala*	21.2	41.2	4.9	4.92	16.5	19.6	2.1	2.08	4.05	2725	60	52	3.49
22	*Liriodendron tulipifera*	17.9	38.4	4.5	8.63	17.8	23.1	1.5	5.02	4.69	1411	75	71	1.32
23	*Vernicia fordii, Machilus pauhoi*	22.3	37.2	2	9.11	13.7	22.6	3.7	4.44	7.65	1542	75	29	3.55
24	*Cunninghamia lanceolata, Schima superba, Paulownia fortunei*	19.4	37.1	4.5	4.55	15.6	24.3	2.9	2.98	5.78	2419	85	95	1.70
25	*Cunninghamia lanceolata, Schima superba, Machilus pauhoi*	19.22	38.1	1.5	7.62	22.2	20.8	2.7	3.49	4.15	1547	75	52	1.98
26	*Cunninghamia lanceolata*	27.4	39.1	2.7	7.12	18.8	17.7	3.2	3.34	3.24	2213	75	71	2.08
27	*Cunninghamia lanceolata*	14.7	33.5	0.8	8.83	12.8	17.3	1.1	3.96	1.81	15539	90	158	6.56
28	*Schima superba, Cinnamomum camphora, Vernicia fordii*	15.2	38.3	1.3	7.58	12.1	22.7	1.4	3.57	2.24	6764	85	121	3.73
29	*Myrica rubra*	9.8	33.2	1.6	7.36	4.2	15.4	0.9	1.98	4.55	6344	65	60	7.05

2.3. Data Analysis

2.3.1. Construction and Decomposition of Three-Dimensional Tree Model

At present, there are many tools for tree modeling [44–46]. We used SpeedTree three-dimensional modeling software, combined with the forest morphological and structural parameters of the main common tree species collected at the Shanxia Forest Farm, including H, DBH, UBH, CW, CH, azimuth angle, and other parameters, as well as the morphology and texture photos of the trees. We selected the morphological and structural parameters of the trees with a good growth status and morphological structure for various tree species. We added the trunk, branches, twigs, and leaves to the SpeedTree software; adjusted parameters such as trunk curvature and branch angle; and pasted the texture of the bark and leaves. Eventually, we constructed initial three-dimensional models for 32 tree species, including *Cunninghamia lanceolata*, *Schima superba*, and *Machilus pauhoi*. The model was constructed using backward elimination technology, which only displays faces facing the screen, improving the rendering efficiency of the three-dimensional model of the forest canopy. Finally, 32 three-dimensional models of forest trees were constructed in order to better reflect the different morphological and structural characteristics of trees in forest scenes and ensure that the changes in the various directions of the trees during the growth process conformed to the law of the forest. In this study, on the basis of the morphological characteristics of forest trees, we decomposed the three-dimensional model of forest trees into two models: trunk and crown. Each model was considered an independent unit, reflecting the growth and variation trends of H, DBH, CH, UBH, and CW. Taking *Cunninghamia lanceolata* as an example, the three-dimensional model and decomposition model of *Cunninghamia lanceolata* are shown in Figure 2.

Figure 2. (**a**) Representation of the texture of the bark and leaves of *Cunninghamia lanceolata* trees. (**b**) The initial three-dimensional model of *Cunninghamia lanceolata* created using (**a**,**b**) and real measurement data. (**c**) Representation of the hierarchical model of *Cunninghamia lanceolata*.

2.3.2. Tree Hierarchical Growth Model

The forest growth model is a mathematical model that describes the growth process of trees. It is generally used to describe the relationship between tree volume and growth age [47]. In order to simulate the polymorphism of forest trees more realistically during the growth process, we used DBH, H, UBH, CH, and CW as the main parameters of forest tree morphology and structure in this study. The DBH growth model and the UBH growth model were used for the tree trunk, while the CW growth model and the CH growth model were used for the tree crown. According to the results of on-site inspections on Shanxia Forest Farm, 12 main distributed tree species were selected for growth simulation. We calculated the values of the tree DBH, H, CW, UBH, and CH after years of growth on the basis of the current age of the trees using other studies and the growth models provided by local forest farms for each tree species. Taking *Cunninghamia lanceolata* as an example, the growth models are shown in Table 2.

Table 2. Hierarchical growth model of *Cunninghamia lanceolata*.

ID	Model Name	Growth Model	Description	References
1	DBH growth model	$D = 21.7106 * \left(1 - e^{-0.0538*age}\right)^{0.7626}$	*D* represents diameter at breast height of tree, age represents time of growth	[48]
2	H growth model	$H = 20.3793 * \left(1 - e^{-0.0684*age}\right)^{1.1943}$	*H* represents height of tree, age represents time of growth	[48]
3	CW growth model	$CW = 0.660 + 0.037 \times age + 0.160 \times Ph$	*CW* represents crown width of tree, *Ph* represents horizontal structural parameters (minimum distance affected by adjacent trees)	[49]
4	CH growth model	$CH = -3.036 + 0.469 \times age + 2.690 \times Pv$	*CH* represents crown height of tree, *Pv* represents vertical spatial structural parameters (the average relative height of adjusted trees)	[49]
5	UBH growth model	$UBH = -3.020 + 0.409 \times age + 2.601 \times Pv$	*UBH* represents height under the branches of tree, *Pv* represents vertical spatial structural parameters (the average relative height of adjacent trees)	[49]

2.3.3. Visualization Simulation of Large-Scale Forest Scene Growth Dynamics Based on Forest Hierarchical Model

In traditional large-scale forest scenes, trees are often scaled proportionally using an integrated model. Integrated models consider the trees as a whole and calculate the actual values of UBH, CW, and CW on the basis of the growth models of H and DBH. This is unreasonable. The hierarchical model divides the model into the trunk and the crown, which can be calculated according to the growth models of different parts, making the large-scale forest landscape more diverse.

Using Unreal Engine 4 (UE4, https://www.unrealengine.com/zh-CN, accessed on 25 May 2023), which is an efficient and versatile game development engine launched by Epic Games, we can directly preview the three-dimensional simulation development effect [44]. We connected to the MySQL database with UE4 and read the stand data table. The three-dimensional model of a tree is divided into two parts: the trunk and the crown. The original point of the coordinates of the trunk and the crown of the tree hierarchical model was assumed to be at the bottom. The initial DBH of the trees in the forest is $D_{initial}$, the height of the trees is $H_{initial}$, the crown width is $CW_{initial}$, the height under branches is $UBH_{initial}$, and the crown height is $CH_{current}$. The current DBH of the trees in the scene is $D_{current}$, the height of the trees is $H_{current}$, the crown width is $CW_{current}$, the height under branches is $UBH_{current}$, and the crown height is $CH_{current}$. The DBH of the trees after growth is D_{grow}, the height of the trees is H_{grow}, the crown width is CW_{grow}, the height under branches is UBH_{grow}, and the crown height is CH_{grow}.

According to the timeline animation event, the forest growth model library (which used the growth model that conformed to the growth of the dominant species provided by the forest farm locally), and the blueprint function script (which is a visual script that provides an intuitive, node-based interface and realizes code programming by dragging and dropping nodes), a set of forest growth blueprint function scripts was built. The scaling coefficient of the initial scene tree was calculated through a function script, and the scaling coefficient after tree growth was calculated according to the annual growth of each part of the tree in real time.

Using LOD technology, the forest hierarchical model and field measurement data were used to construct an initial scene of a real Shanxia Experimental Forest Farm. When the viewpoint is far away, the trees are displayed in the form of billboards. As the viewpoint approaches, a mixture of graphics and images is gradually adopted. When observing the tree morphology at close range, full three-dimensional graphics are used to display the details of the trees. The dynamic visualization simulation process of large-scale forest scene growth is shown in Figure 3.

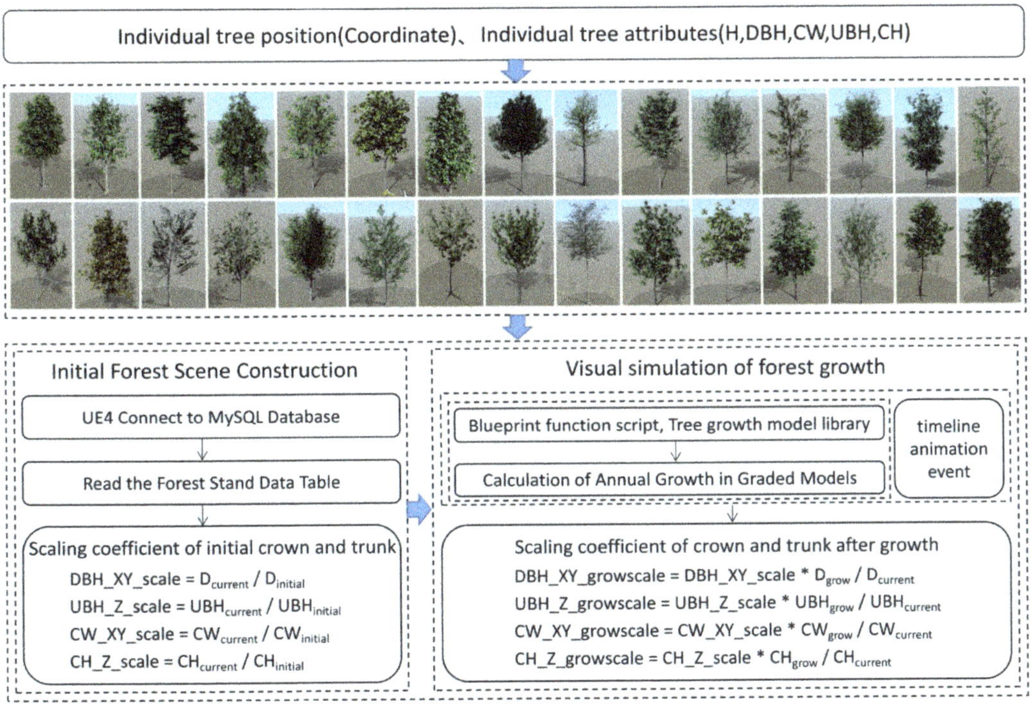

Figure 3. The dynamic visualization simulation process of large-scale forest scene growth.

2.3.4. Evaluation Indicators for Large-Scale Forest Scene Construction

Frame Rate

We compared the performance of scenes using two indicators: frame rate and frame rate fluctuation value (F).

The number of frames refreshed per second was measured in frames per second (fps). Frame rate is an important evaluation indicator for rendering rate in virtual scenes, requiring a frame rate of no less than 30 fps. The higher the frame rate, the smoother the visual display, and the smoother the screen image. Its size determines the smoothness that the user feels when rendering. The frame rate calculation formula is shown in (1).

$$Frame\ rate = frameNum/elapsedTime \tag{1}$$

Frame rate fluctuation value can be obtained by calculating the standard deviation of the frame rates. When calculating the frame rate variability, it is necessary to record the frame rate of each frame during a certain period of time and then calculate the standard deviation of these frame rates. The smaller the frame rate fluctuation value, the more stable the scene runs. The frame rate fluctuation value calculation formula is shown in (2).

$$F = \sqrt{\frac{\sum_{i=1}^{n}\left(Frame\ rate_i - \overline{\sum_{i=1}^{n} Frame\ rate_i}\right)^2}{n}} \tag{2}$$

where $Frame\ rate_i$ represents the frame rate of the ith frame, and F represents the frame rate fluctuation value.

Forest form Evaluation Method

The evaluation of scene quality involves a subjective judgment process. In this study, we adopted a manual evaluation method. We recruited 20 graduate students, including 9 males and 11 females, to participate in the assessment. Males accounted for 45% of the selected sample, while females accounted for 55%. We invited them to conduct experiential simulations of the large-scale forest scene that we constructed and asked them to rate the similarity between the real scenes and the virtual scenes. Accoridng to the recorded scoring results, we quantitatively analyzed the authenticity and reliability of our constructed scene.

3. Results

3.1. Forest Hierarchical Model

Taking 4092 trees in subplot 1 as an example, Table 3 shows the distribution of tree numbers within the ranges of 0–15 cm, 15–30 cm, 30–45 cm, 45–60 cm, 60–75 cm, 75–90 cm, and >90 cm for constructing large-scale forest scenes using hierarchical and integrated models. It can be seen that the hierarchical model used each attribute factor of the tree with a wide distribution of values, such as H, CH, DBH, CW, and UBH, solving the problem of traditional large-scale forests exhibiting a uniform appearance for thousands of trees. The integrated model used two parameters, H and DBH, which can only reflect the H and DBH of each tree. The CW, CH, and UBH increased with the increase in DBH, and the forest form was single.

Table 3. The distribution of tree attributes within the ranges of 0–15 cm, 15–30 cm, 30–45 cm, 45–60 cm, 60–75 cm, 75–90 cm, and >90 cm in the hierarchical model and integrated model.

Tree Attributes	0–15 (cm)		15–30 (cm)		30–45 (cm)		45–60 (cm)		60–75 (cm)		75–90 (cm)		>90 (cm)	
	Hierarchical Model	Integrated Model	Hierarchical Model	Integrated Model	Hierarchical Model	Integrated Model	Hierarchical Model	Integrated Model	Hierarchical Model	Integrated Model	Hierarchical Model	Integrated Model	Hierarchical Model	Integrated Model
DBH (trees)	1907	1907	1620	1620	539	539	26	26	0		0	0	0	0
CW (trees)	112	-	1276	-	1773	-	635	-	208		67	-	21	-
H (trees)	2	2	4	4	49	49	207	207	389		487	487	2954	2954
CH (trees)	35	-	242	-	477	-	611	-	605		530	-	1592	-
UBH (trees)	329	-	1263	-	913	-	534	-	393		318	-	342	-

We conducted a visual simulation of the hierarchical model and the integrated model before and after growth, respectively. The morphological changes in hierarchical and integrated models currently and after 10 and 20 years of growth are shown in Figure 4. The results showed that as the trees grow, the changes in tree DBH and UBH can be clearly seen, as shown in Figure 4a–c. However, Figure 4e,f only shows a proportional increase in tree H and CW, and the DBH and CW are scaled according to the same scaling factor, resulting in morphological structure parameters of the forest that are inconsistent with the actual growth pattern.

Figure 4. The morphological changes in hierarchical and integrated models after 10 and 20 years of growth. (**a**–**c**) Morphological changes of hierarchical model currently and after 10 and 20 years of growth. (**d**–**f**) Morphological changes of integrated model currently and after 10 and 20 years of growth.

3.2. *Visual Simulation Result Evaluation*

Taking the data of the tree information in the Shanxia Forest Farm as an example, a hierarchical model and an integrated model were used to simulate and analyze the spatial distribution of forests in different growth stages. The method of the hierarchical model was to calculate the UBH and CW under the branches using the growth model of UBH and the growth model of CW. The method of the integrated model was to calculate the corresponding UBH and CW proportionally on the basis of the H growth model and the DBH growth model. Due to the long growth cycle of trees, we compared the growth simulation results with sample plots with similar site conditions and tree ages in terms of the branch height and crown width in this paper. The simulation results and actual measurement results of UBH at different ages using the two methods are shown in Table 4, and the simulation results of CW are shown in Table 5. At 40 and 50 years, the average predicted values and errors of the UBH and CW of the hierarchical model were smaller than those of the integrated model. The standard deviation of the predicted values of the hierarchical model was smaller than that of the integrated model, indicating that the values calculated by our method were closer to the actual values.

Table 4. The actual value and growth value of the UBH in the hierarchical model and the integrated model.

Origins		40 Years Old					50 Years Old				
Age (Year)	UBH (m)	Measurements	Hierarchical Model		Integrated Model		Measurements	Hierarchical Model		Integrated Model	
			Forecasts	Error	Forecasts	Error		Forecasts	Error	Forecasts	Error
15	1.6	7.8	8.3	0.5	8.7	0.9	9.4	9.6	0.2	9.8	0.4
15	1.4	7.6	8.8	0.8	6.5	1.1	9.2	10.4	1.2	10.5	1.3
20	2.9	8.2	9	0.8	7.4	0.8	10.1	10.5	0.4	9.4	0.7
20	3.1	8.4	8.7	0.3	8.9	0.5	10.3	11.3	1.1	11.7	1.4
25	4.9	8.1	8.5	0.4	8.7	0.6	9.8	10.1	0.3	9.2	0.6
25	5.8	9.3	10	0.7	8.5	0.8	11.5	11.7	0.2	11	0.5
30	7.1	9	7.3	1.7	11.1	2.1	10.6	11.2	0.6	9.7	0.9
Mean value	3.83	8.34	8.66	0.74	8.54	0.97	10.13	10.69	0.57	10.19	0.83
Standard deviation	5.29	1.51	1.98	1.14	3.49	1.31	1.91	1.82	1.03	2.24	0.96

Table 5. The actual value and growth value of the CW in the hierarchical model and the integrated model.

Origins		40 Years Old					50 Years Old				
Age (Year)	CW (m)	Measurements	Hierarchical Model		Integrated Model		Measurements	Hierarchical Model		Integrated Model	
			Forecasts	Error	Forecasts	Error		Forecasts	Error	Forecasts	Error
15	2.9	6.2	7.1	0.9	7.6	1.4	8.1	8.8	0.7	9	0.9
15	2.7	5.9	7.2	1.3	7.4	1.5	7.9	8.2	0.3	7.3	0.6
20	3.3	6.3	7.4	1.1	7.4	1.1	7.8	8.4	0.6	7	0.8
20	3.4	6.7	7.4	0.7	7.6	0.9	7.6	8.8	1.2	9.1	1.5
25	4.1	5.8	6.4	0.6	7	1.2	8.2	9	0.8	9.4	1.2
25	4.3	6.1	6.9	0.8	7.7	1.6	8.5	9.4	0.9	7.2	1.3
30	5.2	6.9	7.4	0.5	6.1	0.8	8.9	9.3	0.4	8	0.9
Mean value	3.7	6.27	7.11	0.84	7.26	1.21	8.14	8.84	0.70	8.14	1.03
Standard deviation	2.16	0.99	0.9	0.69	1.37	0.74	1.09	1.08	0.75	2.48	0.77

We invited 20 students to compare and rate the constructed large-scale forest scene with the real DigitalOrthophoto Map (DOM) image of the Shanxia Experimental Forest Farm (Figure 5). The evaluation results were mainly decided from four forest scene indicators: scene similarity, scene immersion (whether one feels the authenticity of the environment and the sense of immersive participation), scene fluency, and forest polymorphism. The results of four forest scene indicators are shown in Table 6. In Figure 5a,e, it can be seen that the distribution of the 29 subplots was basically consistent with the DOM. Among them, 19 people thought that the scene authenticity was very high, and 1 person thought that the scene authenticity was general, with a satisfaction evaluation rate of 95%.

Twenty people thought that the scene immersion was good, with a satisfaction evaluation rate of 100%. Eighteen people thought that the scene fluency is good, and two people thought that it is general, with a satisfaction evaluation rate of 90%. In addition, 19 people thought that the scene reflected the polymorphism of the forest very well, while one person thought that the scene did not reflect the polymorphism of the forest very well, with a satisfaction evaluation rate of 95%.

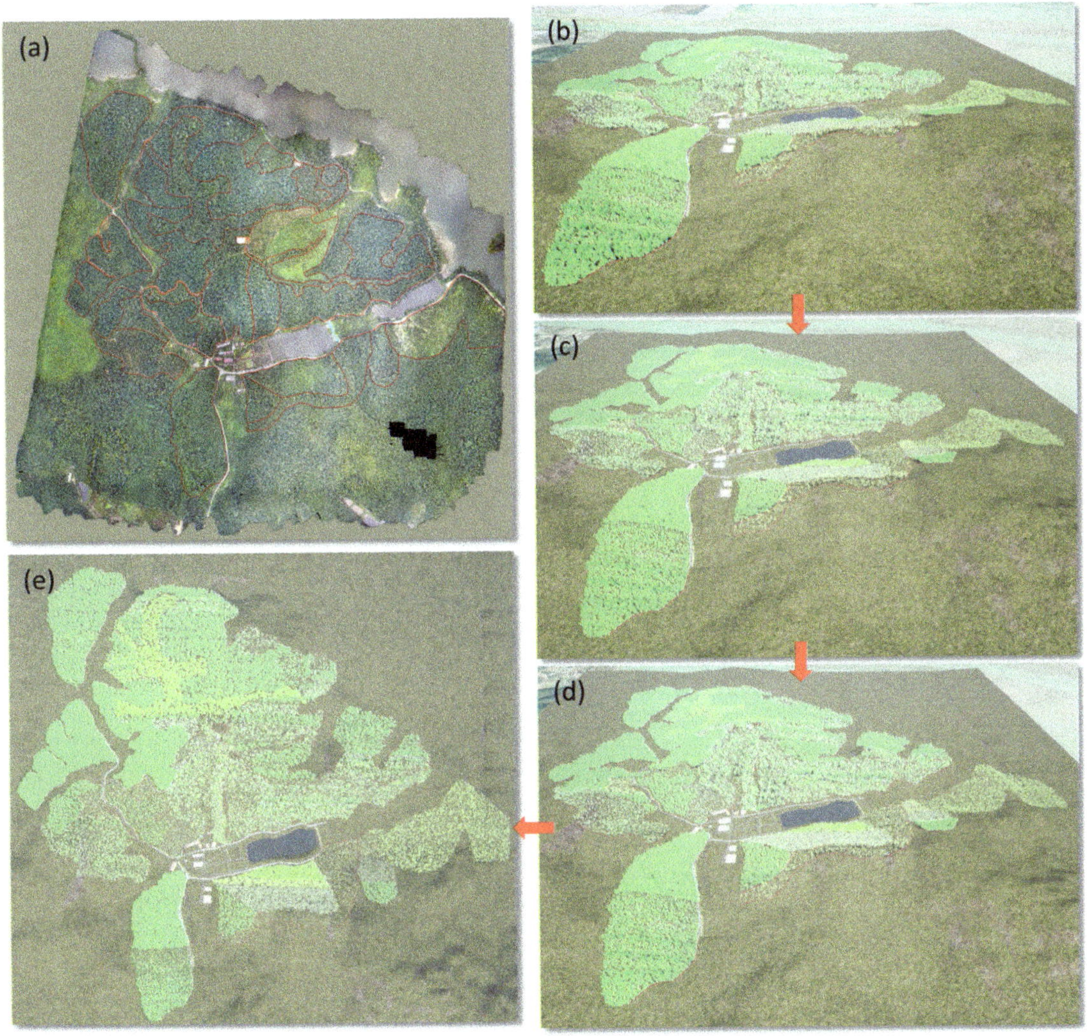

Figure 5. Comparison between real images and forest scene simulations. (**a**) Representation of the DOM of Shanxia Experimental Forest Farm. (**b–e**) Representation of the visual simulation scenes of Shanxia Experimental Forest Farm. Red arrow indicates the forest scene from different angles of the camera.

Table 6. Evaluation of forest scene indicators.

Evaluation Index	Good	General	Poor
Scene similarity	19	1	0
Scene immersion	20	0	0
Scene fluency	18	2	0
Forest polymorphism	19	1	0

The results showed that (1) the constructed large-scale forest scene had high similarity and strong immersion with the real forest farm, and the satisfaction evaluation rate was 95%. Therefore, the construction of large-scale forest scenes using the forest hierarchical model can more vividly and intuitively express and analyze the forest. (2) The construction of large-scale forest scenes using the forest hierarchical model has good running fluency, and trees of different shapes and sizes improve the visual effect of traditional large-scale forest scenes.

3.3. Visualization Simulation of Large-Scale Forest Scene Growth Dynamics

The three-dimensional models constructed with 32 kinds of trees using by SpeedTree software were added to the three-dimensional scene as a class, and we read the coordinates in the stands data table through the inaccurate transform longitude latitude height to the unreal node under the CesiumGeoreference reference in the Unreal Engine 4 rendering engine [50]. The hierarchical growth model and traditional integrated model scaling method were used to draw forest scenes of 10 trees, 100 trees, 1000 trees, 10,000 trees, and 100,000 trees in the Shanxia Forest Farm. The horizontal changes of the forest scene from different height perspectives are shown in Figure 6. The spatial structure characteristics of the forest, such as the diversity of tree species, the size difference, the spatial distribution pattern of trees, and the compression between tree crowns, can be clearly seen. As shown in Figure 6g,h, we could directly observe the uneven distribution or poor growth of trees caused by competition.

The average frame rate and system average frame rate fluctuations before and after growth are shown in Table 7. It can be seen that both two methods achieved a frame rate of over 50 fps when drawing 10, 100, and 1000 trees in small-scale forest scenes. When drawing large-scale forest scenes, such as scenes with 10,000 and 100,000 trees, the method proposed in this study based on a hierarchical model achieved a minimum frame rate of 30.63 fps for forest scenes, which meets people's needs for scene rendering. The minimum frame rate of the integrated model was 30.74 fps, which was 0.11 fps higher than the frame rate of the proposed hierarchical model. The frame rate of the integrated model after growth was 29.68 fps, which was 0.96 fps higher than the frame rate of the hierarchical model proposed in this study. In general, the integrated model ran at a high frame rate, but the constructed forest scene had the problem of "a thousand trees are all similar".

Table 7. The average frame rate and system average frame rate fluctuations before and after growth using hierarchical and integrated model methods.

Number of Trees	Before Growth				After Growth			
	Hierarchical Model		Integrated Model		Hierarchical Model		Integrated Model	
	Average fps	Frame Rate Fluctuation Value	Average fps	Frame Rate Fluctuation Value	Average fps	Frame Rate Fluctuation Value	Average fps	Frame Rate Fluctuation Value
10	120	0.32	120	0.37	120	0.44	120	0.45
100	120	0.39	120	0.45	120	0.47	120	0.44
1000	64.37	0.42	65.47	0.48	64.37	0.53	65.47	0.56
10,000	39.65	0.54	42.43	0.52	39.65	0.66	42.43	0.65
100,000	30.63	0.61	30.74	0.64	29.68	0.69	30.64	0.69

Figure 6. Sequential frames from (**a**–**h**) visualize the horizontal of the large-scale forest scene changes. The camera height is set at 300 m above the terrain, and the camera moved forward from (**a**–**h**).

From the perspective of the frame rate fluctuation value, the smaller the frame rate fluctuation value, the more stable the scene operation [51]. It could be seen that as the number of trees increased, the frame rate fluctuation value also increased. The average frame rate fluctuation value of the scene generated before growth by the hierarchical model was 0.46, the average frame rate fluctuation value of the integrated model was 0.49, and the average frame rate fluctuation value of both after growth was 0.56. In general, the average frame rate fluctuation of the hierarchical model had a relatively small change, and the scene ran more stably than the integrated model's scene.

According to the growth model of the forest hierarchical model, the visual simulation of the growth dynamics in Shanxia Experimental Forest Farm is shown in Figure 7. Figure 7a–e represent the growth status of the forest after 10 years, and Figure 7f–i represent the growth status of the forest after 20 years. The forest scene we constructed through visualization tools will allow forest managers to conduct logging, selective logging, planting, and other management operations and view the approximate spatial structure of the forest. Combining forest growth simulation with forest grading models changes the visual effect of large-scale forest scenes and improves the usability of forests.

Figure 7. Visual simulation of growth dynamics in Shanxia Forest Farm. (**a–e**) Representation of the growth status of the forest after 10 years. (**f–j**) Representation of the growth status of the forest after 20 years.

4. Discussion

Visualization is considered one of the most direct ways to explain the results of a forest growth simulation [26,40]. The visualization of large-scale forest scenes serves as a valuable tool for understanding the composition and dynamics of forests. It allows for the representation of natural growth processes and the impacts of disturbances, such as forest harvesting, providing practical guidance for forest management and further research on forest dynamics. However, previous studies have often overlooked the heterogeneity of individual trees in the representation of large-scale forest scenes, resulting in a lack of diversity in the rendered forests. In this study, we proposed a strategy for visualizing large-scale forest scenes by combining a tree hierarchical model with a growth visualization simulation. Using Unreal Engine 4 and LOD (Level of Detail) technology, we constructed initial large-scale forest scenes and simulated forest growth. The results indicated that our proposed method improves the visual landscape of large-scale forest scenes, achieves dynamic growth of large-scale forest scenes, and has good rendering efficiency.

We used the hierarchical model and the integrated model for visual simulation before and after growth. The results showed that the changes in DBH and UBH of the hierarchical model after growth were significant. The integrated model can only reflect the proportion of tree H and DBH, and the UBH, CH, and CW are calculated proportionally on the basis of their size and the calculated H and DBH, which is unreasonable. The crown shape of the hierarchical model is more diverse than the integrated model, which can better explain the community structure characteristics of forests. However, our method does not completely represent the morphological changes in the forest. We improved the issue of tree uniformity and traditional static expression of forest landscapes at the level of large-scale forest landscapes. In addition, we used the forest hierarchical model to construct a large-scale initial forest scene. The average frame rate of the initial forest scene was 30.63 fps. We calculated the growth changes of trees at different ages using the hierarchical model and scaled the hierarchical model according to the values after growth, achieving the dynamic simulation of large-scale forest landscapes. The results showed that the average frame rate of the scene during the growth process of the classification model was 29.68 fps, which met the basic scene rendering requirements. People can intuitively determine the spatial distribution of trees. Furthermore, we invited 20 graduate students to evaluate the simulated scenes by comparing them with real images, resulting in a satisfaction rate of 95%.

This study provides a reasonable and efficient method for constructing diverse and dynamic large-scale forest scenes. By combining Unreal Engine 4 with field survey data on forest inventories, we constructed a three-dimensional model of each tree species and restored a large-scale forest scene of the Shanxia Forest Farm on the basis of the spatial distribution of the tree positions and species composition. In addition, individual trees or stands can be highlighted according to user-defined operations, allowing forest managers to observe the forest at multiple scales, including stands and landscapes. The three-dimensional exploration of the spatial structure of forests and landscapes contributes to evaluating the development and succession of forest ecosystems and optimizes alternative solutions for forest management. This is essential for assessing forest ecosystem development and succession, optimizing forest management alternatives, and promoting a better understanding of natural and human disturbances among managers and researchers. The visualization of dynamic growth in large-scale forest scenes is crucial for evaluating future forest changes and the impacts of forest management measures. However, for large-scale and long-term forest growth simulation, tree mortality may occur. This study lacked an exploration of tree death. In the future, we will consider changes in forest spatial structure caused by factors such as tree death, fire, wind, and logging, and we will conduct a visual simulation, enabling managers and researchers to better understand natural and human interference.

However, accurately simulating forest growth and representing the branching structure of trees remain challenging due to the wide range of growth variation and driving mechanisms involved. Constrained by existing computational resources and power, this study strikes a balance between frame rate and scene simulation effects by reducing the number of polygons while still capturing the growth status and rendering effects of trees. Our method mainly displays the polymorphism of forests visually. Forest growth does not completely conform to the growth patterns of trees. The system achieves the interpretation and understanding of the output results of the growth model. However, a growth model that is only related to age is too simple. In general, a 4-year-old tree has a smaller crown but also fewer branches, leaves, etc., than a 20-year-old tree. Therefore, we will consider integrating multiple influencing factors, such as stand density, competition index, site index, and environment into the growth model to ensure that the continuous growth of the forest at each stage is in line with reality.

5. Conclusions

In this study, we considered the simplicity and static expression of constructing large-scale forest scenes at present. Using Unreal Engine 4, we utilized three-dimensional visualization simulation technology and LOD technology combined with a forest hierarchical growth model to construct a large-scale forest landscape of 100,000 trees, improving the visual effect of dynamic changes in diverse large-scale forest scenes. The numerical results of tree DBH, height, CW, CH, and UBH were distributed widely before and after growth, which indicated the polymorphism of the forest. However, the integrated model could only be scaled on the basis of the true tree height and DBH. Other factors, such as CW, CH, and UBH, did not conform to the actual growth mode, which was prone to deformation and unreality. In addition, the growth value of the hierarchical growth model calculated by the blueprint function was used to simulate the growth of the forest after 40 and 50 years. The results showed that the average predicted values, average errors, and standard deviations of the UBH and CW of the hierarchical model at 40 and 50 years were smaller than those of the integrated model. The evaluation results of 20 people on the constructed large-scale forest scene showed that the forest scene constructed using this method had a high similarity with the real scene, a strong immersion experience, and a good evaluation rate of 95%. The average frame rate during the initial scene operation was 30.96 fps, the average frame rate during the growth process was 29.68 fps, and the average frame rate fluctuation value of the system operation was 0.46, which ensured the stability of the scene operation. The method proposed in this study for constructing large-scale forest scenes based on hierarchical growth models is applicable to other regions and tree species. It improves the diversity and dynamic change process of large-scale forest scenes visually. In order to analyze the spatial structure characteristics of forests more comprehensively and quantitatively, it is necessary to conduct complex research on growth models for each tree species.

Author Contributions: Conceptualization, H.Z. and K.L.; methodology, K.L.; software, K.L.; formal analysis, Y.L., H.Q., and T.Y.; resources, J.Z. and X.H.; writing—original draft preparation, K.L.; writing—review and editing, K.L., H.Z., and H.Q.; visualization, H.Z. and Z.C.; supervision, H.Z. and Y.L. All authors have read and agreed to the published version of the manuscript.

Funding: This research was funded by the National Natural Science Foundation of China, grant number 32271877, 32071681 and the Foundation Research Funds of IFRIT, grant number CAFYBB2021ZE005-2.

Data Availability Statement: Not applicable.

Conflicts of Interest: The authors declare that there are no conflict of interest.

References

1. Franklin, J.F.; Spies, T.A.; Van Pelt, R.; Carey, A.B.; Thornburgh, D.A.; Berg, D.R.; Lindenmayer, D.B.; Harmon, M.E.; Keeton, W.S.; Shaw, D.C.; et al. Disturbances and Structural Development of Natural Forest Ecosystems with Silvicultural Implications, Using Douglas-Fir Forests as an Example. *For. Ecol. Manag.* **2002**, *155*, 399–423. [CrossRef]
2. Hossein, A.; Djamali, M.; Ghorbanalizadeh, A.; And, E.R. Plant Biodiversity of Hyrcanian Relict Forests, N Iran: An Overview of the Flora, Vegetation, Palaeoecology and Conservation. *Pak. J. Bot.* **2010**, *42*, 231–258.

1. Hof, J.G.; Joyce, L.A. Spatial Optimization for Wildlife and Timber in Managed Forest Ecosystems. *For. Sci.* **1992**, *38*, 489–508. [CrossRef]
2. Huang, J.; Lucash, M.S.; Scheller, R.M.; Klippel, A. Walking through the Forests of the Future: Using Data-Driven Virtual Reality to Visualize Forests under Climate Change. *Int. J. Geogr. Inf. Sci.* **2021**, *35*, 1155–1178. [CrossRef]
3. Lange, E. Integration of Computerized Visual Simulation and Visual Assessment in Environmental Planning. *Landsc. Urban Plan.* **1994**, *30*, 99–112. [CrossRef]
4. Griffon, S.; Coligny, F. de AMAPstudio: A Software Suite for Plants Architecture Modelling. In Proceedings of the 2012 IEEE 4th International Symposium on Plant Growth Modeling, Simulation, Visualization and Applications, Shanghai, China, 31 October–3 November 2012; pp. 141–147.
5. Li, S.; Zhang, H.; Li, Y.; Yang, T.; Shen, K. Three-Dimensional Visualization Simulation of Chinese Fir Stand Growth Based on Unity3D. In Proceedings of the 2018 3rd International Conference on Mechanical, Control and Computer Engineering (ICMCCE), Huhhot, China, 14–16 September 2018; IEEE: Piscataway, NJ, USA, 2018; pp. 530–534.
6. Schroeder, W.J.; Zarge, J.A.; Lorensen, W.E. Decimation of Triangle Meshes. In Proceedings of the 19th Annual Conference on Computer Graphics and Interactive Techniques, Chicago, IL, USA, 27–31 July 1992; pp. 65–70.
7. Kalvin, A.D.; Taylor, R.H. Superfaces: Polygonal Mesh Simplification with Bounded Error. *IEEE Comput. Graph. Appl.* **1996**, *16*, 64–77. [CrossRef]
8. Lluch, J.; Camahort, E.; Vivó, R. An Image-Based Multiresolution Model for Interactive Foliage Rendering. *J. WSCG* **2004**, *12*, 507–514.
9. Max, N. Optical Models for Direct Volume Rendering. *IEEE Trans. Vis. Comput. Graph.* **1995**, *1*, 99–108. [CrossRef]
10. Feng, W.; Gang, W.; Deji, P.; Yuan, L.; Liuzhong, Y.; Hongbo, W. A Parallel Algorithm for Viewshed Analysis in Three-Dimensional Digital Earth. *Comput. Geosci.* **2015**, *75*, 57–65. [CrossRef]
11. Weber, J.; Penn, J. Creation and Rendering of Realistic Trees. In Proceedings of the 22nd Annual Conference on Computer Graphics and Interactive Techniques, Los Angeles, CA, USA, 6–11 August 1995; pp. 119–128.
12. Remolar, I.; Chover, M.; Belmonte, O.; Ribelles, J.; Rebollo, C. Geometric Simplification of Foliage. In Proceedings of the Eurographics (Short Presentations), Saarbrücken, Germany, 2–6 September 2002.
13. Lluch, J.; Vicent, M.J.; Fernandez, S.; Monserrat, C.; Vivo, R. The Modelling of Branched Structures Using a Single Polygonal Mesh. In Proceedings of the VIIP, Marbella, Spain, 3–5 September 2001; pp. 203–207.
14. Perbet, F.; Cani, M.-P. Animating Prairies in Real-Time. In Proceedings of the 2001 Symposium on Interactive 3D Graphics, Chapel Hill, NC, USA, 26–29 March 2001; pp. 103–110.
15. Marshall, D.; Fussell, D.S.; Campbell III, A.T. Multiresolution Rendering of Complex Botanical Scenes. In Proceedings of the Graphics Interface, Kelowna, BC, Canada, 21–23 May 1997; Volume 97, pp. 97–104.
16. Meyer, A.; Neyret, F.; Poulin, P. Interactive Rendering of Trees with Shading and Shadows. In Proceedings of the Rendering Techniques 2001: Proceedings of the Eurographics Workshop, London, UK, 25–27 June 2001; Springer: Berlin/Heidelberg, Germany, 2001; pp. 183–196.
17. Lim, E.-M.; Honjo, T. Three-Dimensional Visualization Forest of Landscapes by VRML. *Landsc. Urban Plan.* **2003**, *63*, 175–186. [CrossRef]
18. Decaudin, P.; Neyret, F. Rendering Forest Scenes in Real-Time. In Proceedings of the EGSR04: 15th Eurographics Symposium on Rendering, Norrköping, Sweden, 21–23 June 2004; pp. 93–102.
19. Larsen, D.R.; Scott, I. The Sylview Graphical Interface to the SYLVAN STAND STRUCTURE Model with Examples from Southern Bottomland Hardwood Forests. In Proceedings of the 14th Biennial Southern Silvicultural Research Conference, Gen Tech. Rep SRS-121, Asheville, NC, USA, 1 January 2010; Department of Agriculture, Forest Service, Southern Research Station: Asheville, NC, USA, 2010.
20. Li, Z.; Zhang, Y.-D. 3D Reconstruction Method of Forest Landscape Based on Virtual Reality. *Multimed. Tools Appl.* **2020**, *79*, 16369–16383. [CrossRef]
21. Tianyang, D.; Jiajia, X.; Jing, F.; Ling, Z. Fast Generation of Large-Scale Forest Scene Based on Spatial Similarity. *Int. J. Digit. Content Technol. Appl.* **2012**, *6*, 13.
22. Schaufler, G.; Stürzlinger, W. A Three Dimensional Image Cache for Virtual Reality. *Comput. Graph. Forum* **1996**, *15*, 227–236. [CrossRef]
23. Bai, Z.Y.; Huang, X.Y. Plum Tree Visualization Based on Speed Tree. *Key Eng. Mater.* **2011**, *474*, 511–516. [CrossRef]
24. Li, S.; Zhang, H.; Li, Y.; Yang, T.; He, J.; Ma, Z.; Shen, K. Dynamic Visual Simulation of Chinese Fir Stand Growth Based on Sample Library. *For. Res.* **2019**, *32*, 21–30. (In Chinese)
25. Li, Y.; He, J.; Yu, S.; Wang, H.; Ye, S. Spatial Structures of Different-Sized Tree Species in a Secondary Forest in the Early Succession Stage. *Eur. J. For. Res.* **2020**, *139*, 709–719. [CrossRef]
26. Tang, L.; Peng, X.; Chen, C.; Huang, H.; Lin, D. Three-Dimensional Forest Growth Simulation in Virtual Geographic Environments. *Earth Sci. Inform.* **2019**, *12*, 31–41. [CrossRef]
27. Zamuda, A.; Brest, J. Vectorized Procedural Models for Animated Trees Reconstruction Using Differential Evolution. *Inf. Sci. (NY)* **2014**, *278*, 1–21. [CrossRef]
28. Tang, L.; Chen, C.; Huang, H.; Lin, D. An Integrated System for 3D Tree Modeling and Growth Simulation. *Environ. Earth Sci.* **2015**, *74*, 7015–7028. [CrossRef]

31. Ijiri, T.; Owada, S.; Igarashi, T. The Sketch L-System: Global Control of Tree Modeling Using Free-Form Strokes. In Proceedings o the Smart Graphics: 6th International Symposium, SG 2006, Vancouver, BC, Canada, 23–25 July 2006; Proceedings 6. Springer Berlin/Heidelberg, Germany, 2006; pp. 138–146.
32. Peng, W.; Pukkala, T.; Jin, X.; Li, F. Optimal Management of Larch (Larix Olgensis A. Henry) Plantations in Northeast China When Timber Production and Carbon Stock Are Considered. *Ann. For. Sci.* **2018**, *75*, 63. [CrossRef]
33. Xie, L.; Widagdo, F.R.A.; Dong, L.; Li, F. Modeling Height–Diameter Relationships for Mixed-Species Plantations of Fraxinus mandshurica Rupr. and Larix olgensis Henry in Northeastern China. *Forests* **2020**, *11*, 610. [CrossRef]
34. Miina, J.; Pukkala, T. Using numerical optimization for specifying individual-tree competition models. *For. Sci.* **2000**, *46*, 277–283
35. Dong, L.; Pukkala, T.; Li, F.; Jin, X. Developing Distance-Dependent Growth Models from Irregularly Measured Sample Plot Data—A Case for Larix Olgensis in Northeast China. *For. Ecol. Manag.* **2021**, *486*, 118965. [CrossRef]
36. Pukkala, T. Methods to Describe the Competition Process in a Tree Stand. *Scand. J. For. Res.* **1989**, *4*, 187–202. [CrossRef]
37. Packalen, P.; Pukkala, T.; Pascual, A. Combining Spatial and Economic Criteria in Tree-Level Harvest Planning. *For. Ecosyst.* **2020**, *7*, 18. [CrossRef]
38. Anastacio, F.; Prusinkiewicz, P.; Sousa, M.C. Sketch-Based Parameterization of L-Systems Using Illustration-Inspired Construction Lines and Depth Modulation. *Comput. Graph.* **2009**, *33*, 440–451. [CrossRef]
39. Letort, V.; Cournède, P.-H.; Mathieu, A.; De Reffye, P.; Constant, T. Parametric Identification of a Functional–structural Tree Growth Model and Application to Beech Trees (Fagus sylvatica). *Funct. Plant Biol.* **2008**, *35*, 951–963. [CrossRef]
40. Bailey, R.L.; Dell, T.R. Quantifying Diameter Distributions with the Weibull Function. *For. Sci.* **1973**, *19*, 97–104.
41. Bailey, R.L.; Cieszewski, C.J. Development of a Well-Behaved Site-Index Equation: Jack Pine in North-Cental Ontario: Comment. *Can. J. For. Res.* **2000**, *30*, 1667. [CrossRef]
42. Feng, L.; De Reffye, P.; Dreyfus, P.; Auclair, D. Connecting an Architectural Plant Model to a Forest Stand Dynamics Model—Application to Austrian Black Pine Stand Visualization. *Ann. For. Sci.* **2012**, *69*, 245–255. [CrossRef]
43. De Reffye, P.; Edelin, C.; Françon, J.; Jaeger, M.; Puech, C. Plant Models Faithful to Botanical Structure and Development. *ACM Siggraph Comput. Graph.* **1988**, *22*, 151–158. [CrossRef]
44. Akarsu, E.; Fox, G.; Haupt, T.; Kalinichenko, A.; Kim, K.-S.; Sheethaalnath, P.; Youn, C.-H. Using Gateway System to Provide a Desktop Access to High Performance Computational Resources. In Proceedings of the Eighth International Symposium on High Performance Distributed Computing (Cat. No. 99TH8469), Redondo Beach, CA, USA, 6 August 1999; IEEE: Piscataway, NJ, USA, 1999; pp. 294–298.
45. Castel, T.; Caraglio, Y.; Beaudoin, A.; Borne, F. Using SIR-C SAR Data and the AMAP Model for Forest Attributes Retrieval and 3-D Stand Simulation. *Remote Sens. Environ.* **2001**, *75*, 279–290. [CrossRef]
46. Madureira, L.; Nunes, L.C.; Borges, J.G.; Falcão, A.O. Assessing Forest Management Strategies Using a Contingent Valuation Approach and Advanced Visualisation Techniques: A Portuguese Case Study. *J. For. Econ.* **2011**, *17*, 399–414. [CrossRef]
47. Meng, X.Y. *Forest Mensuration*; Chinese Forestry Publishing House: Beijing, China, 2006.
48. Luo, J.; Dai, C.; Tian, Y.; Peng, P.; Ma, F.; Zeng, Z.; Zhou, X.; Zhang, M. The tree height and DBH growth model of establishment of main constructive species for Project Forests of Carbon Sink in Hunan Province. *Hunan For. Sci. Technol.* **2016**, *43*, 46–50. (In Chinese)
49. Ma, Z.; Zhang, H.; Li, Y.; Yang, T.; Chen, Z.; Li, S. Visual Simulation of Chinese Fir Crown Growth Based on Spatial Structure. *For. Res.* **2018**, *31*, 150–157. (In Chinese)
50. Lei, K.; Zhang, H.; Qiu, H.; Liu, Y.; Hu, X.; Wang, J.; Cui, Z.; Zuo, Y. Comprehensive Decision Index of Logging (CDIL) and Visual Simulation Based on Horizontal and Vertical Structure Parameters. *Forests* **2023**, *14*, 277. [CrossRef]
51. Garg, K.; Nayar, S. Photorealistic Rendering of Rain Streaks. *ACM Trans. Graph.* **2006**, *25*, 996–1002. [CrossRef]

MDPI AG
Grosspeteranlage 5
4052 Basel
Switzerland
Tel.: +41 61 683 77 34

MDPI Books Editorial Office
E-mail: books@mdpi.com
www.mdpi.com/books

www.ingramcontent.com/pod-product-compliance
Lightning Source LLC
Chambersburg PA
CBHW061813210926
51850CB00006B/687
* 9 7 8 3 7 2 5 8 4 7 2 1 1 *